£24.99

# MASTERING AUDIO

## the art and the science
## second edition

# Mastering Audio

## the art and the science

second edition

Bob Katz

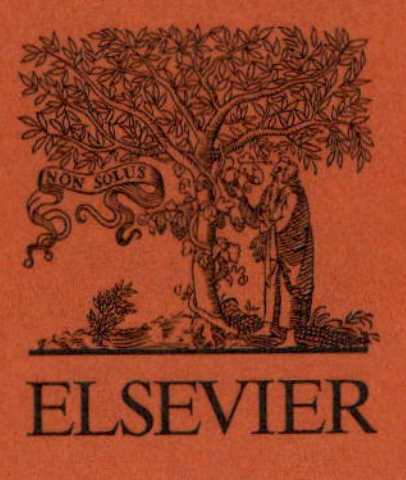


AMSTERDAM • BOSTON • HEIDELBERG • LONDON
NEW YORK • OXFORD • PARIS • SAN DIEGO
SAN FRANCISCO • SINGAPORE • SYDNEY • TOKYO
Focal Press is an imprint of Elsevier

Focal Press is an imprint of Elsevier
30 Corporate Drive, Suite 400, Burlington, MA 01803, USA
Linacre House, Jordan Hill, Oxford OX2 8DP, UK



**Library of Congress Cataloging-in-Publication Data**
Application submitted.

**British Library Cataloguing-in-Publication Data**
A catalogue record for this book is available from the British Library.

ISBN: 978-0-240-80837-6

For information on all Focal Press publications
visit our website at www.books.elsevier.com

07 08 09 10 10 9 8 7 6 5 4 3 2 1

Printed in Canada

# DEDICATION

This book is dedicated to **Mary Kent**, my best friend (and wife) since 1984, without whose love and support I would be absolutely nowhere today!

I also dedicate this book to those mastering engineers whose work I especially admire, **Bernie Grundman, Ted Jensen, Bob Ludwig, Glenn Meadows, Bob Olhsson**, and **Doug Sax**. Your fine work has brought great pleasure into my life through the records you have mastered.

To the many quality-concious engineers who bought the first edition: I hope you will find this edition to be even wider and deeper!

Next, a dedication to the next generation of mastering engineers... who will be working with tools that we have barely dreamed of. If you're young, remember to protect your hearing, because that precious gift can never be returned once it has been taken away.

— B. K.

Edited by
**Eric James, PhD**

Foreword
**Roger Nichols**

**Second Edition**

Contributing Engineers—Surround Chapter
**Dave Glasser**
**Bob Ludwig**
**Rich Tozzoli**
**Jonathan Wyner**

Fact Check
**Dick Pierce**
**Jim Johnston** (Ch. 4)

Layout, Cover Design and Photography
**Mary Kent**

Production Associate
**Robin Reumers**

Graphics Associate
**Matt Stanford**

Production Assistants
**Laurent Cohen**
**Gail Kent**
**Brandon Shope**
**Matt Stanford**
**Jessica Stephenson**
**Tanner Upthegrove**

## CREDITS AND THANKS TO...

A million and one thanks to Britain's **Dr. Eric James**, whose collaborative, empathic, and musical editing helped make this second edition cogent, vital and tight—but still airy and light! Eric can be reached on the web, as can all contacts listed in the book, by visiting *www.digido.com/links/*.

Thanks to Hanover, Massachusetts' Dick Pierce for exemplary fact-checking critical chapters in the second edition, and Seattle, Washington's Jim Johnston for additional contributions to Chapter 4 as well as fact-checking the first edition. I am the only one to blame for any er3ors that may remain.

Thanks to my production team. Mary has outdone herself with the beautiful graphic layout and cover. The very talented Robin "Eagle Eyes" Reumers coordinated logistics, scheduling, collected elements and participated in every part of the book production, from proofreading to sweeping floors. Matt Stanford created our logistical spreadsheet and his exacting illustrative ability can be found in several figures—bless you, Matt!

Thanks to Miami's Roger Nichols for providing (in the Foreword) a philosophy of good audio life! Thanks to Orlando's **Gail Kent** for inventing the punny title! And to Orlando Jazz Producer **Charlie Bertini**, for finding and preserving a perfect print of the Carnegie Hall chart found inside the front cover of this book. It has found its way to many studio walls.

Thanks to **Konrad Strauss** of Indiana, who reviewed the first edition and made helpful suggestions. Also to **Bob Ludwig** of Maine's Gateway Mastering, one of the busiest and nicest guys in audio, who read the first edition manuscript, provided valued input and this time around forms part of the surround mastering roundtable. Thanks to **Dave Glasser** of Colorado's Airshow, **Rich Tozzoli**, mixing whiz, and **Jonathan Wyner** of Cambridge, Massachusetts' M-Works for your surround sound wisdom.

Thanks to New Zealand's **Richard Hulse** for refining the parallel compression technique in the analog domain, and bringing it to my attention, after which I converted it to the digital domain, described in Chapter 11. Thanks to Germany's **Ralph Kessler** of Pinguin Audio, for helping to rigidly specify the attack and decay constants of the K-System meters, and to the many other manufacturers who have also implemented this public domain metering standard.

Thanks to San Leandro, California's **Bob Orban** and **Frank Foti**, radio processing gurus, for providing the text of their excellent article, in Appendix 1. Thanks to San Francisco's **Tardon Feathered** and

**Marvin Humphrey** of Mr. Toads, for producing the **What is Hot** CD, collaborating and organizing the **What is Hot** competition on the mastering webboard, and for locating Bob Orban to answer that essential question.

Thanks to **B.J.** and **Stu Buchalter** of Metric Halo Labs, for devising **SpectraFoo,** one of the most powerful and economical audio analysis tools in the universe. B.J., unsung audio genius, taught me enough about FFT to wade through it, but I could never reach his level of Einsteinian gestalt. Thanks to Tennessee's **Seva**, formerly of **Waves**, for helping to develop their products to true usefulness and inviting serious user feedback. Several images in the chapters are adapted from screenshots of Waves products. Thanks to **Christian G. Frandsen, Paul Frindle**, **Eelco Grimm**, **Dan Lavry**, **Bruno Putzeys**, and **Bob Stuart** for technical advice.

Thanks to Nashville's Producer/Engineer/Designer **George Massenburg** for his inspiration, consistent search for excellence, and endless knowledge of audio history, music and technology. George has produced some of the best-sounding albums. I'm proud that we share the same feelings about what makes good sound. Thanks to New York's **Al Grundy**, founder of the Institute of Audio Research; Boulder, Colorado's **Ray Rayburn**; Hartford, Connecticut's **Steve Washburn**; and New York City's **Noel Smith**, for his friendship and continued dedication to audio education. Similarly, Britain's **Michael Gerzon**, California's **Deane Jensen** and Britain's **Julian Dunn** were short-lived geniuses who contributed so much to this audio world, and liberally shared their thoughts with me over the years.

I thank Denmark's **Thomas Lund** of TC Electronic, a good friend, one of the most dedicated and artistic engineers around. Thomas edited earlier versions of the manuscripts that evolved into a couple of these chapters. He also provided some of the images and research that went into Chapters 5 and 14.

Thanks to **Catharine Steers** and the team at Focal Press, who exemplify as much care and attention to detail in publishing as we do in audio!

Finally, I must thank all my internet comrades, too numerous to mention, who participate in the Mastering and Gearslutz webboards. Your evocative adages, printed with your permission, will be found periodically throughout this book.

— B. K.

**First Edition**

Layout and Inside Graphics Concept
**Toni González** and **Thuan Nguyen** of Art Tested Graphics

Cover Concept
**José Pacheco**

Fact Check
**Jim Johnston**

Production Assistants
**Mark Corbin**
**Dale Drumheller**
**Debbie Dunkle**
**David Hudzik**

Some product photos provided by the manufacturers

# contents

# CONTENTS

# Foreword

By Roger Nichols

When a recording artist I produced heard a great song on the radio he would turn to me and say, "I was going to write that song!" After reading this book my reaction was, "I was going to write this book!" Well, I am glad Bob beat me to it because it looks like he did a much better job than I could have.

What places this book head and shoulders above the rest is the attention to useful detail. Instead of some hyperbole, the reader can actually put these methods to good use. The descriptions of how to perform a task are augmented with the reason that you should perform the task. Not just how downward compressors work, but when and why you would want to use them. Science is meaningless without art.

How do I tell if the digital signal is 16 bit or 24 bit? What does noise shaping do? Should I mix at 96 kHz? How do you make something 3 dB louder when it is already lighting up the over lights? Should I mix to analog or digital? How do I set up my speakers for mixing surround? Which weighs more, a pound of gold or a pound of feathers? These are some of the questions that Bob answers in a clear and concise style.

Bob enters each mastering session with his eyes wide open. Each project is unique, and each mastering session will require a unique approach to bring out the very best results. Bob's musical background helps him select the proper tools for the job. Knowing that a string quartet record does not require the same approach as the Back Street Boys record is half the battle.

Every day clients ask for louder and louder CDs when they come to a mastering session. It is very hard to find Hi-Fidelity CDs these days. Now that you can do your own recording to a digital workstation, buy your own multi-band compressors and burn your own CDs, who needs mastering? My answer is that if you record your own projects at home, you need mastering more than the producer who works with the top engineers in the top studios. The key is outside reference. No, I don't mean that your neighbor came over and said, "Hey, that sounds really great!" I mean reference to other projects, and reference to other engineers who have worked on great sounding CDs.

Bob does an excellent job of dispelling the myth that the louder you make your CD, the louder it will be on the radio. Read this part more than once. Once the reality sinks in, then maybe we will have more viable candidates for a Best Engineered CD Grammy, instead of having to choose a CD for the Least Offensive Engineering award.

The professional mastering engineer works on material from all corners of the music business. This is the last stop before the CD hits the radio and the record stores. The smartest thing any mixing engineer can do is leave the final loudness tools to the loudness professional.

Limiters and compressors should be treated just like firearms. There should be guides for the proper use and classes you must take before you can own one. That class is here in this book. After you read this "audio firearms manual" you will have a much better understanding of the mastering process. You will know when and how to use these tools yourself and when to leave it to the professional. Treat every compressor/limiter as a loaded weapon, and don't point it at anyone unless you intend to use it. It's the **LAW!**

I get e-mail quite often from independent artists who are recording their music at home and want to know what gear to buy to help them mix before they send it to me for mastering. I tell them that the first piece of equipment they should buy is Bob Katz's **Mastering Audio, The Art and the Science**.

Roger Nichols

Miami, August 2002

# Introduction

**On Language**

Sex is good! And being sexy can be fun! I feel that language should be sexy, too, and our centuries-old male-centric language must be rather wearying to the women in our society. It's time to put some vitality back into our syntax. Thus, you will find that in one chapter of this book, the Mastering Engineer may be a female, and in another, male! Vive la différence!

**What Is Mastering?**

Mastering is the last creative step in the audio production process, the bridge between mixing and replication (or distribution). It is the last opportunity to enhance sound or repair problems within an acoustically-designed room—under an audio microscope. Mastering engineers lend an objective, experienced ear; we are familiar with what can go wrong technically and esthetically. Sometimes all we do is—nothing! The simple act of approval means the mix is ready for pressing. Other times we may help the producer work on the problem song they just couldn't get right in the mix, or add the final touch that makes a record sound finished and playable on a wide variety of systems.

**The Approach of this Book**

In the mastering studio we use the scientific tools of audio engineering to illuminate musical art. So this book constantly integrates art with science. Students may inquire why they have to learn all this technical stuff: While you can't get very far without talent, you can get much further with both talent and technical knowledge. In the days of analog processing and analog tape, a practicing audio engineer could get along without a rounded technical education, but digital audio requires far more technical knowledge as well as computer skill.

Technological change has a large effect on society. As late as 1995 no one had an inkling that the public would soon decisively turn from consuming physical music and video product to downloading digital files. In April 2003 the iTunes™ Music Store opened and in less than three years had sold its one billionth song! The full effect of this paradigm shift on our industry has yet to be evaluated.

But regardless of the form in which product is sold, our job as mastering engineers remains: we help music to be presented in the best possible way. This requires old-fashioned craftsmanship and attention to detail, values which never go out of style. The artistic and technical information provided herein will always be precious to students of the art of audio.

**Attention Gearheads**

This book is designed to help you learn to make informed decisions on your own; how audio equipment works, and what happens when you turn the knobs. Just about every day I get a letter like this one from engineers asking me to bless their particular list of equipment...

> Dear Bob, I always master with a Sis-boom-bah brand compressor and equalizer, then I follow it off with a touch of a Franifras enhancer. On the next pass I use a Caramba tool to maximize the sound and then Whosizats dither before going to CD. Please tell me what you think of my choices?
>
> Sincerely, Gearhead.

I usually reply, politely,

> Dear Gearhead, your equipment list sounds pretty extensive, but much more important is how you use it. For example, some of the gear you describe would be entirely inappropriate for some kinds of music....

Mastering is not about "processing" per se, some masters leave the studio with no mastering processing at all. Perhaps the most essential piece of information we can learn is this aphorism written by master engineer Glenn Meadows.

> *"There is no magic silver bullet. There is no one magic anything that will be 'best' in all situations. The ability of the operator to determine what it is that needs to be done and pick the best combination of tools is more important than what tools are used."*
> — Glenn Meadows

Glenn's statement also applies to the amount or setting of each knob or control within our equipment. There is no magic threshold, or EQ setting, or ratio, or preset that will turn ordinary sound into magic. Sonic magic comes from the hard work we put into using our tools (musical magic can only come from the music itself). The truth is that in a typical mastering session, each tool makes only an incremental improvement, and the final result comes from the synergistic totality of the tools working together. In these days of mass-gear-marketing by competitive manufacturers, there is too much emphasis on the glitz, fashion and style of the gear rather than its sound quality and principles of operation. While this book is definitely for gearheads (in the sense that it has lots of glitzy pictures and descriptions of gear designed to produce good sound), serious engineers who want to improve their techniques will also find out how their devices function. Audio principles never go out of style, but gear models fade away.

I have carefully chosen the equipment we discuss as suitable for high-quality audio mastering. Regardless, there are far more models available than I have personally experienced, and their exclusion does not mean they cannot perform a good job.

The theories and background covered here are what I consider to be the minimum necessary to become a competent audio engineer in this digital age. Complex mathematics is not required. There are plenty of good foundational basics for beginners, yet the most experienced digital design engineer will find useful information. I include practical examples at every stage; but if the going gets difficult at any point, simply move to the next section. As you grow in experience, when you revisit those sections you may have skipped, everything will seem less abstract. I try to define special terms the first time we meet them; terms can also be found in a greatly expanded glossary (Appendix 13) and in the index.

### A Taste of This Book

This edition is replete with new, up-to-date information. Just like a well-sequenced record album, these chapters tell a story in a logical, flowing order.

**MYTH:**
*Digital Audio requires less technical skill to use than analog.*

## Part I: Preparation

The mastering engineer has tremendous power, and with that power comes great responsibility. Although it is possible to turn an ordinary mix into a glorious-sounding production, it is also possible to ruin a piece of delicate music by applying the wrong approach.

**Chapter 1: No Mastering Engineer is an Island**, begins by forecasting the 2010 decade, where disc players are becoming obsolete and every room in the home plays music via a central server. We then outline the steps taken in producing a record album, our mastering philosophy, workflow and quality control procedures. We explain what distinguishes a mastering-quality DAW from an ordinary one, how to prepare master files for digital delivery, CD Text, ISRC, error checking, logging, backups and much more.

**Chapter 2: Connecting It All Together**, presents a block diagram of the "ideal" mastering studio, describes mastering consoles, how analog and digital routers function, and metering and analysis gear used in mastering.

**Chapter 3: An Earientation Session**, suggests extended ear training exercises to help develop our listening skills and the vocabulary of audio mastering.

**Chapter 4** makes **Wordlengths and Dither** easy to understand, explains one of digital audio's technical mysteries with lots of examples and a section on dithering at high sample rates.

**Chapter 5: Decibels: Not for Dummies**, is a must-read for working audio professionals. We clarify formerly slippery audio definitions, describe how level meters work, including the latest advances in oversampling (true-peak) meters. We cover proper peak level practices, the world of floating point, the myths of normalization, the deleterious effects of clipping, and how to interface and gain stage analog and digital equipment chains.

**Chapter 6: Monitor Quality**, defines the elements of a high-resolution monitor system, and demonstrates the need for accurate monitoring and proper room acoustics. We describe the methods we use to help a master translate over a wide variety of systems, from Hi-Fis to iPods and the car.

## Part II: Mastering Techniques

**Chapter 7** dignifies the art and science of **Putting The Album Together**. We describe the process of sequencing and spacing from both the esthetic and technical views, PQ coding, editing, and leveling an album.

**Chapter 8: Equalization**, differentiates EQ practice for mastering from that used in tracking or mixing, presents three techniques of *focusing* an equalizer, the relationship of equalization to perceived loudness, and the problems with bass boosts. We focus on the pros and cons of linear phase equalization, and the sonic differences among different equalizer types, and introduce dynamic equalizers.

Next comes our *dynamics trilogy*: **How To Manipulate Dynamic Range For Fun And Profit, Chapters 9-11. Chapter 9** defines the four types of dynamics modification and introduces the ways in which we manually manipulate dynamics.

**Chapter 10** presents *downward* compressors and limiters, the features and use of several analog and digital models, including multiband. We cover linking, sidechains, crest factor, detector characteristics and more.

**Chapter 11: The Lost Processes** describes two often-forgotten but very important approaches to dynamics processing: parallel compression and upward expansion. We show how to use parallel compression *transparently* or for *attitude* and *punch*; and we propose that *upward expansion*, properly employed, can be as powerful a tool as downward compression.

**Chapter 12: Noise Reduction**, explains the various types of noise, methods of reducing it, and most importantly, suggests when we should and should not reduce noise.

**Chapter 13: Top Processors**, looks at various models of high quality analog and digital processors suitable for mastering.

## Part III: Advanced Theory and Practice

**Chapter 14: How To Make Better Recordings in the 21st Century** is a must-read for all concerned audio citizens. We identify *the true culprit for the accelerated digital loudness race*, and methods broadcasters and media are employing to undo the consequences of the race and restore sound quality. We introduce the concepts of integrated monitoring and metering and calibrated monitor level control.

**Chapter 15: Monitor Setup**. No more strings! We use a laser pointer, angle chart and time delay measurement to accurately position the loudspeakers. Then we calibrate the levels and frequency response of the 5.1 system with or without bass management, and provide example acoustical measurements and techniques. We also explain why a single (mono) subwoofer is not enough, and always a compromise.

**Chapter 16: Additional Mastering Techniques**, takes us from required basics to advanced techniques such as methods of *fattening* sound, obtaining *punch*, using stems, and M-S mastering. We then explain how we make a "louder" master when this is what the client requires. Please study Chapters 5, 9-11 and 14 first.

**Chapter 17: Analog And Digital Signal Processing**, describes some of the analytical tools we use to look at sound. We investigate the non-linear relationship between equipment measurements and auditory perception, subjective and objective differences between analog and digital recording, and analog simulators. We also supply the definitive answer to the controversial question: "Which sounds better, analog or digital summing?"

**Chapter 18: How To Achieve Depth and Dimension in Recording, Mixing and Mastering**, teaches us to think beyond the panpot. We learn about the three types of auditory masking, use of delays, early reflections and reverberation.

**Chapter 19: Surround Mastering**, is a roundtable discussion with four guest engineers, who provide their unique insights into this expanding, technically complex field. Among other discoveries, we learn the practical problems of dialnorm and folddown coefficients, and that directional unmasking is a double-edged sword. Thanks to Dave Glasser, Bob Ludwig, Rich Tozzoli, and Jonathan Wyner.

**Chapter 20: High Sample Rates, Is This Where It's At?** tells us why a wide-bandwidth system is important even though our ears are only good to 20 kHz (on a good day!). We examine data that may explain why there are audible differences between sample rates.

**Chapter 21: Jitter: Separating the Myths From the Mysteries**, is a direct and definitive explanation of the topic. With clear diagrams that show how to interconnect the clocks in a digital studio.

**Chapter 22: Technical Tips and Tricks**—how to maintain a digital audio studio, including timecode and wordclock management, interface debugging, and schematics for adapters to convert impedances .

## Part IV: In Conclusion

**Chapter 23: Education, Education, Education**, is where we get to preach what we practice! We start with suggestions on how to approach your first mastering job when your tools may not yet be the best in town. We then discuss how to preserve our valuable hearing. And finally a recommendation on how to roughly gauge the level of a master when calibrated monitoring is not available.

## Part V: Appendices

- **Radio Ready, The Truth**, largely written by guest authors Bob Orban and Frank Foti, with contribution by Tardon Feathered, shows how radio processing severely affects our sound and debunks for all time the myth that super-hot recordings sound better over the FM, satellite or digital radio.
- Up-to-date review of existing **audio file formats.**
- **Premastering for Vinyl**
- **How to prepare** tapes and files for mastering
- **Conversion charts** of decibels, and equalization bandwidth.
- **I Feel The Need For Speed**, a comparison of playback, transfer and network speeds including the fastest economical server technology.
- **Recommended Reading** and test CDs.
- **Glossary**

Visit the **digido.com** website for an online companion to this book, an **Honor Roll** of dynamic Pop CDs to emulate when mixing or mastering, and **links** to websites with resources mentioned herein.

Now that we've had a taste, let's begin **Mastering Audio!**

# PART I: PREPARATION

"Getting ready is half the job."

—Anon

Chapter 1

# No Mastering Engineer Is An Island

## Introduction: The Decline and Rebirth of Hi-Fi

This chapter is about the mastering engineer's approach to audio. Audio engineers are aware that the consumer listening experience is constantly evolving. By the 1990s consumer listening reached a sophisticated level with an appreciation of high fidelity sound but around 1995, when a mass audience discovered that computers could play music, and since they were spending a large proportion of their day in front of computers, it became easy for them to adjust to the idea of music of lesser fidelity coming out of a puny computer speaker. The novelty seemed to be mesmerizing, and though mp3 sound improved as internet bandwidth increased, it generally was played over execrable computer sound systems.

**2000-2010**. Consumers are slowly moving away from CD, and have largely bypassed the higher resolution SACD and DVD-A, making this the decade of the portable digital music player, digital downloads and low-fidelity computer playback. Within the home, the listening experience is deteriorating, but outside, a glimmer of hope can be found in the iPod™, which takes portable listening to a high ergonomic and sound quality. Car sound advances as well, in many cases beyond the quality found in the typical home.

**2010-?**. The next revolution in domestic listening is born from a new appliance—the high fidelity music server.[1] This device lets consumers play downloads, internet radio, and music files on hi-fi and home theater systems (independent of the computer). Each room can have its own wireless connection to the central server, and users play their music collection on demand in any room with a simple remote control. By the end of the second decade, home servers become the dominant playback mechanism, bringing good sound back to the home. **Physical product** becomes less important to anyone except collectors of fine music, though we hope that it will remain an important segment of the music market and the standard-bearer of high fidelity.

Mastering engineers know how to best present audio to the public through a variety of evolving formats. Although we are in a period that emphasizes the lowest common denominator, the internet and disposable singles, we must strive for high quality mastering. We seek to preserve sound quality, reducing it only when the delivery format requires it. And so we urge program producers to create and archive high quality masters....

## I. Technical Steps from Concept to Finished Disc

### Compact Disc Production

The compact disc is the most successful high fidelity music medium in history, with a long life beginning in 1980, and is vital to the music industry. We begin by reviewing the place of audio mastering in the overall scheme of producing and manufacturing a CD. A finished compact disc can

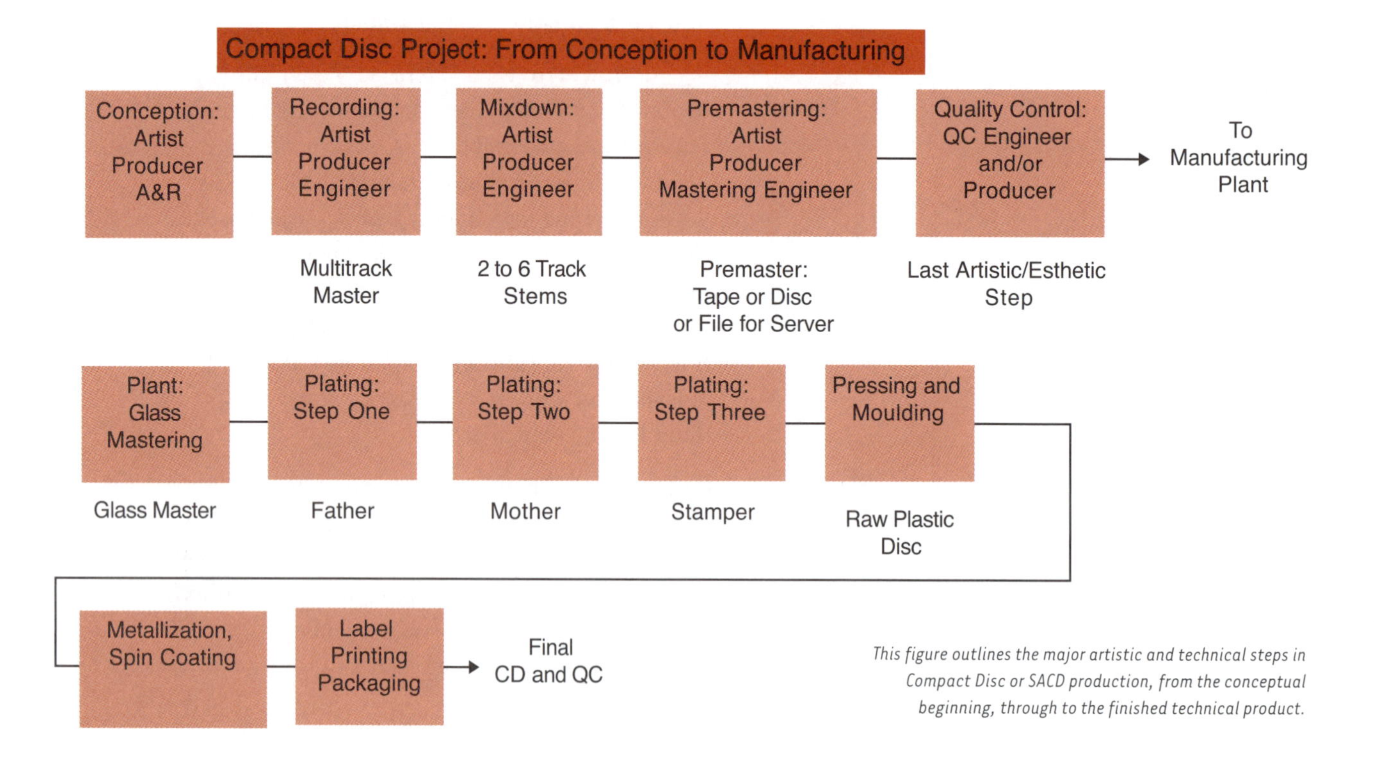


*This figure outlines the major artistic and technical steps in Compact Disc or SACD production, from the conceptual beginning, through to the finished technical product.*

usually be premastered by a specialist in a single day, including the esthetic and technical portion of the premastering.

Prior to the premastering stage, the song composition and album conception can gestate for years, with contributions from the artist, the producer, the record company A&R or all three. Arrangements are written, musicians hired, and the artists go into the recording studio or on location for the recording to multitrack. The prime medium for multitrack recording is the computer hard disk, with analog tape used for some high-end projects. A primary or secondary hard disk may reside on an internet-based server which all participants have access to—containing a database, musical arrangements, performance tracks, mixes, and later, masters.

### Mixing

After the tracking is complete, the producer, artist and mixing engineer produce the mix of each song or section of the work. If mixing to stereo, the mix goes to two tracks, but even then it may be divided into several 2-track *stems* in order to produce TV, performance tracks or instrumentals. The mastering engineer may need stems to tweak the interrelationship between leads and rhythm. If mixing for surround, the mix may go to six or more tracks; and if divided in stems, the surround stems could take up 18 or more tracks!

### Editing and Premastering

The next step, *editing and sequencing (putting the album in order)*, may be carried out at either the mixing studio or the mastering house. Usually sequencing is performed by the mastering engineer, who receives individual songs or segments and puts them in order with spaces or overlaps. Sequencing is followed by *premastering*, which is the official name of our profession, to distinguish it from the technical mastering that takes place at the plant (though everyone calls us *mastering engineers* for short and we use that terminology throughout this book). Premastering can include the artistic and technical tasks of **dynamics processing**, **leveling**, **equalization**, **noise reduction**, and even some mixing, described in detail in later chapters. Naturally, the output medium of *premastering* is officially called the *premaster*, but we usually label it *master*. After the premaster is approved by the producer it may be sent to the plant physically on optical disk or by secure electronic delivery as a DDP file.

### At the Plant

At the plant, the premaster is used to create the *glass master.* But technically speaking, glass is not the master. The glass is the carrier for an emulsion which is applied to its surface. At many plants, glass mastering is performed in a class 10 clean room (or better) by engineers wearing white "space suits" (affectionately known as monkey suits). But an alternative is to house their LBRs (laser beam recorders) in a self-contained clean room that can be loaded up in the morning by one suited individual and run all day without intervention, just observation through a Plexiglas window. The LBR is a multi-million dollar machine that takes the digital information for the master, encodes it[2] to the proper format and then sends an encoded laser

*Attention to detail: It only takes an extra minute to get it done right, but it takes hours to fix a mistake.*

beam onto the light sensitive emulsion. The on-off laser pattern generates a series of pits and lands after the emulsion is developed. The coated glass disc is then moved to another clean room, where the emulsion is sputtered with a fine nickel alloy in a process called *metallization*. Next, the disc is put in a vat where an electrical charge is applied, allowing the surface to be plated, in a process called *electroforming*. After plating, the metal piece, now called the *father*, is peeled off the glass. The glass surface is then cleaned and can be recoated and reused for a new "glass master".

The *father* is the inverse of the final CD (pits are lands and vice versa). For small runs, the father can be used directly as a stamper. But for any significant quantity, the father is electroformed to create a *mother* (which is the inverse of the father) from which many stampers can be produced. Each stamper goes into a press, where a clear polycarbonate disc is inserted and molded. Afterwards, the disc is metallized with an aluminum reflective layer (gold can be used in specialty pressings) and coated with a protective lacquer. Finally, a silk-screened or offset label is applied to the top of the disc, which is then packaged with booklets into the CD boxes by automated machinery. Every element must be carefully inspected for defects—and the CD itself must meet the proper tests for pit depth and spacing (e.g. jitter and RF output tests). It's an exacting process but....

### DVDs are even more complex

Although the physical DVD or DVD-A is very similar to a CD, it requires a much greater magnitude of precision. Because a one-sided DVD has about 7 times the information density of CD, it costs more to produce in the creative, technical and manufacturing stages. The creative department has to generate the graphics and menu copy and the plan for interactivity, well in advance of the authoring stage; furthermore, all of these elements might be in constant flux until the reference audio track has been firmly edited and mastered. Finally, at the pressing plant, DVDs require much more stringent QC standards than CDs, especially because of the delicate bonding process for a multi-layer DVD. The Blu-Ray and HD-DVD successors to the DVD have even more information density and so much complicated capability that all of the above problems are exponentially greater.

## II. Mastering Philosophy

In mastering, meticulousness and attention to detail are the norm, not the exception. We've always been called upon to keep careful track of a project from the time it arrives until it becomes the final product. Days, weeks, or perhaps years later, if revisions are requested, the client has a reasonable chance of ascertaining which processes were used by consulting with the mastering engineer. At RCA Records, through the 80s, analog tape box labels included "dash numbers" (e.g. -1, -2, -3) for each copy generation, and a card catalog carefully logged the tape's status and which one was the correct *master* to use for LP or cassette duplication. When masters were sent for disc cutting, the cutting engineer inserted a written log indicating the Pultec or other equalizer settings they used, left/right channel gains, and so on.

Today, the situation is far more complicated than simply looking in a tape box for cutting information and marking the box with the generation number. Audio-only projects may arrive in multiple forms, digital tapes, DAW sessions on hard disk, optical disks or analog tapes. Projects may be two channel or multichannel surround; they may arrive as full mixdowns, partial mixdowns (stems) or combinations. The definition of what is the **Master** becomes even more vague, since multimedia projects may be finished at the audio mastering studio, or have authoring added at some studio down the road. Metadata (see Chapter 14) including watermarking may be added during a later authoring stage, further complicating the situation.

But one thing has not changed: it is the responsibility of the mastering engineer to ensure that the audio quality which leaves the mastering studio is the same quality that will be represented on the final medium. We must know the project's destination when it leaves our office, and familiarize the producer with what is necessary to preserve the audio quality.[3]

## III. Mastering Tools and Procedures

### Picking the Right DAW

Mastering engineers depend on their DAWs, which must be powerful, reliable, and have very high data (audio) integrity. Sonic Solutions pioneered the mastering Digital Audio Workstation (DAW) and introduced the Source-to-Destination editing model and interactive crossfade editor. To this day, only a few other workstations or software programs have been qualified or dedicated to mastering: Audiocube, Pyramix, SADiE, Sequoia, Wavelab and to a lesser extent, Waveburner and Peak.

Here are some specific advantages of dedicated mastering DAWs:

- High data integrity; the architecture is designed to be bit-transparent except when performing a calculation.
- High resolution (internal calculation precision).
- Multiple playlists (EDLs) can be open and data can be copied and pasted between them.
- Powerful crossfade editor can make "impossible" edits, cutting editing time by over 50% compared to "standard" workstations.
- The project supports multiple sample rates, which switch automatically according to which EDL is opened.
- There is no such thing as a *16-bit* or *24-bit* session. Clips with different wordlengths and file formats (e.g. WAV and AIFF), interleaved and split files can co-exist in the same EDL.
- Conversion is not needed when importing any audio format, except for sample rate conversion.
- Integrated dithering. Separate types of dither and dither wordlengths are available on each output (dither is explained in Chapter 4).

*The last 10% of the job takes 90% of the time.*

*"One challenge in mastering is that half the clients complain if their mixes come back sounding radically different, and the other half complain if their mixes DON'T come back sounding radically different."*

— Jay Frigoletto

· The waveform shows the effect of a fade and optionally a level shift.
· Nearly instant waveform display as each new file is brought into the project.
· Fades are calculated on the fly, a crossfade can be any length.

Other criteria appropriate to picking a DAW include software and hardware reliability and economic stability of the company, a well-maintained support structure, the presence of a user group, and ability to submit feedback. All these measures raise the short-term purchase price of a good workstation, but greatly lower the long-term cost of ownership.

### Mastering Esthetically

Mastering engineers were originally "the men in white coats" who cut the records and were not allowed to have any creativity.[4] Historically, mastering was part of the transfer process, in translating the mix tape so that the record sounded like the mix. Today it is still our goal to present the mix in the best possible way. We should not attempt mixing and mastering at the same time because it defocuses from the goal of mastering.

Our current role falls into one of three basic categories, as outlined by Trevor Sadler:

> 1) The mix is done. The mastering engineer may make modest EQ correction, but nothing that would change the mix. Usually the engineers that bring in these types of mixes are very good and have achieved what they want to hear in the mix process.
>
> 2) The mix is done, but the producer wants something to happen...
>
> 3) The mix ends up not being what the engineer or artist intended, and they are now looking for major changes in mastering.[5]

Every piece of music is unique, and requires an approach that is sympathetic to the needs of that music, the producer and artist. A good mastering engineer is familiar with and comfortable with many styles of music. She knows how acoustic and electric instruments and vocals sound, and she's familiar with the different styles of music recording and mixing that have evolved. In addition, an experienced mastering engineer knows how to take a raw recording destined for duplication, determine what may be lacking, and help make it sound like a polished record. She should also know when to leave a project alone.

By sympathetically listening to, and working with the producer, the engineer can produce a master that is a good combination of her ideas and the producer's intentions, better-sounding than if the engineer had simply mastered on her own. The best masters are produced when both the producer and the engineer consult with each other and are willing to experiment and listen to new ideas. As this book progresses, we'll cover mastering approaches, from the purist to the extreme.

### Mastering Without a Producer Present

To help make the mastering process smoother, we suggest a listen/evaluation prior to the session and a discussion of the producer's goals. Then we can master quite comfortably even without the producer or artist physically present. After the mastering session, we'll send a reference disc for their approval before cutting the master. Usually by that time, we are enough in sync needing perhaps minor changes, so there is no need to produce a second reference.

### The Mastering Workflow

The mastering engineer's workflow comprises critical auditioning, editing, cleanup, leveling, processing, PQ coding, and output to the final medium. When the raw material (usually files of mixes) arrives at the mastering studio, it must be loaded into a mastering DAW, then auditioned and edited/sequenced prior to mastering. Many engineers work with DAWs in very much the same way we worked before there were any DAWs.[6] First, we take the source for each tune (e.g. Analog tape, DAT, CDR, Masterlink, AIFF or WAV file), and process one song at a time. If that source is digital and if analog processing is to be used, we send it to a high-quality DAC, pass it through one or more analog audio processors and possibly control the level, EQ, or fade via a customized analog mastering console. The signal is then passed to a high quality ADC, optionally through various digital processors, dithered to 16 bits for the compact disc, and then recorded into the DAW. We then move on to the next song, resetting processors until the best sound is achieved for that song. And so on as illustrated in **Variation 1 of the figure on the next page.**

In this variation, as all leveling, fading, processing and equalization has already been accomplished, the DAW is only used for assembling and spacing, which is a very efficient approach. When we reach the end of the tune, if it requires a fadeout and we missed it, instead of reloading the entire song, we may back up before the fadeout, do a simple punch-in on the workstation, perform the fade, and then a matched edit. What this means is that, if the client orders any revisions, the engineer must repatch the entire chain, reset the processors, make any processing changes, and re-record/ replace the old destination file or a portion of the destination with a new one. For example, in the figure titled **matched edits** on page 27, mastered track 3 is an assembly that begins with a piece of version 2, followed by version 1, a retouched section of version 1A,[7] and finishes with version 1.

*"The Customer is Always Right."*
— Dale Carnegie

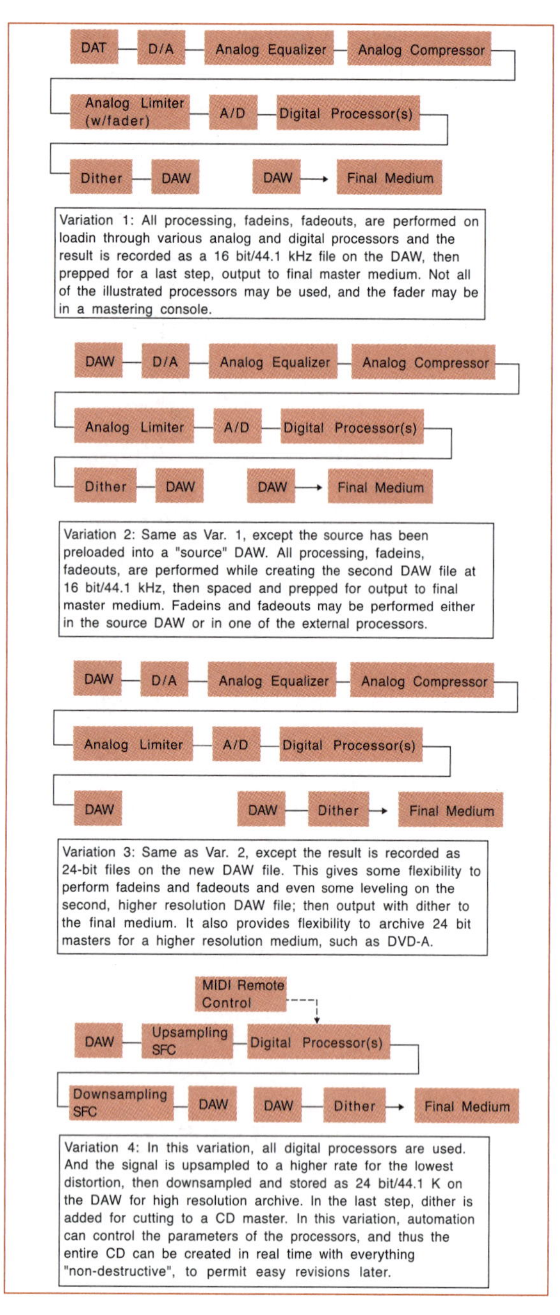

DAT
D/A
Analog Equalizer
Analog Compressor
Analog Limiter (w/fader)
A/D
Digital Processor(s)
Dither
DAW
DAW
Final Medium
Variation 1: All processing, fadeins, fadeouts, are performed on loadin through various analog and digital processors and the result is recorded as a 16 bit/44.1 kHz file on the DAW, then prepped for a last step, output to final master medium. Not all of the illustrated processors may be used, and the fader may be in a mastering console.
DAW
D/A
Analog Equalizer
Analog Compressor
Analog Limiter
A/D
Digital Processor(s)
Dither
DAW
DAW
Final Medium
Variation 2: Same as Var. 1, except the source has been preloaded into a "source" DAW. All processing, fadeins, fadeouts, are performed while creating the second DAW file at 16 bit/44.1 kHz, then spaced and prepped for output to final master medium. Fadeins and fadeouts may be performed either in the source DAW or in one of the external processors.
DAW
D/A
Analog Equalizer
Analog Compressor
Analog Limiter
A/D
Digital Processor(s)
DAW
DAW
Dither
Final Medium
Variation 3: Same as Var. 2, except the result is recorded as 24-bit files on the new DAW file. This gives some flexibility to perform fadeins and fadeouts and even some leveling on the second, higher resolution DAW file; then output with dither to the final medium. It also provides flexibility to archive 24 bit masters for a higher resolution medium, such as DVD-A.
MIDI Remote Control
DAW
Upsampling SFC
Digital Processor(s)
Downsampling SFC
DAW
DAW
Dither
Final Medium
Variation 4: In this variation, all digital processors are used. And the signal is upsampled to a higher rate for the lowest distortion, then downsampled and stored as 24 bit/44.1 K on the DAW for high resolution archive. In the last step, dither is added for cutting to a CD master. In this variation, automation can control the parameters of the processors, and thus the entire CD can be created in real time with everything "non-destructive", to permit easy revisions later.

*Infinite Variations on a Mastering Theme. Four examples of approaches to audio mastering.*

Often there is no real-time load in, since sources may arrive as high resolution or high sample rate computer files on DVD-Rs, CD-ROMs or hard disks, and can be loaded at high speed directly into the workstation (**Var. 2**). The engineer then has to listen to each tune to get the feel of the whole album and check for noises or other problems that may need fixing. She may begin by putting the material in order, cleaning up heads and tails, perform fadeouts and spacing, and then proceed as in Var. 1, except she uses the workstation as the new "source" as well as destination. In **Var. 3**, the mastering engineer waits until the final output to perform dithering, which gives some flexibility to perform fadeins and fadeouts on the final DAW file and perhaps some leveling. With the increasing number of high sample rate projects, another variation is to use two workstations, one to play back high sample rate material, the other to record a sample-rate-converted and dithered output for CD prep. Yet another variation is to use **upsampling** followed by **downsampling** because even if the source material is ready for CD at 44.1 kHz, digital audio processing at a higher rate sounds better (Chapters 17 and 20). The engineer may reproduce the source material at the lower rate, feed an upsampling **sample rate converter** (SRC), then perhaps convert using a high-resolution DAC for analog processing, record the material into a high sample rate ADC for optional further digital processing, then finally downsample and dither (if destined for compact disc, which must be 16-bit). First, the material is stored at 24-bit/44.1 kHz on the new DAW file, then it is dithered in the last step to the 16-bit master file or medium. The source DAW in this figure may not be the same

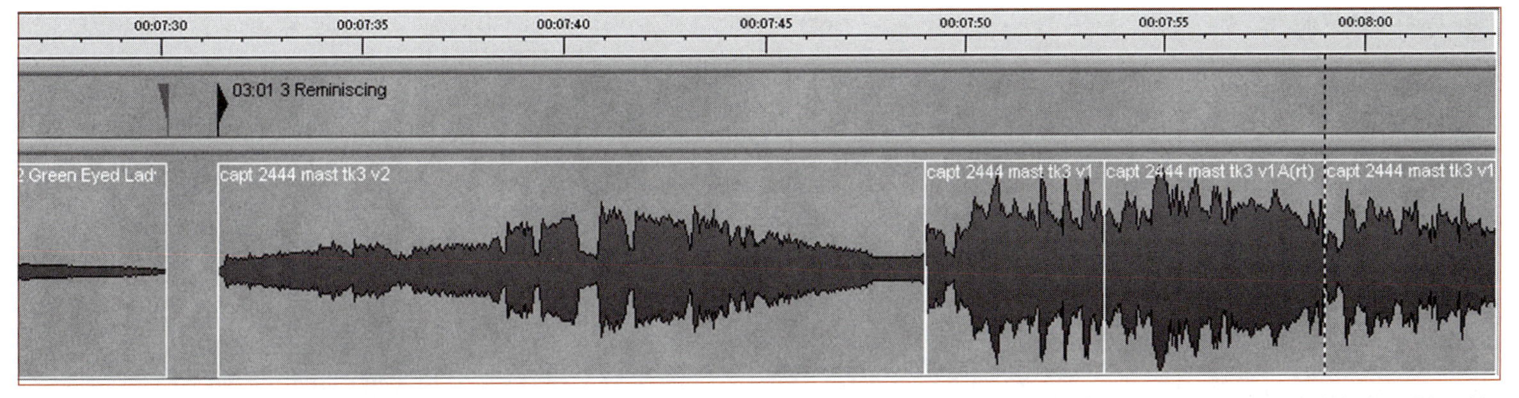


*Matched Edits in an EDL combine multiple revisions and save production time.*

machine as the destination DAW, for if the source material is not at the destination rate (44.1 kHz for CD), then two DAWs are necessary, since most DAWs can only work at a single rate.[8] To save time, I've used as many as three DAWs in a single session, each to capture a different resolution and format.

Material that arrives at multiple sample rates (different songs at different rates) is particularly problematic, often necessitating sample rate conversion of files to a common rate before the mastering can get started.

### Tune by Tune or Fully-Automated?

All of the above descriptions have one thing in common: they follow a tune by tune approach to mastering, resetting the processors, before moving on to the next. Although engineers have been making excellent albums using this method for years, an increasing number of digital audio processors are remote-controllable via MIDI (Var. 4), which permits them to be automated and thus completely integrated with the workflow. Most engineers already use some sort of automation in their work, since advanced workstations provide automated equalization, leveling, fades, dynamics, and even plug-ins. If a later revision is requested, the mastering engineer can recall the previous EDL (edit decision list) and instantly make changes in the amounts or timing of the workstation's internal processing. **The MIDI technique extends this ability to the outboard equipment.** For me this is a revolution—finally I can work with the album in the making in a comfortable, fluid, non-linear manner. I work with a song until it is cooked, save the parameters in the memories of the processors, and then move on to the next song, postponing the capture to the final file until the entire album has been programmed. I then return to near the end of the previous song and play the sequence with the MIDI automation following along, nondestructively. This makes it easy to integrate two dissimilar songs, e.g. if one ends big and the other begins small. We can revisit any portion of the album while the context and the details are in development. For example, we may make a great climax to the album, then recheck the first song in that context and reprocess it if necessary without having to reload or recapture. Full automation also permits special effects—for example, as we approached the climax on one tune, upon the entrance of a big vocal chorus, I created MIDI-automated changes in an outboard processor that gradually increased the spaciousness and depth, producing a gigantic sound in the final chords. After we're satisfied that the album sounds good, we then return to the beginning and cut the master in real time with full automation, except of course for the analog processors, the majority of which must be manually adjusted.

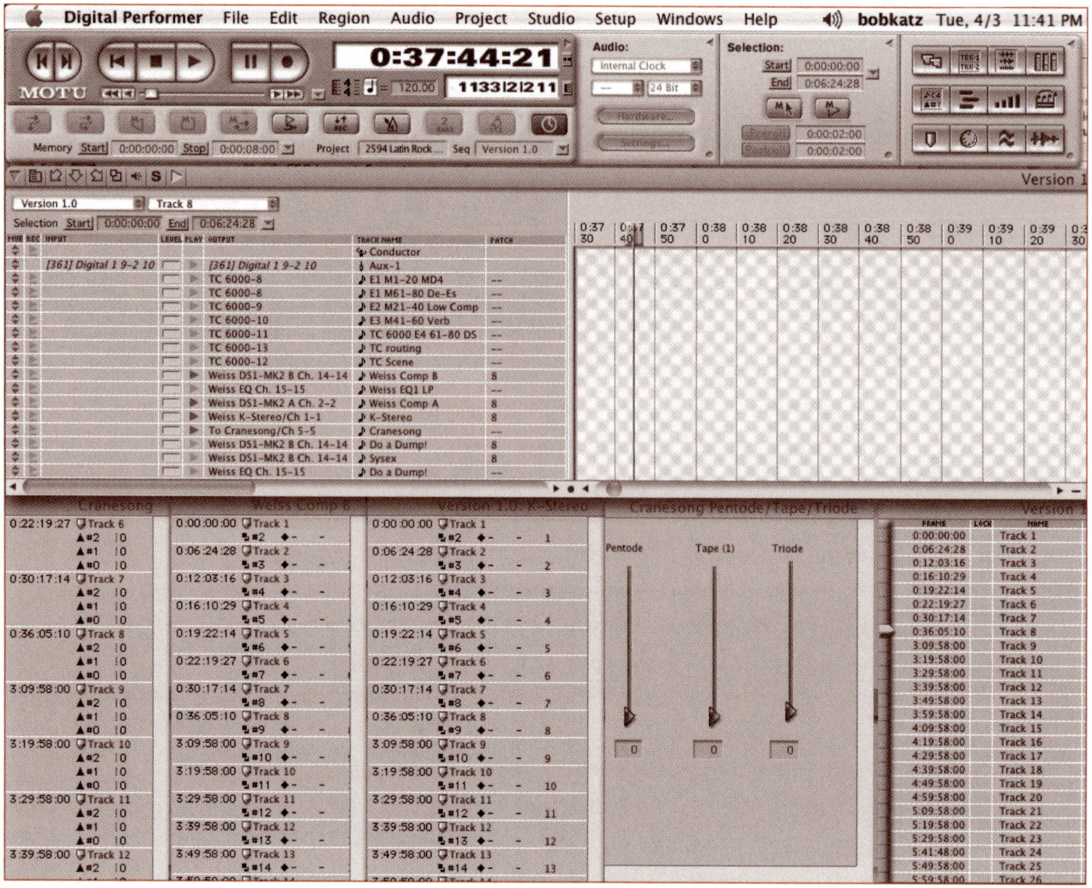

The biggest advantage of full-automation is the ease of revision, especially if we have a critical clientele. Processing is applied in a non-destructive, non-cumulative manner so anything can be undone without going down another generation or forcing a reload. Another advantage of this method is that the raw sources can be immediately compared with the master and demonstrated to the client. This method also has sonic advantages. Since the MIDI automation accomplishes the required processing changes from song to song (except for the analog gear), we can assign any gain changes involved to the most transparent-sounding processor. The biggest disadvantage of this method though is the learning curve required to run a MIDI sequencer and

*Digital Performer (MIDI sequencer) in action, controlling program changes in outboard gear*

Elton John's version of *Your Song* will have a different ISRC code from any cover of the same song. As long as a song is not edited or remastered, it retains the same ISRC code even if the rights are sold to another record label; ISRCs are for tracking, and may not reflect the current owner of the title.

### CD Text[9]

CD Text is a facility that displays song title, artist, album name, and even lyrics on specially-equipped CD players, most often those found in cars. The term is very misleading for some clients who pop their CD reference into iTunes, and expect to see the titles. In fact, iTunes and Windows Media Player get their title and artist data via a database on the Internet and most computer and CD players do not read CD Text from the disc.[10] If the client requires CD Text encoding for those few players that can read it, we require he proofread the CD text on his own player prior to our cutting the master.

Although modern mastering DAWs can cut masters with CD text, the plant should be notified in advance if CD text is incorporated into a project as by default they turn off this facility to prevent spurious characters from being encoded on the pressing.

## V. Media Preparation, Verification, Backups

### Mastering Output Formats for CD-DA

While we can accept sources in nearly any format for mastering, only two formats are suitable for replicating CD-Digital Audio (CD-DA) discs: **CD-DA** (on CDR media), or **DDP** files (Disc Description Protocol image file, sometimes abbreviated DDPi) which can be placed on data disc or uploaded to the

**Label** Emusica Records, LLC  **Date** 1/7/2006
**Title** El Rey Del Bajo  **Source** ◉ Analog ○ Digital
**Artist** Bobby Valentin  **Format** CD Audio
**Cat #** 87731300002
**UPC/EAN** 0877313000023

Mastered by: **Bob Katz**. This master was created on SADiE ver. 5. All levels, fades, & PQ times are client-approved. Please do not alter in any way. Please refer all technical questions to Digital Domain at (407) 831-0233.

```
-----------------------------------------------------------------------------
Tno Ind Start Time        PQ Duration       CD Time          PQ Time
-----------------------------------------------------------------------------
  1 01 Hay Craneo                                  USDBB0600010
      0 00:02:55.67       00:00:02.00       00:00:00.00      00:02:54.67
      1 00:02:57.67       00:03:54.04       00:00:02.00      00:02:56.67
-----------------------------------------------------------------------------
  2 02 Arenas Del Desierto                         USDBB0600011
      0 00:06:50.66       00:00:04.18       00:03:56.04      00:06:50.71
      1 00:06:55.26       00:04:02.17       00:04:00.22      00:06:55.14
-----------------------------------------------------------------------------
  3 03 Guaraguao                                   USDBB0600012
      0 00:10:57.26       00:00:03.37       00:08:02.39      00:10:57.31
      1 00:11:01.05       00:03:29.44       00:08:06.01      00:11:00.68
-----------------------------------------------------------------------------
  4 04 Mi Ritmo Es Bueno                           USDBB0600013
      0 00:14:30.32       00:00:03.17       00:11:35.45      00:14:30.37
      1 00:14:33.66       00:05:42.72       00:11:38.62      00:14:33.54
-----------------------------------------------------------------------------
  5 05 Codazos                                     USDBB0600014
      0 00:20:16.46       00:00:03.26       00:17:21.59      00:20:16.51
      1 00:20:20.14       00:03:58.43       00:17:25.10      00:20:20.02
-----------------------------------------------------------------------------
  6 06 Cuando Te Vea                               USDBB0600015
      0 00:24:18.40       00:00:03.25       00:21:23.53      00:24:18.45
      1 00:24:22.07       00:05:25.39       00:21:27.03      00:24:21.70
-----------------------------------------------------------------------------
  7 07 Esperame En El Cielo                        USDBB0600016
      0 00:29:47.29       00:00:02.15       00:26:52.42      00:29:47.34
      1 00:29:49.61       00:04:44.39       00:26:54.57      00:29:49.49
-----------------------------------------------------------------------------
  8 08 La Vibora                                   USDBB0600017
      0 00:34:34.08       00:00:04.49       00:31:39.21      00:34:34.13
      1 00:34:38.74       00:04:19.14       00:31:43.70      00:34:38.62
-----------------------------------------------------------------------------
  9 09 Aqui No Me Quedo                            USDBB0600018
      0 00:38:57.71       00:00:03.19       00:36:03.09      00:38:58.01
      1 00:39:01.32       00:03:49.64       00:36:06.28      00:39:01.20
-----------------------------------------------------------------------------
 10 10 Coco Seco                                   USDBB0600019
      0 00:42:51.04       00:00:03.25       00:39:56.17      00:42:51.09
      1 00:42:54.46       00:03:36.39       00:39:59.42      00:42:54.34
-----------------------------------------------------------------------------
 AA   1 00:46:28.73       T00:43:36.06      00:43:36.06      00:46:30.73
-----------------------------------------------------------------------------
```

*PQ Listing showing engineer's comments, track times, ISRC codes and other information*

> *"Most computer and CD players do not read CD Text from the disc."*

plant via FTP.[11] DDP is the more reliable format (nearly foolproof), because it is file-based and files can be compared against a master file for 100% data accuracy. Less reliable is CD-DA, first because it has less robust error correction than a data disc, second because there is no easy way to verify that a CD-DA copy matches the source, and third because clients can play CDR masters (though they shouldn't) and mishandle them or leave fingerprints. Our procedure is to "seal" the CD-DA master in a plastic bag marked "to be opened only by plant personnel".

There are usually 4 files in a DDP fileset, the most critical being the audio image file **image.DAT.** Auxiliary files **ddpid**, **ddpms**, and **sd** carry the PQ codes, version and ancillary information. Since DDP is file-based, it is possible to include a checksum along with the file set. One popular type of checksum is called an MD5. The procedure is to calculate an MD5 as soon as the master is made, save that information in a small text file, and pass that file along with the master wherever it is copied. A verification program then compares the MD5 listing against its calculation of the copy, and if they match, then the data must be identical.[12] It also reports if any of the files are missing.

The md5 file contains four checksums and the names of the files they apply to:

```
SADiE v5.6.0 pre-rel 2
14150E4A23C243909B1D211754FF6E84BED7F  sd
4BC7D638FF80DB8046F5CD806C3B16E2  ddpid
57D7A40C2AB2441A6BDC90B35CCFAD7B  ddpms
EFA51D04A1A863C93D32C65850140715  image.dat
```

Here is a listing of a DDP fileset ready to be sent for replication.[13] As a convenience to the plant, I include four additional files—the MD5 listing, the verification program provided by SADiE to enable the plant to quickly verify their copy is accurate, a read me file with instructions, and a PQ listing.

| Name | Date Modified |
|---|---|
| ddp_checksum.md5 | Mar 22, 2007, 1:39 PM |
| ddpid | Mar 22, 2007, 1:31 PM |
| ddpms | Mar 22, 2007, 1:31 PM |
| image.DAT | Mar 22, 2007, 1:38 PM |
| md5_validate.exe | Feb 11, 2007, 4:32 PM |
| pqprint.txt | Mar 22, 2007, 1:31 PM |
| read me Phil Ware Trio CD master.rtf | Mar 22, 2007, 1:54 PM |
| sd | Mar 22, 2007, 1:31 PM |

The master cannot be edited; it must be recorded in one continuous pass, under the control of a computer. Some recording engineers attempt to deliver "masters" on CDRs recorded on a stand-alone CD audio recorder, but this is usually unsatisfactory because of the inaccuracy of the track points, inability to put separate track end marks (which creates extra-long track times), and E32 errors introduced every time the recorder stops its laser (breaking the "one continuous pass" rule). More masters are now being sent to the factory via high-speed internet lines. I recommend zipping the file collection before uploading, as this guarantees the entire group will be received intact.

### Listening Quality Control

At the end of the project, quality control testing may be performed by a separate engineer, who must have musical/artistic ears, technical prowess, and also a lot of common sense: since the project has already been auditioned by the mastering engineer and producer, presumably all the noises were accepted, perhaps even welcomed as "part of the music." But if a single unacceptable tic or noise is discovered anywhere in a master, the entire master has to be remade and listened to/evaluated. There is no shortcut. During the QC listen, which is done with headphones, he may hear noises or problems that were not picked up in the mastering studio. For example, small dropouts on one channel are often masked in loudspeakers. He notes the timecode of each offending noise, and if it is suspicious, compares it with the original source to see if the problem was introduced during mastering, then brings questionable noises to the attention of the mastering engineer. Bob Ludwig suggests that headphone listening becomes essential when the number of channels multiplies. Potentially embarrassing noises or glitches hidden in the surround channel when auditioned on loudspeakers become quite audible when that channel is isolated in a pair of headphones. To complicate the situation even further, one consumer may be listening to all channels using surround headphones while others will be hearing stereo reductions (folddowns). Clearly, a surround master requires much greater attention to detail, and costly time to evaluate, requiring several hours to QC an hour program, including any extra passes necessary to check a fold down!

QC includes ensuring that the songs are in the proper place, based on client-supplied lists of the song lengths, lyric sheets, etc, and that the correct master goes out for duplication. We must be especially wary of misidentifying individual CDs of a multiple CD set.

The responsibility for QC must be accepted by someone. There is usually no press proof except when very large quantities are involved. Pressing plants used to have rooms where masters were critically listened to, prior to glass mastering. But now, when the master arrives at the replication plant in physical or electronic form, it will likely be copied at high speed to the factory's central server, so no one at all listens during glass mastering. The day has come when the home consumer is the first person to audition the product![14]

### Objective Media Verification/Error check

Digital media are susceptible to data dropouts which cause errors, which is why all the digital audio storage formats utilize error correction algorithms.[15] Uncorrected errors result in glitches, clicks, or mutes. Normally, when playing a digital tape or disc, we do not know how much error correction is going on. It can sound great, but the tape or disc could be near to dying! If the error correction system is

> *"If a single unacceptable tic or noise is discovered anywhere in a master, the entire full-length master has to be remade and listened to/evaluated. There is no shortcut."*

*Myth: An audio load-back/null test shows the integrity of a CD Master.*[18]

working very hard, the next time that disc is played, a speck of dust or head alignment problem, or simply wear and tear, could cause a signal dropout during playback. Our job is to look behind the scenes using specialized measurement tools. Listening alone is like having a doctor look at the patient without doing any clinical tests. So media verification is a thorough internal examination.

There is also the issue of *error concealment*, which is the last defense mechanism in digital playback. If there is an uncorrectable error of fairly short duration, instead of muting, the playback machine interpolates between the audio level before and after the error. Short bursts of error concealment can be virtually inaudible or smooth the sound pleasantly, but professionals never use a master medium that is so degraded. So we verify our media with evaluators like the standalone Clover System or Plextools Pro, which runs on a PC and requires a Plextor brand writer.

We call correctable errors **soft errors**, and uncorrectable errors which would mute or interpolate on playback, **hard errors**. Soft errors on CD-DA are correctable in two layers of defense: **C1** and **C2**. Hard errors are known as **CU**. If the C1 correction fails, C2 takes over, and if that fails, a CU error occurs and the player goes into error concealment. If error concealment fails, then the player will mute for a period of time. For replication masters, we do not allow any C2 or CU errors, though we may accept a reference CD which has an occasional C2. And for further comfort, we count the average number of C1 errors per second, also known as **BLER**. CD plants permit BLER values up to 200, but our in-house standard is no more than 50, which allows a CDR to age and deteriorate with a margin of safety. Another conservative mastering house accepts BLER up to 100. As we can see from the Plextools error report on the next page, this master has a remarkable BLER level of 0.2. A very good CD can have a BLER lower than 10, yet CDs will still play with BLERs of 1000 or even above—which illustrates how robust the error correction system is for CD-DA. CD ROMs use an additional layer of error correction since data cannot be interpolated like audio.[16]

When the CD-DA master reaches the plant, it will be error-tested again and copied to hard disk. However, there is no error testing during the copy to hard disk, which I have already noted is the fundamental difference in reliability between CD-DA and DDP.[17] The plant simply assumes that a CD with a low error rate will transfer dependably in a reader that's in good condition. Millions of CDs have been successfully mastered from CD-DA masters with no problems. But an error test is no substitute for listening back to the master, since when cutting the master, there are many electrical components in the chain after the audition point. You're always one generation behind; if you listen while making a copy, you've only proofed the generation in front of it.

By the way, every computer CD-DA copy is effectively an original as **soft errors do not accumulate when copying**; the C1 and C2 errors from the source are corrected and the new disc will have its own error count. If, however, the source disc has a hard (uncorrectable) error, a mute or a glitch **will** turn up on a computer copy. This is a subtle difference from CD audio copies performed via SPDIF or AES/EBU, where hard errors on the source which are successfully interpolated will not produce an audible glitch.

### Mastering for the Internet

No special preparation is needed for masters destined for digital download, though level and sound translation should be considered as we will discuss in other chapters. Original masters are high resolution files which should then be downsampled or data-rate-reduced to the final format. There is an advantage to coding an mp3 from a 24-bit file as it will have subtly better sound quality than from a 16-bit source.

### Backups/Archives

After a project is finished, we wait until the client has approved the master (usually by listening to a copy of the master). We then may wipe the audio material from our main hard disks, but not before making an in-house backup of the audio and EDLs on hard disk. This backup is mostly in case a revision is requested. Some record labels require full backups of the masters, often on DVD-R, hard disk, or some other acceptable archive format.

The critical difference between a **backup** and an **archive** is that an archive is made to a medium which is supposed to last a long time (30 years or more). However, will the equipment still be around to read them ten years later? Will the DAW software be able to reproduce automation moves? Perhaps the idea of full data recovery is truly an illusion.

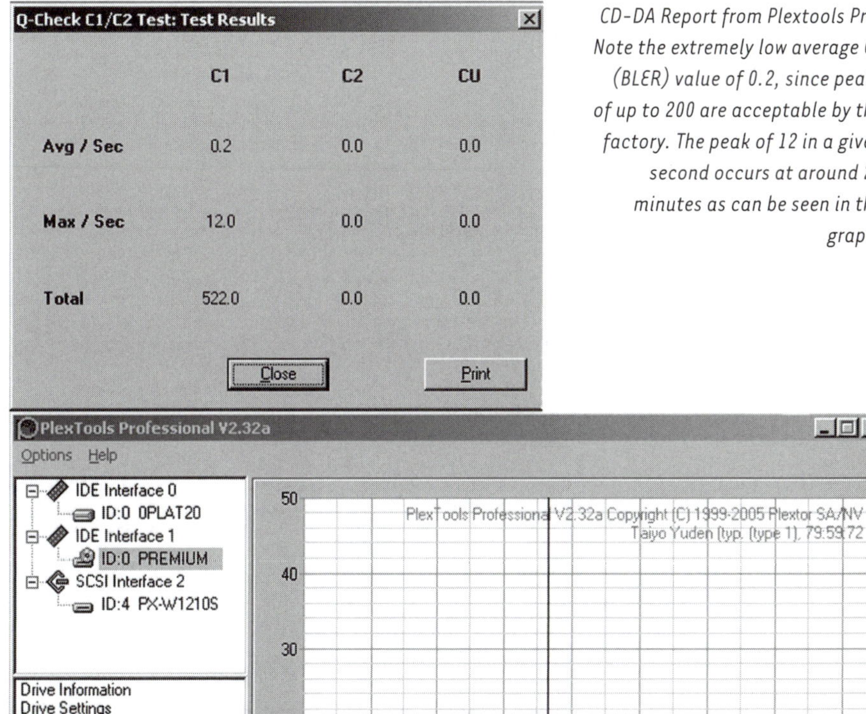


*CD-DA Report from Plextools Pro. Note the extremely low average C1 (BLER) value of 0.2, since peaks of up to 200 are acceptable by the factory. The peak of 12 in a given second occurs at around 25 minutes as can be seen in the graph.*

1 The Slim Devices Squeezebox, introduced November 2003.

2 The encoding includes EFM modulation and error correction information. The exact nature of compact disc and DVD encoding is beyond the scope of this book. Further references can be found in Appendix 10.

3 These are real horror stories from the trenches: One mastering engineer reported a situation where another house added the CD-ROM portion to an extended CD, and somehow in the process, changed the audio quality of the audio portion. Never assume that everything will be fine when the master goes out the door, even to the extent of (on critical projects) approving and testing the final product. It is possible to do **null tests** or **bit for bit comparisons** which compare the original audio master against the final pressing, assuring that the audio data had not been altered after it left the mastering house.

In another situation, a less than reputable plant copied **all** incoming masters using a consumer-based program which automatically shortens tracks to the end marks, and then puts 2-second silent gaps between all the tracks. So the final pressing of a beautifully-engineered live concert sounded like it was edited with an axe!

4 Emerick, Geoff & Massey, Howard (2006) *Here, There and Everywhere: My Life Recording the Music of the Beatles*.

5 Trevor Sadler, via email and webboard, 2005.

6 Well, this is true for CD mastering. But if you go back to the age of LP, the engineer was forced to cut an entire record in one continuous pass. If he stopped, he created a locked groove, which you call yesterday's E32 error. A sophisticated LP cutting engineer would note settings for each tune and manually change her processors during the banding between each track. Equalizers were developed with A and B settings, allowing her to press one switch during the intertrack gap, and then leisurely preset the opposite equalizer for the next track. Primitive, but roughly equivalent to the fully-automated process described here.

7 **Retouched** refers to Cedar Retouch, a specialized noise reduction or restoration processor (see Chapter 12).

8 One workstation (Sequoia) can work at two rates simultaneously by opening two instances, each using different sound cards.

9 Thanks to mastering engineer Jim Rusby for being the original resource on CD Text for the first edition of this book.

10 In iTunes, when a CD is inserted, the Gracenote database is accessed by default (this preference can be turned off). The database counts the number of tracks on the disc, their lengths and spacing to determine the name of the album, which has caused a few embarrassments over the years (such as when a Christian singer discovered her album came up as hip hop). Currently the best solution is to add or subtract even a frame of space anywhere on the disc until it looks unique to Gracenote. Any iTunes user can submit information on a CD to Gracenote, so it is advisable for the record company to beat the consumers to the punch by uploading data before the CD is released. Content owners can apply to become Gracenote partners, which gives them upload priority and allows them to lock consumers out of potentially disturbing the listing of a title. Thanks to Glenn Meadows for this tip.

11 The PCM-1630 and DDP on Exabyte tape are obsolete. The master medium used for DVDs is either a DVD-R disc, or DLT (Digital Linear Tape).

12 There is a very slight mathematical possibility that two files can have the same MD5. MD5 is extremely reliable and I recommend using it when moving large groups of files from server to server and before erasing the source! On the PC, a shareware program called **Advanced Checksum Verifier** is easy and convenient to use.

13 There are two versions of the DDP protocol, version 1.0 and 2.0. Version 2 is probably safe with most plants, but check in advance.

14 Thanks to Mike Collins, *One To One* Magazine, November 2001, and to various discussions on the Mastering Webboard, for inspiring this section.

15 Hard disks generally do not require error-correction, since their error rates are extremely small.

16 Ironically, there is no correlation between a disc's error rate and its readability in a given player, especially delicate players like in the car. The measured RF signal level is a better measure of a disc's readability; unfortunately, Plextools does not measure RF level.

17 Unless the replication plant adds a custom error-reading interface to the CD-ROM reader which is used to rip the CD-DA to hard disk.

18 On the contrary, all the null test proves is that there were no uncorrectable errors, but it is not a measure of media reliability or error-count. The null test is post the error correction. You could be one bit away from failure and not know it. The next time an error-prone disc plays, there could be an interpolation or a mute if the error count is high. Thanks to Glenn Meadows for pointing out these facts.

{ *Backups? We don't need no ba&*9 u.* }

Chapter 2

# Connecting It All Together

## I. Introduction

Unlike mixing studios, mastering studios may change their configuration several times during a busy day. One morning, the mastering engineer might spend an hour auditioning clients' mixes, deciding whether they are ready for mastering or may need some mix revisions; these could be at varying sample rates, wordlengths and formats. Later that morning he might do a two-hour revision to an existing project which requires a complete repatch of the room setup. And from afternoon through the evening, a full-album mastering with yet another setup. With the increasing emphasis on singles, it's not inconceivable to require 4 or 5 separate setups throughout the day. Thus, the mastering engineer is highly dependent on the efficiency of his workflow routines, and the power of the equipment within those routines to produce consistent and repeatable results.

## II. The Modern Mastering Studio

Let's take a close look at the connections in the "ideal" audio mastering studio (pictured next page, signal flows from left to right). Notice that three major devices are under software control: the digital router, the analog router and the monitor control. Digital sources for routing could be DAW, CD, DAT, DVD, hard disk recorders, or the outputs of processors such as compressors, limiters, equalizers, reverbs, noise reduction units. Analog sources might be analog tape LP, the analog outputs of DVD players that do not have digital outputs, or analog processors, typically equalizers or compressors. A digital router is a dedicated digital patchbay which can distribute one source to many destinations and handle several specific impedances—which standard plug and jack patchbays cannot. A relay-switched analog router is a patching system whose advantage over a standard analog patchbay is software control and instant reset. Other critical components of the system are the ADC used to transfer analog tapes to hard disk, the DAC/ADC combination for inserting analog processors in the mastering chain, and of course, the monitor DAC, which should be jitter-immune and of the highest-quality (covered in Chapter 21). Since we are expected to make consistent quality judgments, auditioning all digital sources and pressed media through this single converter guarantees that the monitor gain and sound quality will be exact. Unfortunately, this principle of consistent monitoring has been subverted by the advent of copy-protected media such as DVD-A and SACD, whose players do not have digital outputs.[1] These make it impossible to proof the final product through the same converters that were used during the mastering. In this case the output levels of these and other analog sources have to be calibrated to that of the digital components to ensure consistent monitoring level.

## Digital Routing

The digital router connects digital audio sources and destinations in any combination. A single source

***The ideal mastering studio*** *(block diagram) has all these elements. Routers, manual switches or patchbays interconnect the gear, either manually or under software control.*

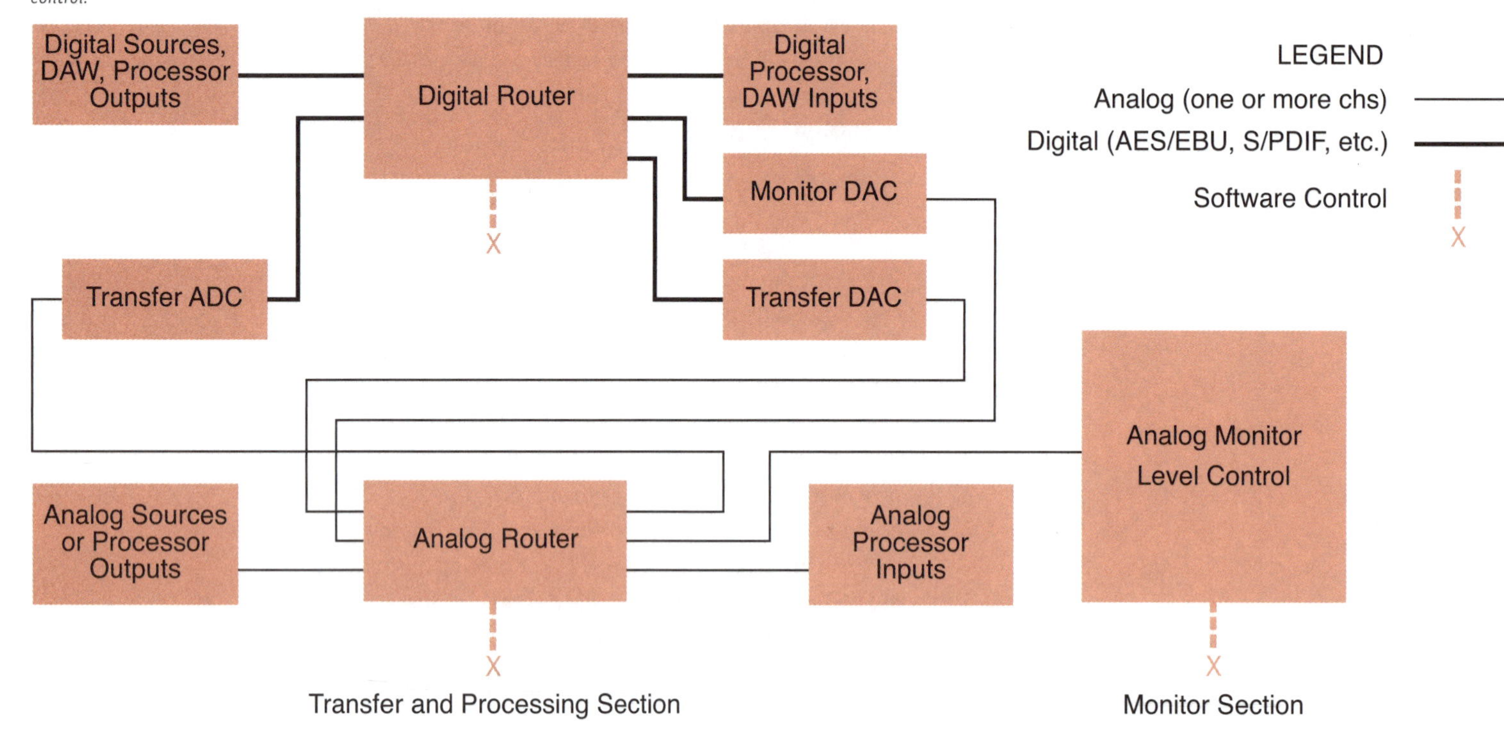

can be distributed to multiple destinations, but multiple sources cannot be routed to a single destination without a digital mixer. A 16x16 router is sufficient for a small mastering studio, but a medium size studio processing stereo and a small amount of multichannel work would require 32x32, and the largest studios need up to 128x128. Since AES/EBU carries 2 channels, a 128x128 router actually switches 256 channels in pairs. There are two basic types of digital routers:

**Asynchronous routers**, which do not require clocking, such as the Z-Systems Detangler or Crookwood models, can switch virtually any type of signal, support multiple sample rates and different synchronizations in the same chassis, and can be configured for different voltage and impedance standards. Thus a single unit could be used to route AES/EBU or S/PDIF (2 channels per connection), Dolby E (8 channels per connector), Dolby Digital (6 or more channels), MADI (multiple channels) or encoded formats such as mp3, even distribute wordclock and some can handle composite video—all at the same time! Routing setups can be saved via external software or hardware controllers.

**Synchronous routers**, which do require clocking (either internal or external), are limited to handling only one type of signal (usually AES/EBU or MADI). All signals have to be the same sample rate and framed to the identical clock. Nonetheless, there are tasks that synchronous routers can perform which asynchronous routers cannot. They can switch signals in the middle of a mastering session without losing clock connections between devices. They can mix multiple sources to a single destination; good models can mix bit-transparently.[2] In some, such as those available from TC Electronic, Metric Halo and RME, individual channels can be split, reversed or shuffled. The RME routing software application comes bundled with their computer interface cards. Note however that a synchronous router cannot deal with a foreign source—one that is not locked to the system clock, or at a different sample rate from the session; nor can it handle two sample rates at once. For the monitoring of foreign sources, a DAC with multiple input selection—fundamentally an asynchronous switcher—is required. Manley and Dangerous Music manufacture economical asynchronous digital switches that can be used to expand a DAC's inputs or do basic routing. The ideal studio will have an asynchronous router, perhaps supplemented by a synchronous one.

### Bit Transparency

Digital routers for mastering should be *bit-transparent*, in other words, the output should be identical to the source. Asynchronous routers are likely bit-transparent by design, because they do not reclock or process the signal format and consist of software-controlled hardware switches. But synchronous routers often contain DSP, dithering or mixing capability, and so must be used with caution if a bit-transparent output is required. Avoid routers containing sample rate converters, as these change every incoming signal in order to lock to a common clock, a process which adds distortion and is not bit-transparent. Though this type of router is not acceptable in a mastering environment, you will find them at broadcast and post houses that require myriads of foreign sources to be accessed at any time and locked to a common (*house*) clock.

### Analog Routing

Pure analog routers are passive switches with no active electronics in the signal path (though of course they do require power to make their connections). Relay-based analog routers with sealed gold contacts are much more reliable than a standard plug and jack patchbay. If analog tape is the source, and optimal processing is in the analog domain prior to digital conversion, this router should be flexible enough to insert analog processors between the tape and the ADC.[3] As shown in the block diagram, digital, analog, or hybrid chains can be created in any desired sequence.

### Mastering Console

Some studios use custom mastering consoles, which provide for source and insert selection, routing and, in some cases, processing modules. For example, Sound Performance Lab (SPL) manufactures an analog mastering console, the MMC1, which integrates semi-automated analog routing and monitoring with high-quality discrete electronics. But underneath the exterior of any mastering console is a series of interconnected modules. The implementation, whether via a series of switches, relays, pushbuttons, a patchbay or an analog router, is a matter of ergonomics and personal preference. Besides these custom-built designs, a mastering "console" can also consist of a set of high-quality analog processors and a few faders in a rack or laid into a desk surface. Every mastering studio has its own variation on these themes but all incorporate something like the ideal studio block diagram depicted on page 38.

*Custom routing software which can be connected to a Z-Systems digital router*

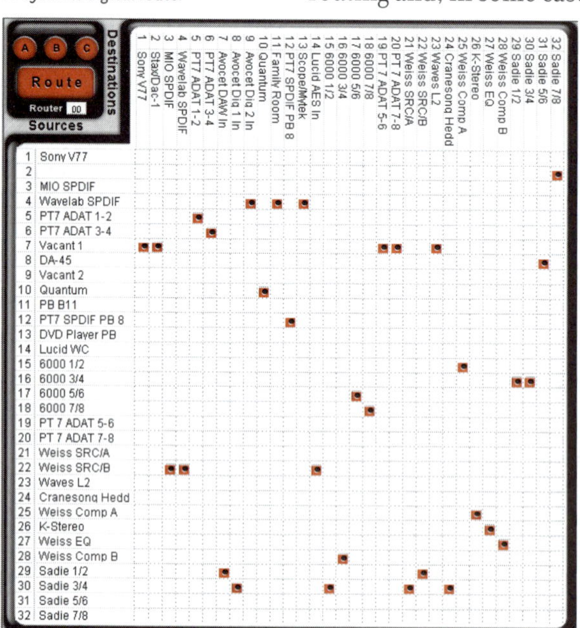


### Monitor Level: DSP or Analog Controlled?

What factors decide the choice between an analog or DSP-based monitor level control? My advice is that a DSP-based monitor controller is cost-effective and the better models sound very good, provided that you are working in the digital domain, not regularly changing sample rates, and have some other means of monitoring foreign signals (such as CD players). Most DSP-based monitor controllers prefer being locked to a consistent master clock and mute awkwardly when rates are changed. The main advantage to an analog monitor attenuator is it permits instantly monitoring a variety of analog and digital sources (as most mastering engineers need).[4] Cranesong, Crookwood, Grace, and SPL make high quality monitor controllers with stepped, calibrated controls changing level in the analog domain.

## III. Software Control

### Basic Software Control

In a basic studio, the digital router is the only device under software control. Complex chains of analog or digital components can be created at the push of a button, since the analog processors are connected to the converters, and the converters are digitally connected to the router. For example, the figure at bottom left (previous page) shows a software-control system I developed that communicates with a Z-Systems 32x32 router.

This matrix display is not very ergonomic, but it does show every digital connection to the router. So the computer screen also displays a chain showing how the signal passes from box to box, as seen below. Analog devices can be inserted into a chain using a discrete DAC/ADC, or a processor with integrated converters like the TC 6000 or Cranesong HEDD. The digital input to the transfer DAC and the digital output from the transfer ADC show up in the display.

Individual setups can be saved and named for each project. In the chain pictured, a stereo loop begins at the DAW: SADiE output 3/4 feeds a TC 6000 engine, followed by a Weiss dynamics processor, K-Stereo, Weiss EQ, a second Weiss Dynamics processor, another TC engine, and returns to SADiE inputs 1/2 for capture at high sample rate, out to a Weiss sample rate converter, to Wavelab for capturing at a lower sample rate, and finally to the Cranesong Avocet for monitoring. The Avocet (pictured top right) can monitor six sources (three digital and three analog), compare the sound before and after mastering or compare the master with a pressed CD.

## Fully-Integrated Software Control

In our studio A, analog processors are manually patched with passive switches, but in studio B, using a Crookwood Igloo system (pictured next page), the entire studio block diagram can be reconfigured at the touch of a button. The Crookwood is an integrated analog and digital transfer console, including an asynchronous digital router, analog router and analog monitor controller. Signal chains can be analog, digital or a mixture. Source and destination selection and the digital and analog processing chains are displayed on an LCD screen on the selector remote. The selector panel at the left shows a processing chain, the top line is the digital portion and the bottom line the analog, with each number representing a different device. An analog chain attenuation of -2.75 dB has also been applied. The selection process is very ergonomic, hence switching between load-in, mixing or mastering modes is possible with very little downtime.

The monitor panel at the right is fully integrated, all the sources can be monitored independently of the processing chain. Like the Avocet, the Crookwood monitor selector can compare any digital or analog source at matched gain, recalling the levels and trims for each source. In this picture, the monitor gain is -17.0 dB with a 0.0 dB trim offset.

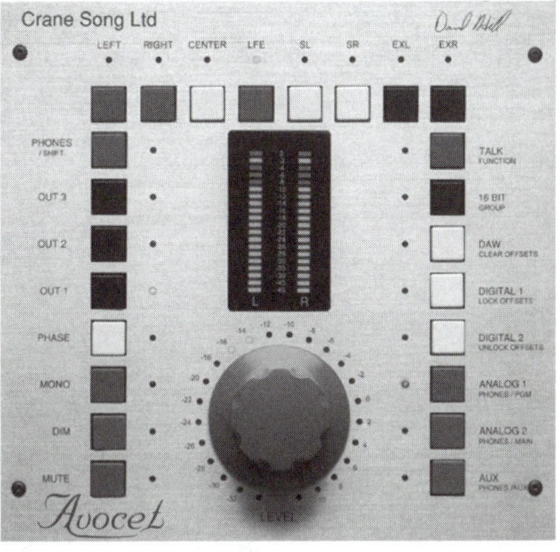


*Cranesong Avocet Monitor Controller*

*Custom routing software displays the signal flow "chain"*

Sadie 3/4 | 6000 1/2 | Weiss Comp A | K-Stereo | Weiss EQ | Weiss Comp B | 6000 3/4 | Sadie 1/2 | Weiss SRC/B | Wavelab SPDIF | Avocet In

Crookwood Selector Panel and Monitor Panel

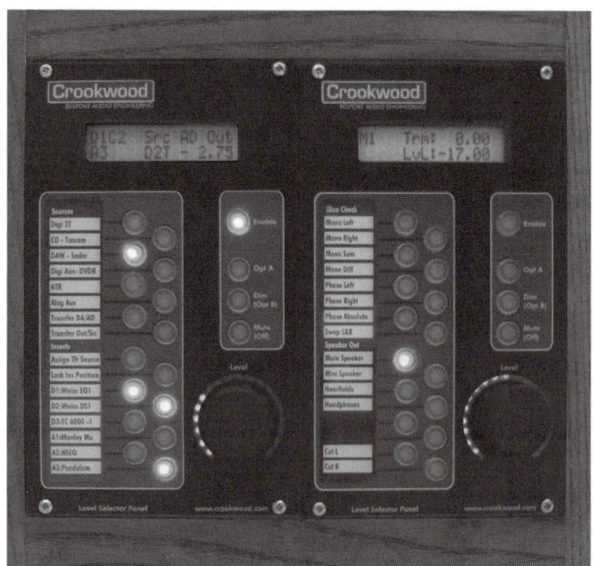


## IV. Other Equipment

Other useful equipment connected to the digital router include:

- an AES/EBU-ADAT-Tascam format converter which allows the connection of client-provided equipment (largely digital recorders).

- The bitscope, pictured at right, which serves to double-check the bit-integrity of the source, confirm that there are no extra bits due to hardware or software bugs, and that the dither appears to be functional (covered in Chapter 4).

- The Mytek digital meter (at left) which measures peak audio level, and the Dorrough meter (at right) which displays average and peak on the same scale (covered in Chapters 5 and 14).

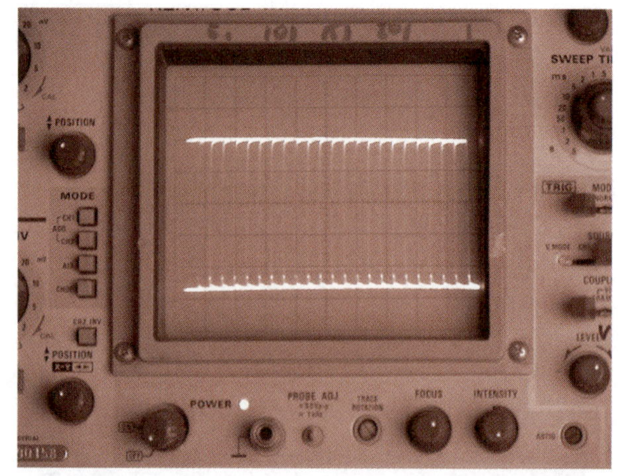


24 Bits active on the bitscope

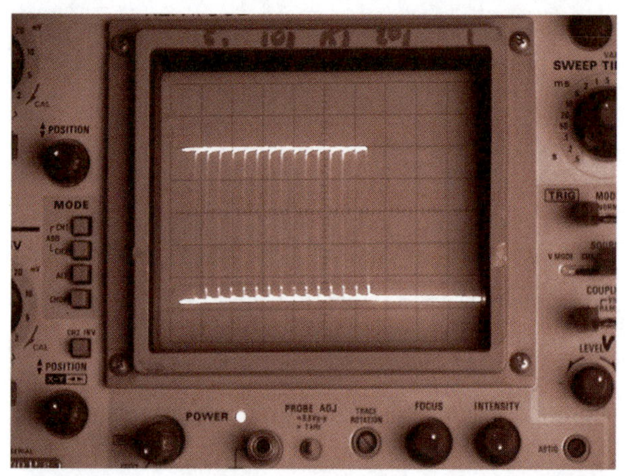


16 Bits active on the bitscope, truncated after the LSB

Dorrough Loudness Meter. It's extremely useful due to the dual-scales, but it does not correlate with loudness any better than a standard VU meter.

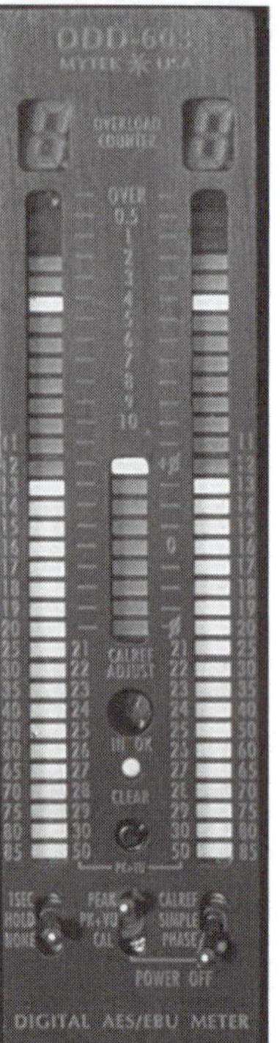


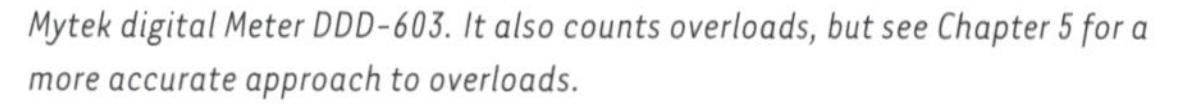

Mytek digital Meter DDD-603. It also counts overloads, but see Chapter 5 for a more accurate approach to overloads.

## V. Block Diagram and Wire Numbers

When constructing a mastering studio, it is best to begin with a detailed block diagram, inserting wire numbers from a separate wire number list. Pictured here is an example, with wire numbers in parentheses.

Finally, proper grounding and wire layout techniques are crucial for minimizing signal interference.[5] Analog gear used for mastering can be customized for minimalist signal path, removing transformers and superfluous active stages, and as a digital mastering studio may contain only a few analog processors, it is easy to put all the analog gear physically together in its own rack, at a distance from other equipment to avoid clock interference.[6]

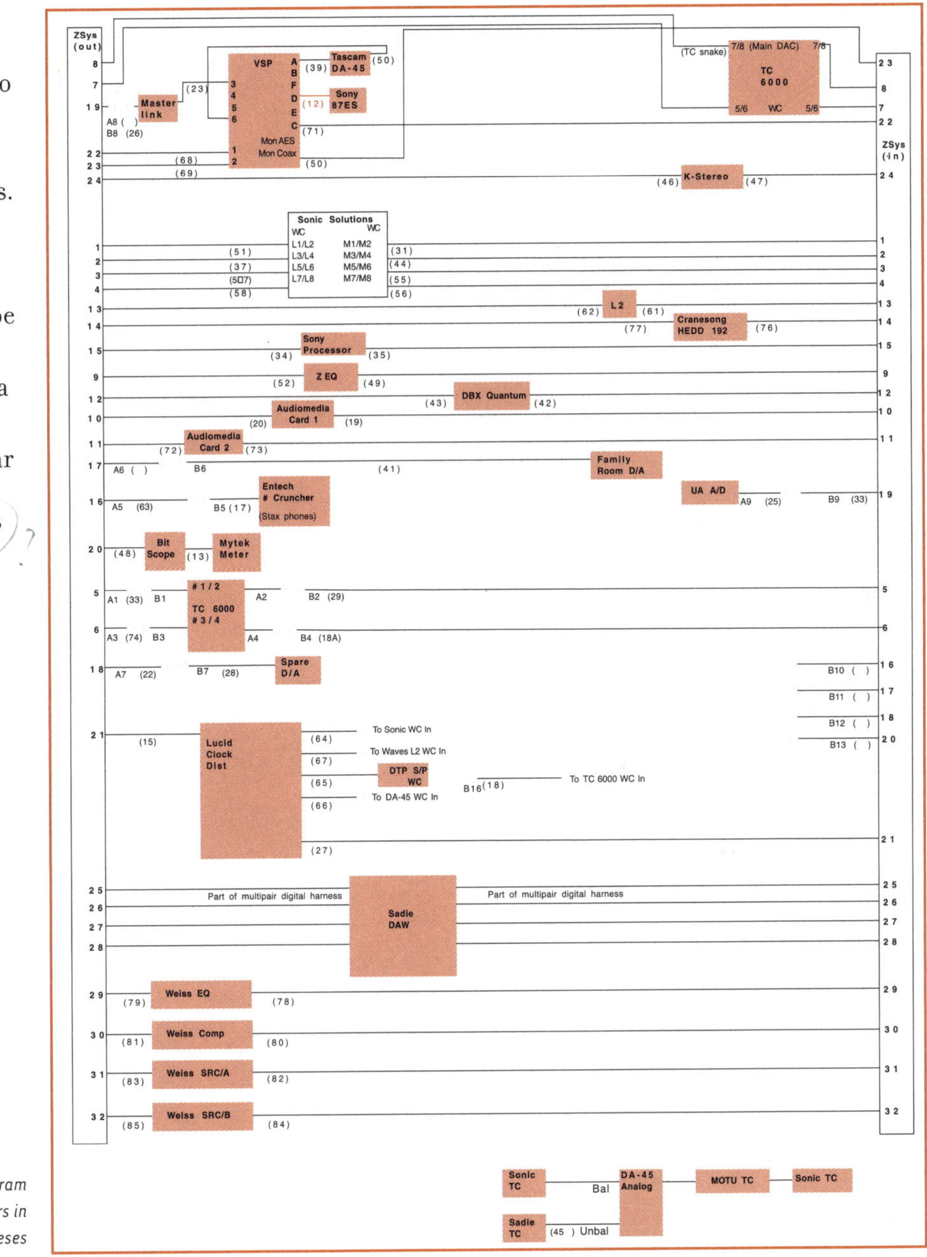


*Mastering Studio block diagram with wire numbers in parentheses*

1 Or whose digital audio outputs are in a copy-protected format which can't be accessed by standard DACs.

2 A well-designed digital mixer can be used as a synchronous router. It can even mix sources and be bit transparent to each source as long as the sum of all the levels does not exceed full scale (0 dBFS). In other words, if you mute all but one of the sources that you are mixing, the output is a perfect copy of the input. In order for a mixer to be bit-transparent, the levels must be set to exactly 0 dB and the pan controls set to produce unity gain. Depending on the brand of mixer, unity pan can be centered or full right or left. Regardless, never turn your back on digital and test the mixer for bit-transparency (Chapter 17 discusses tests for bit-transparency).

3 To commit or not to commit? That is the question. If you are the type who commits on loadin, I recommend keeping on archive an unprocessed safety transfer of the precious analog tape so another transfer would not be required if a different sound is desired later.

4 Some people question the wisdom of using a digital attenuator in a monitor chain, however, there's a lot of latitude in a 24-bit DSP-based monitor attenuator; if doing small amounts of attenuation and relatively little analog gain after the DAC, distortion and noise is probably below the analog noise floor. So there may be no audible resolution loss in a properly-dithered DSP-based monitor level control coupled with proper analog gain staging. However, since mastering engineers depend on making accurate monitoring judgments, an analog-based monitor attenuator, which does not add quantization distortion, makes more sense.

5 See Appendix for recommended reading.

6 If you break the digital-analog proximity rule, use analog gear which is well-shielded and immune from RF interference. Look for interference signatures in a measurement of the noise floor.

CHAPTER 3

# An Earientation Session

## I. Introduction

Ear training is actually mind training, because the appreciation of sound is a learned experience and the more we experience, the more we learn. Although to our modern ears, Edison's acoustic phonograph gave a crude representation of the original, its first listeners felt that its reproduction was indistinguishable from real life. It is only with each advance in the state of the art of sound reproduction that people become aware of the shortcomings of the previous technology. For example, whenever I work at a very high sample rate, and then return to the "standard" (44.1 kHz) version, the lower rate sounds worse, although after a brief settling-in period, it doesn't sound that bad after all (see Chapter 20).

As we bccome more sophisticated in our approach to listening, we develop a greater awareness of the subtleties of sonic and musical reproduction. We can also grow to like a particular sound, and each of us has slightly different preferences, which vary over the years. When I was much younger, I liked a little brighter sound, but from about the age of 20, I've tended to prefer a well-balanced sound and immediately recognize when any area of the spectrum is weak or over present.

A mastering engineer requires the same ear training as a mixing engineer, though the mastering engineer becomes expert in the techniques for improving completed mixes, while the mixing engineer specializes in improving the mix at the level of the individual elements which make up the whole. Mastering from stems (sub mixes) is a hybrid of these two, since the mastering engineer has some

*"Make passive ear training a lifelong activity."*

control over individual instruments or groups. Still, mastering is not the same as mixing and should be separated from it since it's very hard to do two jobs well at the same time.

Ear training can either be a **passive** or a **hands-on** activity. Passive ear training goes on all the time ("what a tinny speaker in that P.A. system"), while active ear-training occurs while your hands are on the controls. Make passive ear training a lifelong activity—as it increases your ability to discriminate fine sonic differences. Practice being more consciously aware of the sounds around you and identifying their characteristics. Acousticians and classical recording engineers can't help judging the reverberation time of every hall they enter.

**Hands-on** ear training is the process of learning how to connect technique with the sound you have in your head; like all skills, developing hand to ear coordination requires practice. Before working on a piece of music, try to visualize (audiolize?) the sound you are trying to achieve; have a definite sonic goal in mind. Sometimes even if we don't know how we're going to solve a problem, a clear goal keeps us from fumbling.

## II. Speaking the Language

The classic chart folded into the front cover was hand-drawn in 1941 by E.J. Quinby of room 801 within the depths of Carnegie Hall.[1] We've reproduced it for the benefit of musicians who want to know the *frequency language* of the engineer, and for engineers who want to speak in a musical language. Sometimes we'll say to a client, "I'm boosting the frequencies around middle C", instead of "... around 250 Hz". Learn a few of the key equivalents, e.g. 262 Hz represents middle C, 440 is A above middle C, and then remember that an octave is a 2x or 1/2x relationship. For example, 220 Hz is the frequency of A below middle C in the equal-tempered scale. The ranges of the various musical instruments will also clue you to the characteristics of sound equalization—next time you boost at around 225 Hz, think of the low end of the English horn or viola.

Although it helps an engineer to have played an instrument and be able to read music, many successful engineers can do neither as they have good pitch perception, can count beats and understand the musical structure (verse, chorus, bridge...).

The chart on the next page is a graphic representation of the subjective terms we use to describe excesses or deficiencies of various frequency ranges. Excess of energy is shown above the bar and a deficit below. The bar is also divided into eight approximate regions though there are no standard terms for these divisions: what some people call the **upper bass**, others call the **lower midrange**; some call the **upper midrange** what others call **lower treble**. Notice that we have more descriptive terms for areas that are boosted as opposed to those which are recessed. This is because the ear hears boosts or resonances more easily than dips or absences.[2]

With an equalizer, the sound can be made **warmer** in two ways: by boosting the range roughly between 200 and 600 Hz; or by dipping the range roughly between 3 and 7 kHz. These two ranges form a yin and yang, which we'll discuss in Chapter 8. Another way to make sound warmer (or its converse, edgier) is to add selective harmonics, as described in Chapter 17. Too much energy, and/or distortion, in the 4 to 7 kHz region can be judged as **edgy**, especially with high brass instruments. Extra energy in the lower midrange, or a strong upper midrange can add presence to a sound, but too much can sound fatiguing or harsh. If the sound is edgy, it can often be made **sweet(er)** by reducing energy in the 2.5 to 8 kHz range. Too much energy in the 300-800 range gives a **boxy** sound; go up another third octave and that excess is often termed **nasal**. A deficiency in the range from roughly 75 to 600 Hz creates a **thin** sound.

## III. Exercises

### Ear Training Exercise #1: Learn to Recognize the Frequency Ranges

This is an exercise in the perfection of pitch perception. To have perfect pitch means that you can identify each note (or feedback frequency) blindfolded. But this ability is not just a trick: if you learn how to identify frequency ranges by ear, this will greatly improve your equalization technique. I used to practice until I could automatically identify each 1/3 octave range blindfolded, but now my absolute pitch perception is between 1/3 and 1/2 octave, which is about what you need to be fast and efficient at equalizing. Start ear training with pink noise and then move to music, boosting each range of a 1/3 octave graphic equalizer until you can recognize the approximate range. Take a blindfold test, with a friend boosting EQ faders randomly. Don't be dismayed if at first you're only accurate to about an octave, even this will get you close enough to the range of interest to be able to better "focus" the equalizer.

### Ear Training Exercise #2: Learn the Effects of Bandwidth limiting

Less-expensive loudspeakers usually have a narrower bandwidth, as do lower-quality media and low sample rates (e.g. the 22.05 kHz audio files often used in computers). Train your ears to recognize when a program is either naturally extended or bandwidth-limited. It's surprising how much low and high end filtering we can get away with, as can be heard when old films with optical sound tracks are shown on TV. The listener may not notice the voice is very thin-sounding until it's been pointed out because the ear tends to supply missing bass fundamentals when it hears the harmonics. We can take advantage of this in mastering, (e.g. reduce

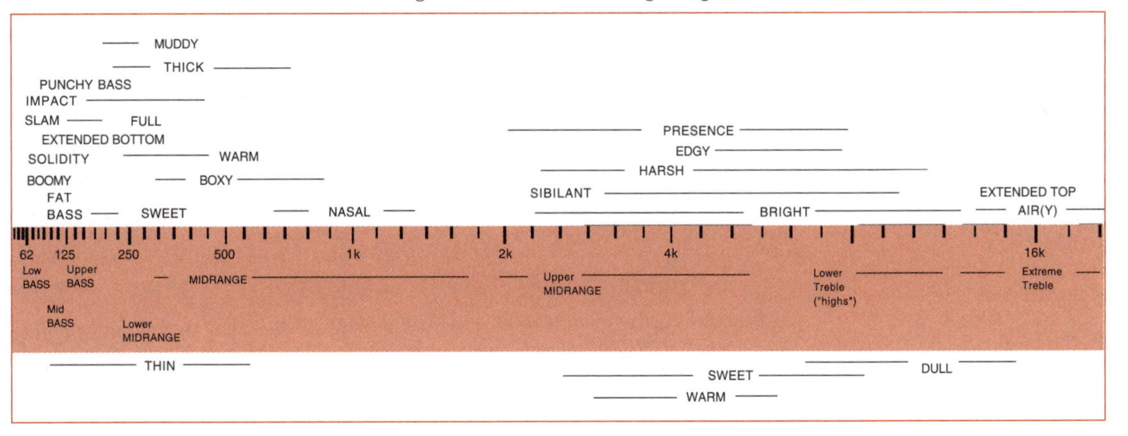


*Subjective Terms we use to describe Excess or Deficiency of the various Frequency Ranges.*

low frequencies to obtain a higher level) but this is an audible compromise and the best productions are always the ones with full bandwidth.

Most musical information is safely tucked away in the midrange, the only frequencies that remain in an analog telephone connection, but a 5 kHz bandwidth takes away the life and clarity of the sound, even if all the informational content is there. Practice learning to identify the effects of bandwidth limiting using high and low pass filters on various musical examples. Another way to study the contribution of the low bass range is to turn subwoofers on and off.

### Ear Training Exercise #3: Learn to Identify Comb Filtering

The only advantage of the English system of measurement is that the speed of sound translates neatly to about one foot per millisecond. When a single sound source is picked up by two spaced microphones, and those microphones are combined into a single channel, unwelcome audible comb filtering will result if

- the gain of each microphone is about the same and the microphones are identical or similar models.
- the relative mike distance from the source is in the critical area from about 1/2 foot (~150 mm) through about 5 feet (~1.5 m). At 5 feet, the more distant mike's signal is lower in level, also reducing the combing effect.

Comb filtering can occur anytime a source and its delayed replica are mixed to a single channel; when one source's gain is reduced at least 10 dB, the comb filtering becomes audibly insignificant. The figure (next page) shows the frequency response resulting when the source and the delay are at equal gain. The vertical divisions are 3 dB. From top to bottom—a delay of 3 ms (about a 3 feet/1m path difference), 1 ms, and 2 ms. In real life, reflections will be diffused and somewhat attenuated, with less obvious effect.

The reflections from a singer's music stand are one important source of this problem, but this cannot be fixed (as some think) simply by adding a piece of carpet because carpet has no meaningful effect in the range below about 5 kHz, which, as we can see from the figure, is where the major problems occur. The ear really begins to notice comb filtering when the delay is changing, for example, the classic *flanging* effect when an artist sways in front of a reflecting music stand. That's why the best music stand is none at all; open-wire stands are second-best and careful placement does the rest. A related kind of comb filtering occurs when the sound from an instrument reaches the microphone both directly and also via reflections from the floor.

Television and film soundtracks provide excellent laboratory exercises in learning how comb-filtering can mutilate sound, since the proper operation of a lavalier microphone depends on indirect sound, which can include nasty reflections from nearby surfaces. Listen to the TV weather report blindfolded and try to identify position by the sound: "Now she's crossed her hands on her chest, about 3 inches below the lavalier microphone. Now she's turned around to face the blue screen, about 2 feet away. She's sitting down at the anchor desk and you

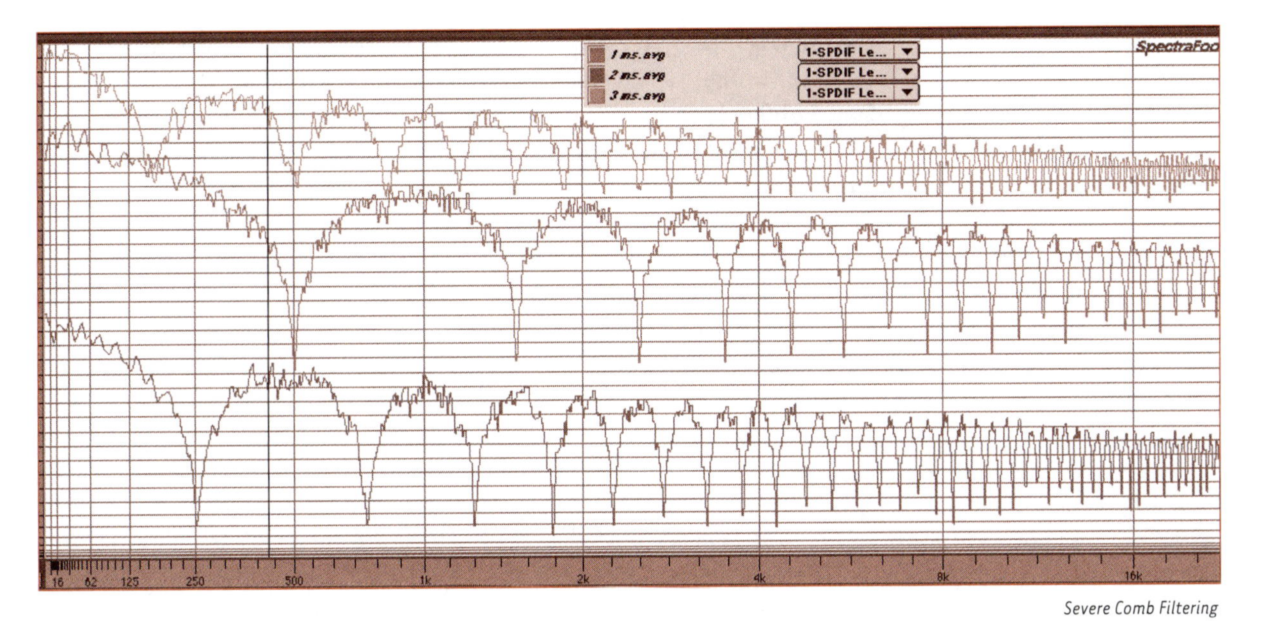


*Severe Comb Filtering*

can hear from the hollow dip at 500 Hz that her mike is about a foot above the desk."

### Ear Training Exercise #4: The Sound of Great Recordings well-reproduced. Perception of Dynamics, Space and Depth

Because mastering engineers may be called on to work on a wide variety of music, train your ears to recognize good recorded sound in each genre. Start by becoming familiar with great recordings made with purist mike techniques, and with little or no equalization or compression. Learn what wide dynamic range and clear transients sound like, so as to recognize more quickly when dynamic range has been limited. Listen to live music; the percussive impact of a real live big band, or the clear transients of a classical piano, which provide a standard that can never be bettered. Compare the depth of a live recital which can be captured with simple miking techniques, versus how much of this is lost when multiple miking is used.

When comparing a master to the mix, try concentrating on one instrument or quality at a time as you switch and confirm that each stage of the mastering process has made things better and not worse.

### Ear Training Exercise #5: The Proximity Effect Game

Most recorded pop vocals have greater lower midrange and presence than real life. The trick is aided by the recording engineers' use of the proximity effect: an increase in bass response when a directional microphone is moved closer to the source. Learn to recognize when a vocalist was recorded too closely, overemphasizing those frequencies.

"Did you know that wearing a hat with a brim puts a notch in your hearing at around 2 kHz?"

**Ear Training Exercise #6: The Sound of Overload**

When solid-state amplifiers overload, the round part of their output waveform starts to square off. We use the term **clipping** or **clipped** to describe when this overload is severe and the waveform is squared-off. Some amplifiers overload drastically, producing high odd harmonic distortion, others (particularly tube amps) overload more gracefully, a characteristic which turns them into a form of compressor, fattening sounds when pushed past their linear region. Learn to identify the sound of overload in all its forms: analog tape reaching or into severe saturation, overdriven power amplifiers with intermodulation distortion, optical film distortion (as in classic 1930's talkies), etc. As a training exercise, study the saturation on peaks of a classical or pop recording made from analog tape as compared to a modern all-digital recording. Learn the characteristics of each piece of equipment; soon you'll discover some rare digital processors that overload more gently than others.

**Ear Training Exercise #7: Identify the Sound Quality of Different Reverb Chambers**

Artificial reverb chambers have progressed tremendously over the years. Become familiar with the artifacts of different models of reverbs. Some exhibit extreme *flutter echo*, some sound very flat, while others produce an excellent simulation of depth. We'll learn how they accomplish this in Chapter 18.

**Ear Training Exercise #8: The differences between sampled pianos and the real thing.**

Sampled pianos are sounding better all the time. Sometimes we get fooled! Practice your fine perception until you don't get fooled very often.

**Ear Training Exercise #9: Mono, Weak Stereo, Good Stereo**

Train your ears to distinguish a good stereo recording from one which has little separation or depth. Distinguishing a mono recording from a stereo is not as simple as looking at the level meters and saying, "oh, one is moving a little differently", which could be a mono recording with a gain difference between channels. The phase meter on a true stereo recording should show a variation in phase shift with the dynamic movement. An imperfect monitoring environment can give the false impression of stereo information, check on headphones when in doubt.

**Ear Training Exercise #10: Listening Acuity—Identifying Tiny Differences.**

Make a test master with 0.5 dB difference in equalization of one band. Can you hear the difference in a blind test? 0.5 dB is probably the threshold below which we work on the feeling level and above which we work on the assurance level. Don't underestimate the importance of audio voodoo; what we believe to be true has a power of its own.[3] However, unless certain, don't be fooled into thinking a difference is truly perceivable. And test Katz's law: The length of silence between two successive plays is proportional to the number of incorrect conclusions.

### Things to Recognize

Experienced mastering engineers learn to recognize...

- dropouts (digital mutes and analog types), especially audible in headphones
- space monkeys (twitters and glunges, artifacts of lossy coders)
- skewed analog tape
- compression pumping
- hiss
- different frequency ranges of sibilance
- IM gurgle from bad bias in analog tape deck
- phasing (which sounds like varying comb filtering)
- noise reduction misalignment causing pumping
- electrical noises (ticks, clicks, pops)
- phonograph associated noises (tracing/tracking distortion, rolloff, swishes, inner groove distortion, non-fill)[4]

**Bad Edits.** An experienced mastering engineer should be able to recognize a bad edit where the ambience or the sound is partially cut off, or the sound partially drops out. Practice making edits and bring them to the attention of an experienced engineer for critical analysis.

**Wow and Flutter.** Wow and flutter are caused by speed variations in recordings, and are no longer a problem with digital recording. But mastering engineers are sometimes called upon to restore older analog recordings. So to enhance perceptual acuity, make a cassette recording of a solo piano, and compare it side by side with a digital recording of the same instrument.

**Polarity problems**. Learn to recognize when the left channel of a recording is out of polarity with the right. Reverse the polarity of the wires to one loudspeaker and become familiar with the characteristic sound of the error: thin sound, with a hole in the middle of the image. This will also help you to recognize when some instruments in a mix are out and others are in polarity. A good engineer can recognize polarity problems, by the vagueness of the stereo image, even without switching to mono.

**Recognize the hum frequencies.** Hum at the fundamental of the power line (50 Hz in Europe, 60 Hz in the U.S.) usually means a bad shield, an open mike line, or a ground loop. Hum at the second harmonic (100 Hz in Europe, 120 Hz in the U.S.) usually means a bad power supply filter capacitor. Hum at the third harmonic (150 Hz/180 Hz) can indicate induction into an audio cable from a power transformer or a ground loop between chassis.

### In Summary

*Earientation* should be a lifelong activity and no one can become an expert overnight. These exercises will help start the process.

> *"The length of silence between two successive plays is proportional to the number of incorrect conclusions."*
>
> — Katz's Law

1 I've never visited that room, but it would be an interesting archeological voyage to find his lair. Internet references indicate he was a renaissance man, from subway trains to music.

2 Jim Johnston (in correspondence) points out that peaks change the partial loudness more than dips. It's psychoacoustic!

3 Thanks to Andrew Hamilton for that piece of philosophy.

4 Thanks to Jim Rabchuk for this list of items which should be part of the engineer's listening skill set.

CHAPTER 4

# Wordlengths and Dither

**I. Introduction**

Although audio engineers must learn how to deal with and take advantage of wordlengths and proper dithering practices, we must also keep our problems in perspective. If the mix isn't good, or the music is not working, then dither probably doesn't matter much at all. But if everything else in a project is right, and we want to maintain the sound quality, then proper dithering is very important.

## I. Dither in the Analog Domain

In an analog system, the signal is *continuous*[1], but in a PCM digital system, the amplitude of the output signal is limited to one of a set of fixed values or numbers. This process is called **quantization**. Each coded value is a discrete step. For example, therc are exactly 65,536 discrete steps, or *values* available in 16-bit audio, and 16,777,216 discrete steps in 24-bit audio. The approximate codable dynamic range of any PCM system is calculated by multiplying the wordlength by 6; e.g. 8 x 6 = 48 dB for an 8-bit system. So the lowest value that can be coded by 16-bit is 96 dB down from the top; in 24-bit it is 144 dB. If a signal is quantized, this will induce a **distortion** related to the original input signal which can introduce harmonics, harmonics aliased down to lower frequencies (see Chapter 20), intermodulation, or any of a set of highly undesirable kinds of distortion, perceived as a buzz, grit, harshness, coldness and/or loss of depth in the sound.

Pictured on the next page is a portion of a sine wave which has been quantized.[2] Since there is no resolution below the level of each quantization step, the result is a stepped, distorted waveform. Low level

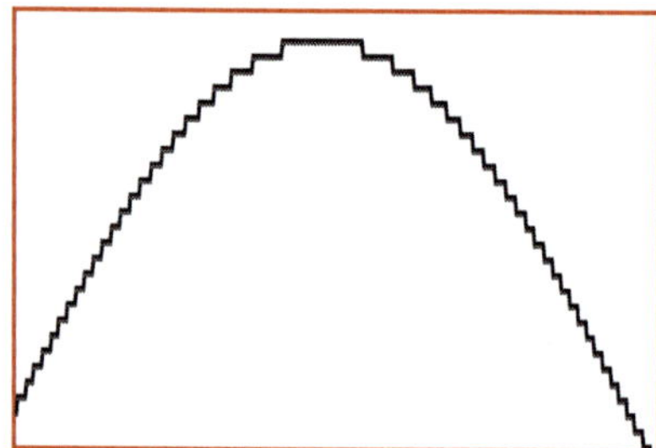

*Section of a sinewave quantized without dither ("stepping").*

information is completely lost. To prevent this kind of distortion, we use dither, which is a process that mathematically removes the highly undesirable distortions entirely, and replaces them with a fixed noise level.

Here's a simple *thought experiment* that explains why dither is necessary and how it works.[3] Let's create a basic ADC. We'll make it sensitive to DC, and bipolar, so it responds to both positive and negative analog inputs, and we'll give it a very big LSB size of 1 volt to make the numbers easy, and LSB threshold at 0 volt. LSB means **least significant bit**, the bit with the smallest (lowest) analog value in the PCM system. We'll construct our ADC so that an analog source in the range between -1 volt and 0 volt produces a digital output word of 0, and an analog source in the range between 0 volt and +1 volt produces a digital output word of 1. If, without applying any dither, we present a -0.25 volt DC (continuous) signal to the input of the ADC, the output of the ADC will be a string of zeros. Any information below the LSB threshold is completely lost, as illustrated at left. If we remove the -0.25 volt signal and apply dither to the input of the ADC in the form of a completely random signal (i.e., noise) centered around 0 volt, its peak amplitude will randomly toggle the LSB of the ADC (pictured above right).

Now the output of the ADC is a stream of very small random values whose **average is zero volt** (there is an equal number of 0's and 1's).

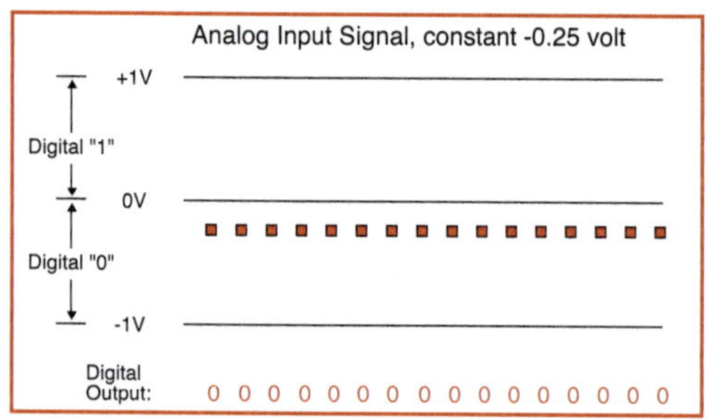


*Graph of a hypothetical ADC whose LSB threshold is 0 volt. Each sampled analog input is represented by a small orange square; in this example, the analog source is held at a continuous -0.25 volt. Note that any input between -1 volt and 0 volt will be lost, because it is below the threshold of the LSB, producing a string of zeros. Because it is below threshold, a DC signal held continuously at -0.25 volts will not be detected.*

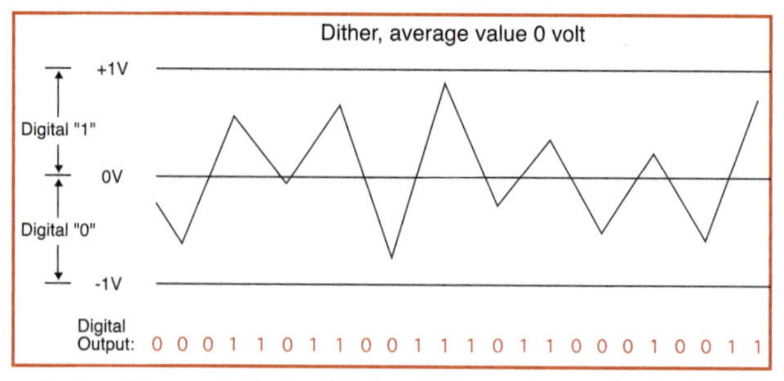


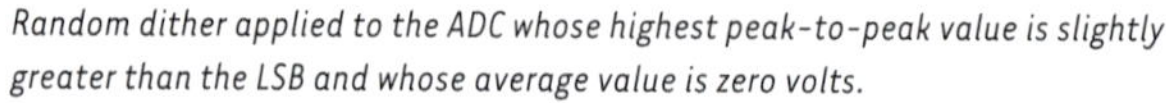

*Random dither applied to the ADC whose highest peak-to-peak value is slightly greater than the LSB and whose average value is zero volts.*

Leaving the dither on, let's apply our -0.25 volt signal again (pictured below). At each sample point (in time), the -0.25 value of our analog signal is added to the random dither value.

The output stream now is a stream of numbers whose average is equivalent to -0.25 analog volts.[4] We have thus detected and captured information

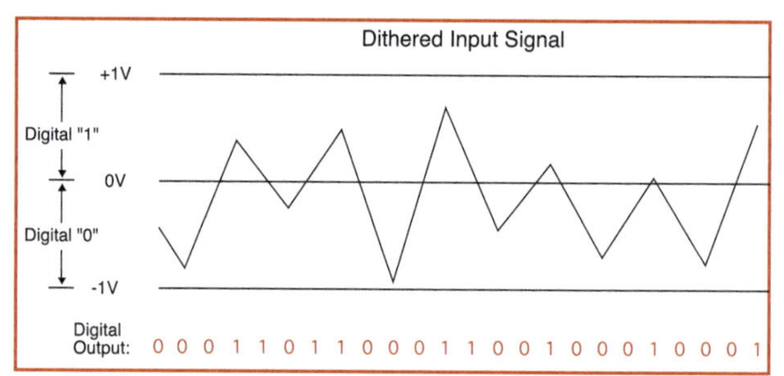


*Dither summed with the input signal produces an output whose average value is -0.25 volt.*

that was previously lost (even though it's buried in "noise"). In other words, our low level resolution has improved. The conversion is still essentially random, but the presence of the -0.25 volt signal biases the randomness. Put another way, the characterization of the system with dither on is transformed from being completely deterministic to one of statistical probability. The periodic alternation of the LSB between the states of 0 and 1 results in encoding a source value that is smaller than the LSB. This time, on the average, the LSB puts out a few more zeros than ones because of our -0.25 volt signal. We say that dither exercises, toggles or modulates the LSB.[5] With the dither on, we can now change the input signal over a continuous range and the average of the ADC output will track it perfectly.[6] An input signal of 0.373476 volts will have an average ADC output of (the binary equivalent of) 0.373476. The same will hold true of inputs going over the 1 threshold: an input of 3.22278 will have an average ADC output of 3.22278. So not only has the dither enhanced the resolution of the system to many decimal places, it has also eliminated "stepping".

Dither's resolution enhancement is a truly physical/mathematical phenomenon, **not** simply a means to fool the ear or "to mask the low level digital breakup". In addition to being able to record and reproduce **all** the analog values at high and medium levels, dither lets us encode low level signals **below the -96 dB limit for 16-bit!**[7] We can digitally measure undistorted test tones lower than -130 dBFS in a 16-bit dithered system (see figures next page). These results—increased resolution and the elimination of quantization distortion—cannot be achieved by adding noise **after** the A/D conversion. So dither must be added at the proper point in the circuit: adding noise after quantization is not the same as dithering and is as effective as locking the stable door after the horse has escaped.

*"We use the term* **resolution** *to indicate whether a source signal of a given level will be represented in the output."*

The perceived (audible) dynamic range of the dithered system is greater than its codability because human beings are able to hear signals in the presence of noise of greater energy than the signal. Although its noise is raised to about -91 dBFS with the addition of dither,[8] we can hear discrete tone signals about as low as -115 dBFS in a 16-bit recording.

With 24-bit conversion and storage, dither is probably not necessary during the original analog encoding because the inherent thermal noise on their inputs tends to self-dither around the 20th bit. Since this thermal noise may not be perfectly random, some manufacturers may add a bit of their own dither to guarantee lowest distortion.

Similarly, a transfer from analog tape theoretically may have enough hiss to self-dither a transfer to 16-bits, but as its noise may not be sufficiently random for dithering I would advise encoding analog tapes to 24 bits for processing and later redither with a proper generator. There is no resolution advantage in recording the output of any ADC to a 32-bit floating point file, and it uses more storage. Floating point is meant for processing, it can encode dynamic ranges far greater than real life, but this is pointless when your source is real life! (More about this in Chapter 5). The dynamic range of an ADC at any frequency can be measured with an accurate test tone generator and a low-noise headphone amplifier with

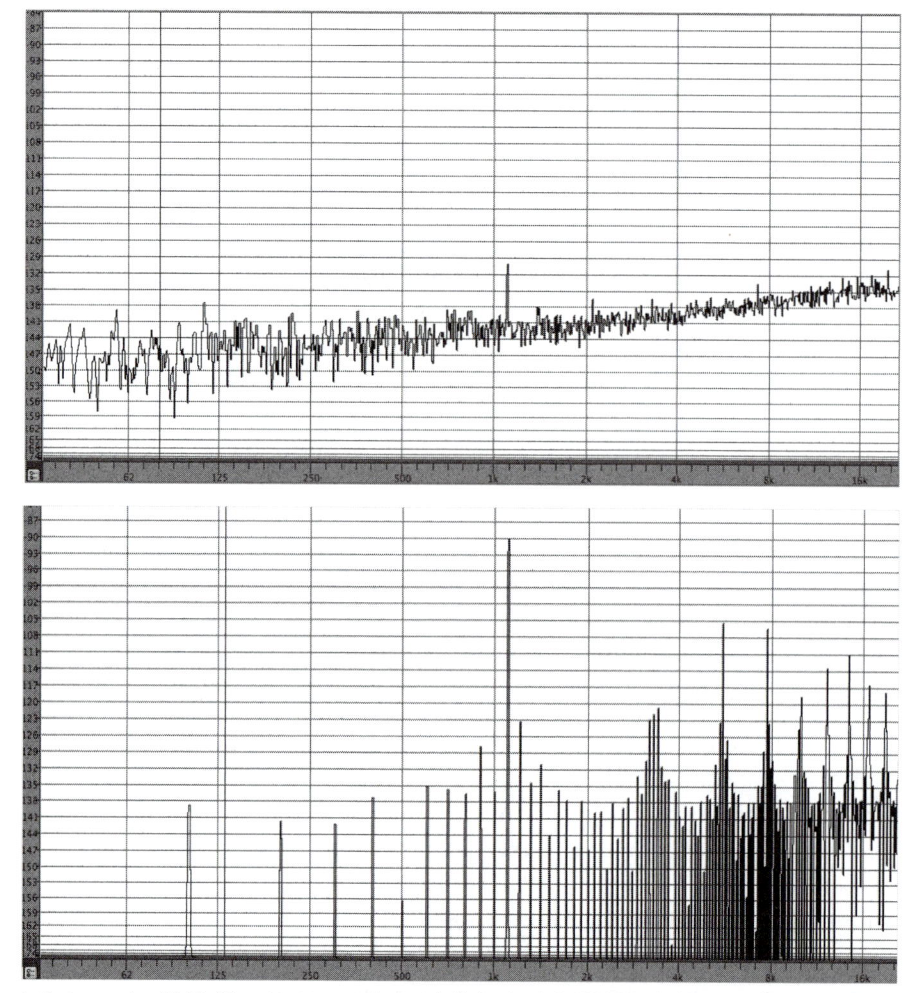


*In the top graph, a 16-bit dithered test tone at the remarkably low level of -130 dBFS, which is measurable but inaudible. Since the world's best DAC has a noise floor around -120 dBFS, the lowest tone in 16-bit we can practically hear is about -115 dBFS. The noise components in each bin in the -147 dBFS range add up to an RMS total of about -91 dBFS. However (bottom graph) an undithered 16-bit test tone at -90 dBFS yields severe audible distortion. Many of these distortion spikes are audible above a 16-bit noise floor so noise added after truncation is ineffective.*

sufficient gain. To conduct the test, simply listen to the analog output while lowering the level of the tone and find when it disappears (use a high quality DAC for this test). Another important test is to attenuate music in a workstation (about 40 dB) and listen to the output of the system with headphones. Listen for ambience and reverberation; a good system will still reveal ambience, even at that low level.

## II. The Need for (re)Dither in the Digital Domain

### The First Secret of Digital Audio: How Wordlengths Expand

Let's face the music: as soon as we transform audio by changing its level, equalizing, compressing, or any other calculation that requires multiplication, **its wordlength increases.** For example, after processing, a 16-bit source grows to 24 bits **or more**. This means sound quality will deteriorate if we simply truncate that product to a shorter wordlength. Let's see why this occurs.[9]

Digital audio is all arithmetic, but the accuracy of that arithmetic, and how the engineer (or the workstation) deals with the arithmetical product, can make a tremendous difference to the final sound. All DSPs (Digital Signal Processors) tackle digital audio on a sample by sample basis. At 44.1 kHz, there are 44,100 samples in a second (88,200 stereo samples). When changing gain, the DSP looks at the first sample, performs a multiplication, producing a new number, and then moves on to the next sample. It's that simple.

To avoid unnecessarily complicated mathematics, I'm going to invent the term *digital dollars.* Suppose that the value of our first digital audio sample is expressed in dollars instead of volts, for example $1.51. And suppose we want to reduce it by 6 dB. 6 dB is half the original value.[10] So, to attenuate our $1.51 sample, we divide it by 2.

This creates a problem: $1.51 divided by 2 equals 75 1/2 cents, or $0.755. So, what should we do with the extra decimal place we've just gained? It turns out that being able to deal effectively with extra places is what good digital audio is all about. If we just drop the extra five, we've lost just half a penny—but in the audio world that 'half a penny' contains a great deal of the natural ambience, reverberation, decay, warmth, and stereo separation that was present in the original $1.51 sample. Multiplications generally result in a longer wordlength than we started with and the wordlength can increase, up to the precision of the DSP. For example, a 1 dB gain involves multiplying by 1.122018454 (to 9 place accuracy). $1.51 multiplied by 1.122018454 equals $1.694247866. Although the lower decimal places may seem insignificant, remember that DSPs perform **repeated** precision calculations for filtering, equalization, and compression, so *unless adequate precision is maintained,* the end number may not resemble the right product, yielding distortion. The higher the precision (up to a reasonable limit), the better the resolution and the cleaner the digital audio.

While DSPs are capable of performing higher precision arithmetic, the math slows them down, takes up more cycles, and that means fewer plugins can run on your computer or fewer operations in an outboard processor, so ironically, most manufacturers are fearful of a consumer backlash if they do things right. Inside a digital mixing console (or workstation), the mix bus must be much longer than 24 bits, because adding two (or more) 24-bit samples together, then multiplying by a coefficient (the level of the master fader is one such coefficient) can result in a 48-bit (or larger) sample, with every little bit significant.[11] Since the AES/EBU standard can carry up to 24-bits, external processors calculate at the highest possible precision, then bring this long word down to 24 bits and send the result to the outside world, which could be a 24-bit storage device (or another processor). The next processor in line also must reduce its internal long wordlength back to 24 bits for AES/EBU transmission. The result is a slowly cumulating error in the least significant bit(s) from process to process, which is distortion. If the cumulative distortion of the processors is around the 24-bit level (around 144 dB down), it will be inaudible. DAWs handle wordlength in different ways; Pro Tools HD calculates at 48-bit precision, but truncates data from 48 to 24 on the way to and from the plugins; most native DAWs use either 32 or 64 bits floating point (see Chapter 5).

### How Dither Works in the Digital Domain

Since truncation is so bad, what about rounding? In our digital dollar example, we ended up with an extra 1/2 cent. In school, they taught us to round numbers according to the rule "even numbers...round up, odd...round down". But rounding is little better than plain truncation, still

**Resolution, Wordlength and Precision**

**Resolution** is an overused term that we must define to make it effective. We use the term resolution to indicate whether a source signal of a given level will be represented in the output. This can be expressed as a number of equivalent bits.

We define the term **wordlength** as the number of discrete data bits employed to transfer a digital value from the source to a destination.[16]

**Precision** is defined as the internal data wordlength within the algorithm. The internal precision of a processor must be significantly greater than its output wordlength because of the great number of steps used to perform a complex process (such as equalization). It is not easy for someone on the outside to determine the internal precision of a processor. We have to depend on the skill and reputation of the programmer and the sound quality of the result.

adding lots of quantization distortion. So, when dealing with more numerical precision and small numbers that are significant, we still have to use dither to bring the information from the LSBs into the bits we intend to use.

When processing digitally, we add dither similar to the analog approach, except that the processor must generate the dither as a series of random numbers. This technique is often called *redithering*, because the signal may have been already dithered during the encoding (recording) process, giving rise to a misconception that additional dither is unnecessary. **There is no such thing as self-dithering:** When processing digitally , no matter how much hiss or noise exists in a source, the noise from the original dither becomes irrelevant. Like the program material, the source noise has now been distributed amongst a longer wordlength due to the processing. We must add new dither to preserve resolution before truncation, even when going 24-bit to 24-bit (because the intermediate wordlength is longer than 24). In the analog example, we learned that the dithered result contains all the low level information below the LSB, because we **added** the analog dither to the analog signal. Similarly, in the digital domain, we add two digital numbers together, one of which is the digitally-generated dither.

To do this, we calculate random numbers and add a different random number to every sample. Then, cut it off at 16 bits (or whatever shorter wordlength we desire). The random numbers must also be different for left and right samples, or else stereo separation will be compromised.

For example, starting with a 24-bit word:

```
                  ---Upper 16 bits---  -Lower 8-
Original 24-bit   MXXX XXXX XXXX XXXW  YYYY YYYY
Add random number                      ZZZZ ZZZZ
```

The result of the addition of the Z's with the Y's gets carried over into the new least significant bit of the 16-bit word (LSB, letter W), and possibly higher bits if we have to carry. Just as in the analog example, the random number sequence combines with the original lower bit information, modulating the LSB. The result is that much of the resolution and sound quality of the long word is carried up into the shorter word. Random numbers such as these translate to random noise (hiss) when converted to analog and this hiss is audible only if listening carefully with headphones.

**Some Tests for Linearity**

To verify whether a digital audio workstation truncates digital words or does other nasty things, the only measurement instruments we need, are headphones with sufficient gain. Track 42 of *Best of Chesky Classics and Jazz and Audiophile Test Disc, Vol. III*[12] is a fade to noise test without dither, demonstrating quantization distortion and loss of resolution. Track 43 is a fade to noise with flat dither, and track 44 uses noise-shaped dither (to be explained). Using Track 43 as the test source, it is possible to hear smooth and distortion-free signal down to about -115 dB. Track 44 shows how much better it can sound. If we then process track 43 with digital processing (with and without dither) we can hear what it does to the sound, especially when reduced to 16 bits. If the workstation is not up to par, the result can be quite shocking.

### The Effect of Masking

In ***Color Plate Figure C4-1***, we compare the levels of 16, 20, and 24-bit flat dithered noise. -91 dB seems like so little noise, but there is a tradeoff between its benefits (distortion removal and ambience recovery) and the masking effect of the noise. Though the noise is not directly audible at normal listening levels, critical listening demonstrates that sometimes dither noise masks or obscures the very ambience and spatiality we are trying to recover! This is especially true with flat dither at the 16-bit level, which to my ears sometimes adds a slight veil to the sound, narrows the imaging and reduces the depth. It's a tradeoff because dither's benefits outweigh the losses due to masking.

### Improved Dithering Techniques

Although the required amplitude of 16-bit flat dither is about -91 dBFS, it's possible to shape (equalize) the dither to minimize this masking effect. Noise-shaping techniques re-equalize the spectrum of the dither while retaining its average power, effectively moving the noise away from the areas where the ear is most sensitive (circa 3 kHz), and into the high frequency region (10-22 kHz). The best of these processors yield 19-20 bit performance on a 16-bit CD, and the dither noise for most listeners at -60 or -70 dB is inaudible.

***Figure C4-2*** in the color plates shows a graph of the amplitude versus frequency (at 44.1 kHz/16-bit) of one of the most successful noise-shaping curves, POW-R dither, type 3 (red trace). For comparison, we can see flat 20-bit dither (orange), and 24-bit dither (green). POW-R 3 uses a very high-order noise-shaping filter, with several dips where human hearing is most sensitive, the inverse of the "F" weighting curve which defines the low-level limit of human hearing. Note that at one critical point in the midrange the POW-R 3 curve is as low as the level of flat 20-bit dither; this is what noise-shaping is all about, dropping the noise where it would be most audible. There are numerous noise-shaping redithering processors on the market, some in hardware, most in plugins or built into DAWs. The curve does not directly equalize the material, however when these were introduced, critical listeners complained that the high frequency rise of the noise-shaping curves changed the tonality of the sound, adding a bit of brightness. But my listening tests indicate that it is masking in the lower midrange which affects the tonality. I found that shapes with a little bump around 200 Hz sounded brighter than those with a flat midrange, and that the early dip in POW-R 3 around 2 kHz seems to unmask low level material in that range, producing a brighter sound. But the results are not 100% repeatable, and depend a lot on the source material, so different shapes are applicable for different recordings. To generalize—depth, transient response and brightness seem to increase with the higher order shapes, but the accuracy of the tonality increases with the lower orders. For example, a warm-sounding 24-bit master which benefits from depth may translate best to 16-bit using a dither with a higher order shape. When in doubt, choose a moderate shape. Many engineers still prefer flat dither, especially for grungy material where a little noise is a good thing!

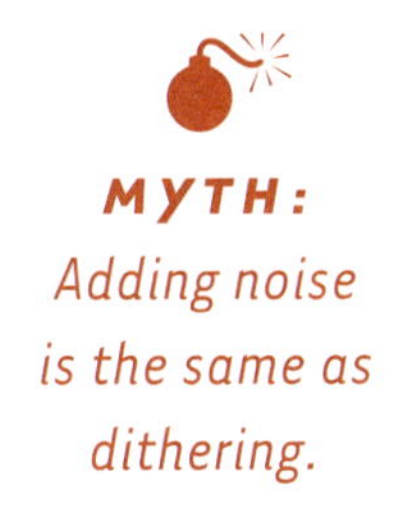

MYTH:
Adding noise
is the same as
dithering.

Noise-shapers on the market include: Lavry Engineering model 3000 Digital Optimizer, Waves L1 and L2 Ultramaximizers (in three flavors), Prism, POW-R, and several others. Apogee Electronics produced the **UV-22** system in response to complaints about the sound of earlier noise-shaping systems, declaring that 16-bit performance is just fine. Apogee does not use the word "dither" (because their noise is periodic, they prefer to call it a "signal") and instead of noise-shaping, UV-22 adds a carefully calculated noise around 22 kHz, with a slight noise improvement in the midband compared to flat 16-bit dither.

We can effectively compare the sound and resolution of these redithering techniques by lowering music (around -40 dB) into the dither unit, and listen to the output of a high-quality DAC on headphones. The sonic differences between high and low quality dithering systems can be shocking: Some will be grainy, some noisy, and some distorted, indicating improper dithering or poor calculation. When making judgments at high gain, try to discount any obvious high frequency noise due to noise-shaping, because noise-shaping is designed to be inaudible at normal gain.

#### The Cost of Cumulative Dithering at 16 bits

As we have already seen, the measured amplitude of 16-bit dither is extremely low, approximately -91 dBFS. But a skilled listener does not have to listen at a very high level to hear the degradation of truncation or improper dithering. At 16 bits, dithering always sounds better than simple truncation.[13] But to avoid a potential sonic veil—**let there be only one generation of 16-bit dither in a CD project, the one-time, final process.**[14] Mix to a long wordlength medium and send that file to the mastering house, which will apply 16-bit dither once, at the last stage. Noise-shaping is fragile, 16-bit noise-shaped material should not be further processed, and cumulative noise-shaped 16-bit dithers can cause artifacts, clicks, ticks, or distortions especially if processed.

#### Diminishing Returns with 24-bit Chains

However, the effect of cumulative dither noise at 24 bits is so low it should not be a concern; a single dithered processor produces -139 dBFS noise, about 20 dB below that of the quietest known converter.[15] Within a native DAW, plugins communicate at 32 bits or longer, so there is an advantage to producing a 32-bit file (floating point, to be explained in Chapter 5). When feeding external digital processors via AES/EBU, the 32-bit word should ideally be dithered to 24 bits. And each processor should redither to 24 bits before feeding the next processor in a chain. But not all external processors provide dithering. Is redithering to 24 bits that important? The answer is in the masking—converter noise is usually enough to mask or cover the distortion products from one or more 24-bit truncations in a row. In my comparative listening tests, using high resolution monitors in a quiet room, it is extremely difficult to hear any difference between 24-bit dither versus truncation. Nevertheless, when possible, dither wherever wordlength is reduced, because accumulated distortion can be audible at the tail of a long series of processes. Bob Stuart, "We have on occasions found errors in dithering at the 22nd bit through listening."

In summary: If your DAW is 32-bit, mix if possible to a 32-bit file, which retains its resolution, or dither to 24 bits, but don't lose any sleep if you have to truncate to 24 bits once in a while.

### The Sound Effects of Defective Digital Processors

Never take a digital processor, or any DAW or computer that processes audio, for granted. For example, when software is changed or updated, we should never assume that the manufacturers have found all the bugs and we should assume that they may have created new ones. We even need to ensure that BYPASS mode, which seems seductively simple, actually does produce true clones in bypass. The illustration at right (courtesy of Jim Johnston) shows a series of FFT plots of a sinewave, illustrating the type of non-linear distortion products generated by truncation without dithering. The top row is an undithered 16-bit sinewave. Note the distortion products (vertical spikes at regular intervals, not harmonically related to the source wave). The second row is that sinewave with uniform dither. Note that the distortion products have disappeared. The bottom row is the formerly dithered sinewave, going through a popular model of digital processor with a defective BYPASS switch, and truncated to 16 bits. This is what would happen if a (16-bit) CD was fed through this processor in so-called BYPASS mode, and dubbed to a CDR! That is why every processor should be tested for bit-transparency before attempting to use them for master-quality work.

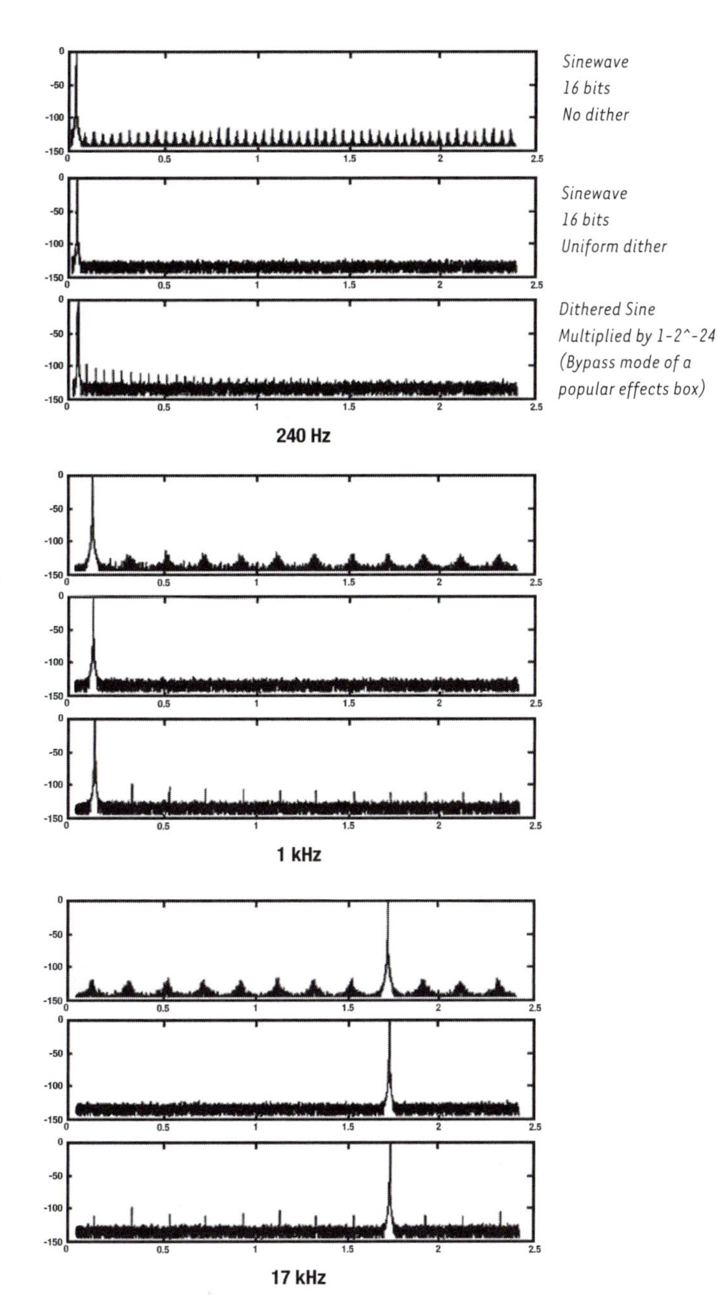

**MYTH:** *You can't mix source wordlengths in a single workstation session.*

## III. Some Practical Dithering Examples and Guidelines

1) When reducing wordlength, add dither. Example: From a 24-bit processor to a 16-bit CD.

2) Avoid dithering to 16 bits more than once on any project. Example: Use 24-bit intermediate storage, do not store intermediate products on 16-bit recorders.

3) Wordlength increases with almost any DSP calculation. Example: The outputs of digital consoles, DAWs and processors will be 24-bit, even if you start with a 16-bit source.

4) Every "flavor" of 16-bit dither and noise-shaping type sounds different and none is as good as the 24-bit original (though some come very close). It is useful to choose the "flavor" of dither which is more appropriate for a given type of music.

5) **In any project, sample rate conversion should be the next-to-the-last operation, and dithering to the shortest wordlength must be last.** Intermediate dithering may occur "behind the scenes", e.g. from 48 to 24 within the processor. While truncation should be avoided, if the SRC does not dither internally, truncation to 24 bits sounds far less bothersome to the ear than to 16 bits.

6) Peak-limited material with high levels may overload with the addition of dither, and some peaks could take it a smidgen past the top, so to be safe, drop the gain of peak-limited material 0.1 dB when dithering.

7) **Most** software adds the dither and produces the shorter length file at the same time. But sometimes that has to be done in two steps. For example, in a DAW like Pro Tools, invoke the dither (usually) in a plugin, and then bounce to a new file and tell it explicitly to make the new file 16 bits.

8) When bouncing tracks, if possible, dither the bus to the wordlength of the multitrack or DAW. If the multitrack is 16-bit digital, that's a violation of #2 above, so try to avoid bounces unless the multitrack is 24-bit.

## IV. Managing DAW Wordlengths

As of this writing, two existing workstations require that a session be defined in terms of wordlength and file format. This leads to user misconceptions when bouncing. For example, when bouncing to a new file in Pro Tools, even if the session is 16-bit, the output file should be 24-bit, or we would be truncating (distortion, loss of ambience). Similarly, consumer programs which purport to let consumers equalize and process CDs to produce new CDs—are **truncating**.

It is not necessary to "expand" the wordlength of the session before mixing: the sound can never get more resolved than what was originally encoded. Regardless of the source sample's wordlength, a workstation will always calculate to its internal precision, effectively adding zeros to the tail of any shorter words to facilitate the calculation (the padded zeros do not change the original value). In other words, 16, 24 and 32-bit samples can coexist in a well-designed workstation, and when calculations take place, all samples are multiplied to the wordlength of the workstation. Plugins should be captured to their full wordlength. However, since Pro Tools and Digital Performer restrict a session to a single wordlength, in order to insert bounced tracks back in the session, it is necessary to convert the session to a longer wordlength. This is an inconvenience because all they do to "expand" files is add padding zeros. Perhaps because of the inconvenience, neither of those workstations is commonly used by mastering engineers, who regularly mix wordlengths (and file formats) in the same session.

### Auto-Dither

We often have to combine previously-mastered and dithered music with new material. If no mastering processing needs to be applied to the previous material, we try to clone it and avoid adding a second generation of dither. There are a couple of ways to accomplish this. The first is by using **auto-dither by source wordlength**, which is a clever facility currently only available in a Prism brand processor. In its absence, we can route the already-mastered material to another DAW stream, direct to the output, bypassing the dither generator. There are other kinds of **auto-dither**, including **auto-black** which turns off the dither if the source audio level goes below a certain threshold for a period of time, useful if the producer insists on total silence between pieces.

## V. Dither At High Sample Rates

Moving to high sample rates automatically provides a signal-to-noise advantage, so 16 bits at 96 kHz is 3.4 dB quieter than at 44.1, sonically equivalent to about 16 1/2 bits. Noise-shaping at high sample rates can allow shorter wordlength files with very low psychoacoustic noise floor—the noise can be made extremely low and flat in the audible band and the shaping moved above 20 kHz. In fact, 16-bit noise-shaped dither at 96 kHz can sound as good as 24-bit/44.1, as I discovered one day when I accidentally left 16-bit dither on while working at 96 kHz. ***Figure C4-3*** in the Color Plates compares dithers, from highest to lowest level. At 16-bit/44.1 kHz: **in yellow**, POW-R 1, which is very flat with an extreme rise above 20 kHz. **In turquoise**, POW-R 3, with its extreme shaping. At 16-bit/96 kHz: **In green**, POW-R 3, whose shaping is postponed to almost 16 kHz, in the midband it is totally flat, but it is 16 dB quieter than its 44.1 kHz counterpart! At 24-bit/96 kHz: **in blue**, flat dither. Is there any further sonic advantage to shaping 24-bit dither? At least it measures quieter: **In orange** POW-R 1, whose rise doesn't occur until well above 32 kHz. **In red**, POW-R 3, whose midband level is below the chart (below -198 dBFS!). It is at least 18 dB below flat 24 bit dither, and since flat 24-bit dither is already considered inaudible....

**MYTH:**
*Expanding the wordlength of the samples from 16 to 24 (or 32) makes the sound better.*

1 **Continuous** does **not** mean that the analog signal has infinite resolution. Its finite resolution is defined as the lowest signal which is not covered by the noise floor.

2 Image courtesy of Jim Johnston.

3 Based on original concept by Mithat Konar, director of engineering, biró technology, with contributions from Robin Reumers and Jim Johnston.

4 As an exercise, count the number of 0's and 1's in this image and take the average. The average of the 1's should result in 9/24, which is 0.375. This relates to -0.25 volt (0.375*2 - 1) on the graph. We only present 24 samples, but since dither is a probabilistic system, an exact measurement would require an infinite number of samples!

5 In practice, it's more than just the LSB which is exercised. It can be all the bits. In base 10, if we add two numbers, and the sum is greater than 9, we have to carry. In base 2, we also have to carry and if the next significant digit to the left is not a zero, we have to keep on carrying until the next digit up is a zero and turn it into a 1. In 2's complement, the addition of dither at the LSB level will affect the values of many digits, including the MSB, as the number changes polarity between negative and positive. Seen on a bitscope, it seems to show two values at once because the numbers are always toggling with the addition of dither.

6 "Perfectly" to the lowest decimal place that can still be accurately determined over the noise floor.

7 Or below the coding floor of any particular wordlength. In other words, if we dither to 20 bits, whose coded range is 120 dB, we can include low level signals below the -120 dB limit. Or if we dither to 8 bits, we can include low level signals below 8-bit's normal limit of -48 dBFS.

8 The noise floor is raised 4.77 dB to be exact. This is the least amount of noise necessary to properly dither a digital audio signal and eliminate all possible distortion. The statistical distribution of the noise must be triangular probability. See Wannamaker, R. A., & Lipshitz, S. P., & Vanderkooy, J. (1992) Quantization and Dither: A theoretical Survey. *J.Audio Eng. Soc.*, Vol 40, No 7, pp.601.

9 Some processor and DAW manufacturers still have not recognized the importance of internal processor precision and the fact that wordlengths expand, a prime reason why some digital devices sound sweeter than others. The situation was once much worse, but by the end of this century's first decade, consumer-level, standalone and introductory DAWs still have not implemented dithering and wordlength management. Regardless, study the menus and preferences of each professional workstation to ensure they are making the right dithering choice.

10 For signals which are correlated, the formula is: dB change = 20 * log (ratio). For example, if we drop the level by a ratio of 1/2, whose log is -.3010, then multiply by 20, the approximate result is -6 dB (6 dB down), to the nearest decibel. Note the use of the word approximate, and yes, the degree of accuracy used in such calculations affects the quality of our audio.

11 To be exact, the low level (ambience) information that was present in the original wordlength is now spread proportionally over a much longer wordlength.

12 Chesky JD111, available at major record chains or through Chesky Records, Box 1268, Radio City Station, New York, NY 10101; 212-586-7799 (I produced this disc). The hard-to-find CBS CD-1, track 20, also contains a fade to noise test.

13 For the inharmonic distortion caused by quantization is very unmusical to the ear. Very different-sounding than turning a Marshall amplifier up to 11.

14 Since analog tape's noise floor is much higher than that of dither, many argue that several generations of 16-bit dither circa -91 dBFS should be insignificant. It depends on the material. Pristine, digitally-recorded material can sound veiled when "over dithered". But some rock and roll sounds better with lots of noise, or with flat dither instead of noise-shaped dither. And the psychoacoustic argument goes on, which is why we have ears to make judgments!

15 For example, six 24-bit dithered processors in a row raise the noise floor by 8 dB to -131, which is more than 11 dB below the noise floor of the quietest known converter.

16 Thanks to Paul Frindle for this most concise definition of wordlength.

CHAPTER 5

# Decibels: Not For Dummies

**Introduction**

So many of us take our meters for granted—after all, recording is simple: *all you do is peak to 0 dB and never go over!* But things only appear that simple until you discover that with the same material, one machine says that it peaks to -1 dB, another machine shows an OVER level, and yet your workstation tells you it just reaches 0 dB! To make things worse, among the expensive digital meters available, only a handful accurately convey the information we really need to know. In this chapter we will explore the different types of meters, the concept of the digital OVER, analog and digital headroom, gain staging, loudness, signal-to-noise ratio and we will also take a fresh look at the common practices of dubbing and level calibration.

## I. Stamp Out Slippery Language

### Bob's Top 10 List of Slippery & Confusing Audio Terms[1]

**10. INTENSITY**... is a measure of energy flow per unit area. For practical purposes, sound intensity is the same as SPL (see below).[2]

**9. LEVEL**... is a measure of intensity, but when used alone, because it can mean almost anything, it means absolutely nothing! To avoid confusion, the *level figure should always be qualified by a 'unit' term, e.g. voltage level, sound pressure level, digital level.* Examples: 40 dB SPL, -20 dBu, -25 dBFS. Each suffix defines the reference. SPL (sound pressure level) is a measure of the amplitude or energy of the physical sound present in the atmosphere. 40 dB SPL and 0.002 Pa (Pascals) are the same pressure, the first expressed in decibels relative to 0 dB SPL, the second in absolute pressure units.

"Level is often confused with Gain!"

**Level** is very often confused with **Gain**. For example, the sound pressure level from your monitor loudspeakers is often confused with the monitor gain. The term *monitor gain* is so slippery that I have started using a more solid term that everyone seems to understand: **MONITOR (control) POSITION.** For example, I say "the monitor control is at the 0 dB position."

**Stamp Out Monitor Controls Marked in SPL!** Some manufacturers think it's sophisticated to label a monitor control in dB SPL, such as 73, 74, 75. But this is confusing. They forget that gain must always be expressed in simple decibels with no suffix. When people use those mislabeled controls, they talk in circles. Speaking of the control position, Betty says: "I always play back at 83 dB." Thinking she meant sound pressure level, Fred replies: "But I measured 90 dB at your loudspeakers." That's why a monitor control should be simply marked from 0 dB to -∞, like any other attenuator.

**8. DECIBEL...** (dB) is a relative quantity; it is always expressed as a ratio, compared to a *reference*. For example, what if every length had to be compared to one centimeter? You'd say, "this piece of string is ten times longer than one centimeter." It's the same thing with decibels, though sometimes the reference is not explicitly stated but just implied. +10 dB means "10 dB more than my reference, which I defined as 0 dB." Decibels are logarithmic ratios, so if we mean "twice as much voltage," we say "6 dB more" [20 * log (2) = 6].

**DBU, DBM, DB SPL, DBFS...** are ratios with predefined references, so they can be converted to absolute values in volts, power, etc. I believe the term **dBu** was coined in the 1960's by the Neve Corporation; it means *decibels unterminated, compared to a voltage reference of 0.775 volts.* **dBm** means *decibels compared to a power reference of one milliwatt.* **dBFS** means *decibels compared to full scale PCM;* 0 dBFS represents the highest digital level we can encode.

**Plurals**. We **do** say "two decibels", but we do **not** pluralize the abbreviation. We do say "two dee bee", but we do not say "two dee bees".

**7. GAIN or AMPLIFICATION...** is a **relative** term expressed in plain decibels with no suffix: it is the ratio of the amplifier's output level to the input level. If an amplifier receives an input of -23 dBu and puts out an output of +4 dBu, it has 27 dB gain (without any suffix). See Sidebar.

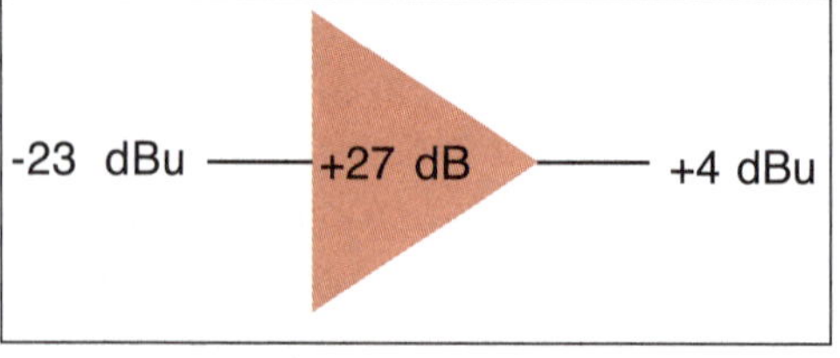


*The meaning of Gain vs. Level. An amplifier with 27 dB gain is fed an input signal whose level is -23 dBu to yield an output level of +4 dBu. The decibels of gain should never need a suffix.*

**6. ATTENUATION...** when expressed in dB is an optional term for negative gain, e.g. a loss. Examples: 20 dB attenuation is the same as -20 dB gain.

**5. SOUND PRESSURE LEVEL** (SPL)... is a measure of sound pressure in dB relative to 0.0002 dyne/$cm^2$ (0 dB SPL).[3] 74 dB SPL is the typical level of spoken word 12 inches away, which increases to 94 dB SPL at one inch distance. While we often see language like *95 dB SPL loud*, this is both inaccurate and ill-defined, as *loud* refers to the user's perception, and *SPL* to the physical intensity.

**4. LOUDNESS...** is used specifically and precisely for *the listener's perception. Loudness* is much more difficult to represent in a metering system, in fact, it's best presented as a series of numbers rather than as one overall figure of "loudness." Two pieces of music that measure the same on a flat level meter can have drastically different *loudness.* A true loudness meter makes a complex calculation using SPL, frequency content, and duration. Exposure time also affects our perception; after a five minute rest, the music seems much louder, but then we get used to it again—another reason why it is wise to have an SPL meter around to keep us from damaging our ears.

**3. INTRINSIC LOUDNESS...** In the first edition I invented a term "absolute loudness", but it wrongly gives the impression that we somehow have control over the consumer's volume control. So I'm replacing that with a new term **intrinsic loudness**, which I define as the loudness of a

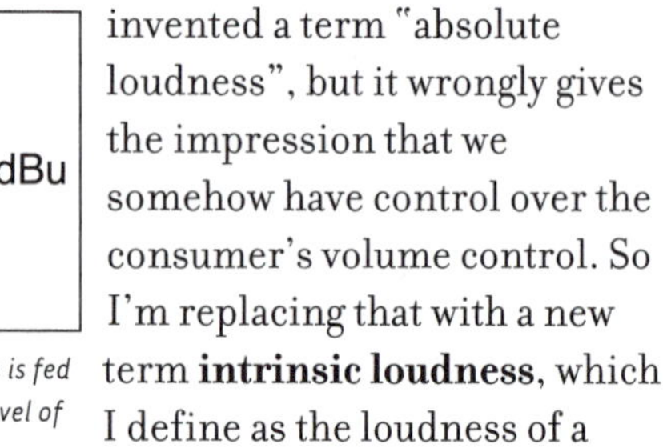

program **before the level is adjusted using the monitor control**. Since there is no SPL reference in a digital file, intrinsic loudness has no absolute units, but the term can be used in a relative way. We can compare two programs' intrinsic loudness by switching between them, adjusting the monitor control until they sound equally loud, and noting the decibel difference in the monitor positions. Then we can say that program 1 has "2 dB more intrinsic loudness than program 2" though, for brevity, I may say program 1 is 2 dB "louder", using quotation marks. When I use the term **hot CD** or **hot master,** I am referring to a recording which has a high intrinsic loudness. Our perception of the program's loudness is also affected by the behavior of the monitor DAC. For if a program is so distorted that its analog reconstructed level would cause a certain DAC to overload, this DAC may appear louder due to the high frequency distortion. **Intrinsic loudness** would not have been a meaningful term in the analog era because analog tapes and LPs do not have a consistent reference, but with digital 0 dBFS is always the same.

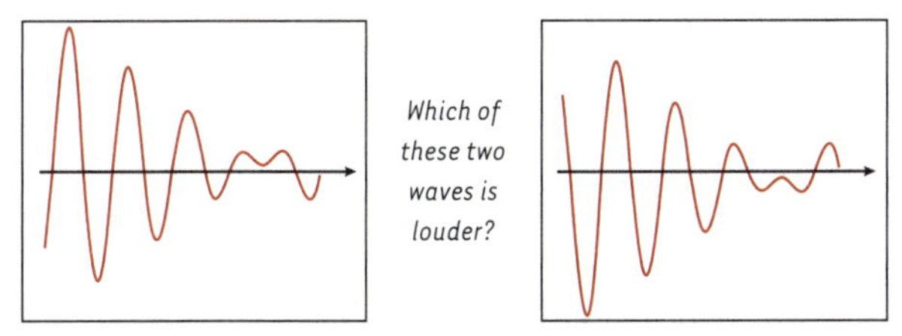

*Which of these two waves is louder?*

**2A. Average vs. Peak Level**... Which of these two waves is louder? The answer is: Both have the same loudness. Both have the same average and maximum peak level. The first wave is identical to the second except its polarity is exactly reversed.[4] Think of this picture as the graph of the movement of a drumhead seen from the front and back side. The terms average and peak are a bit misleading if we consider the up-going direction as *positive* and the down-going as *negative*, because the top wave has a maximum in the positive direction and the bottom has a maximum in the negative; however the **absolute value** of the greatest peak in each wave is the same, so we say each wave has the same maximum peak level. Over a long period of time, the average of the positive and negative numbers must be zero, which is the static pressure of the atmosphere (the position of the drumhead at rest). However, the ear generally ignores polarity when it considers loudness, so it sees this wave **rectified** (pictured), where all the negative segments have been converted to a positive, or more correctly, absolute value.

*Rectified wave*

The average value of this rectified wave over time is somewhere between its peak value and zero; our meters determine this average depending on the method of averaging and the time length of the average. Sometimes we set our SPL meters to read peak level, but generally we set them to read average sound pressure level because this correlates closer to how the ear hears. For proper monitor calibration, a good SPL meter should use the *RMS* averaging method, as opposed to a simple averaging which can produce as much as 2 dB measurement error. RMS ignores issues such as phase shift and tells us the true **energy level** of the recording. The word *average* in this book can refer to the true RMS level or the simple average; when it is important, I will specifically say RMS.

**MYTH:**
*"The red light came on while I was recording, but when I played it back, there weren't any overs, so I thought it was OK."*

*Contributed by Lynn Fuston*

**2B. Crest Factor**, also known as **Peak-To-RMS Ratio**, is the difference between the RMS level of a musical passage and its instantaneous peak level. In practice, any averaging meter may be used, e.g. if a fortissimo passage measures -20 dBFS on the averaging meter and the highest momentary peak is -3 dBFS on the peak meter, it has a crest factor of 17 dB. It is extremely rare to encounter a piece of music with a crest factor greater than 20 dB, so this is the commonly cited maximum. When the dynamic range (the difference between the loudest and softest passages) of a recording has been reduced, we say the material has been **compressed**, and that a compressed recording has a lower crest factor than an uncompressed one.[5]

*And the # 1 most confusing audio term is...*

**1. Volume** ... is **usually associated with an audio level control, but is an imprecise consumer term.** Volume is measured in quarts, liters and cubic meters! The words more properly used in our art are **Level** and **Loudness.** The big problem is that consumers use the term ambiguously, to mean both the loudness they perceive and the position of the "volume control"—a perfect example of confusing gain with level! So in this book I prefer to use the professional term **monitor control.** I rarely use the word *volume* except when speaking informally to clients or consumers, and occasionally succumb to saying "volume control" when referring to a consumer's system.

## II. Meters Meters Meters

**We Won't Get Fooled Again.** Recording engineers rely heavily on their favorite meter, and this book is not intended to change people's favorite. But as practicing engineers, it is prudent to learn the defects and virtues of each meter we encounter.

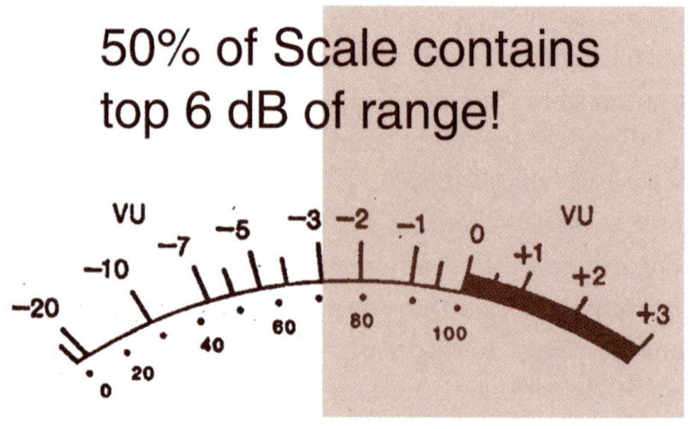


*VU meter operators are often fooled into treating the top and bottom halves of the scale with equal weight, but the top half has only 6 dB of the total dynamic range.*

**The VU Meter.** Relative newcomers to the industry may have never seen a VU meter, and some of them may be using the word "VU" incorrectly to describe peak-reading digital meters. **VU** should only be applied to a true VU meter that meets a certain standard. The first thing we must learn is that *the VU meter is a dreadful liar...* It is an averaging meter, and so it cannot indicate true peaks, nor can it protect us from overload. However the VU does do one thing better than a peak meter—it comes closer to our perception of loudness, but even so, it is a very inaccurate loudness meter because its frequency response gives low frequency information equal weight, and the ear responds less to low frequencies. Another problem is that the VU meter's scale is so non-linear that inexperienced operators think that the greater part of the musical action should live between -6 and +3 VU, but this is wrong. A well-engineered music program has plenty of meaningful life down to about -20 VU, but since the needle hardly moves at that level, it scares the operator into thinking the level is too low. Only highly-processed (dynamically compressed) music can swing in such a narrow range; in other words, the VU scale encourages overcompression (we'll discuss compression of dynamic range in Chapter 9). Hence the VU meter should only be taken as a guide.[6] A much better averaging meter would have a linear-decibel scale, where each decibel has equal division and weight down to -20 dB. We'll discuss the use of averaging meters in more detail in Chapter 14.

### Digital Peak Meters

Digital Peak meters come in three varieties:

1. Cheap and dirty
2. Sample-accurate and sample-counting (but misleading)
3. Reconstruction (oversampling)

**Cheap and Dirty Peak Meters**. Recorder manufacturers pack a lot in a little box, often compromising on meter design to cut production costs. A few machines even have meters which are driven from analog circuitry—a definite source of inaccuracy. Some manufacturers who drive their meters digitally (by the values of the sample numbers) cut costs by putting large gaps on the meter scale (avoiding expensive illuminated segments). The result is that there may be a -3 and a 0 dB point, with a large unhelpful no man's land in between. When recording with a meter that has a wide gap between -3 and 0, it is best practice to stay well below full scale.

### Sample-Accurate and Sample-Counting Meters.

Several manufacturers have produced sample-accurate meters with 1 dB (or smaller) steps, that convert the numeric value of the samples to a representation of the sample value, expressed in dBFS.[7]

**The Paradox of the Digital OVER**. When it comes to playback, a meter cannot tell the difference between a level of 0 dBFS (*FS = Full Scale)* and an OVER. That's because once the digital signal has been recorded, the sample level cannot exceed full scale, as in this figure.

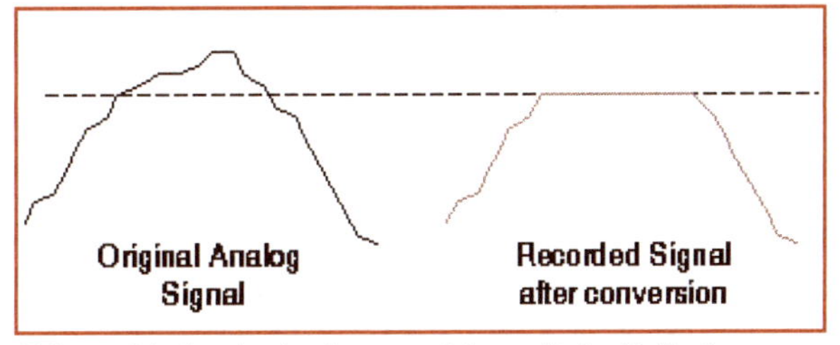


*While an original analog signal can exceed the amplitude of 0 dB, after conversion there will be no level above 0, yielding a distorted square wave. This diagram shows a positive-going signal, but the same is true on the negative-going end.*

We need a means of knowing if the ADC is being overloaded during recording. So we can use an early-warning indicator—an **analog** level sensor prior to A/D conversion—which causes the OVER indicator to illuminate if the analog level is greater than the voltage equivalent to 0 dBFS. If the analog record level is not reduced, then a maximum level of 0 dB will be recorded for the duration of the overload, producing a distorted square wave.

After the signal has been recorded, a standard sample-accurate meter cannot distinguish between full scale and any part of the signal that had gone over during recording, it shows the highest level as 0 dBFS. However, a **sample-counting meter** can analyze a recording to see if the ADC had been overdriven. This meter counts contiguous samples and can actually distinguish between 0 dBFS and an OVER after the recording has been made! The sample-counting digital meter determines an OVER by counting the number of samples in a row at 0 dB. If 3 contiguous samples equal 0 dBFS, the meter signals an OVER, because it's fair to assume that the incoming analog audio level must have exceeded 0 dBFS somewhere between sample number one and three.[8] Three samples at 44.1 kHz is a very conservative standard; on that basis, the recorded distortion would last only 33 **microseconds** and would probably be inaudible. While this type of meter was sophisticated in its day, current thinking is that the sample-counting meter is only suitable for evaluating whether an ADC has overloaded. Authorities now feel that meters which display the digital value of the samples and which count samples to determine an OVER are no longer sufficient for mastering purposes and should be used with caution during mixing. Their place is taken by...

### The Reconstruction Meter: Even More Sophisticated

As long as a signal remains in the digital domain, the sample level of the digital stream is sufficient to tell us if we have an OVER. However, **signals which migrate between domains can exceed 0 dBFS and cause distortion.** This includes any signal that passes through a DAC, a sample rate converter, or is converted through a codec such as mp3 or AC3. During the conversion from PCM digital to analog or mp3, filtering within the converter yields occasional peaks **between the samples** that are higher than the digital stream's measured level, which can be higher than full scale. This next figure shows that contrary to what we might assume, **filtering or dips in an equalizer which we'd imagine would produce a *lower* output can actually produce a higher output level than the source signal**. B.J. Buchalter explains that

> the third harmonic is out of phase with the fundamental at the peak values of the fundamental, so it serves to reduce the overall amplitude of the composite signal. By introducing the filter, you have removed this canceling effect between the two harmonics, and as a result the signal amplitude increases. Another reason for the phenomenon is that all filters resonate, and generally speaking, the sharper the filter, the greater the resonance.[9]

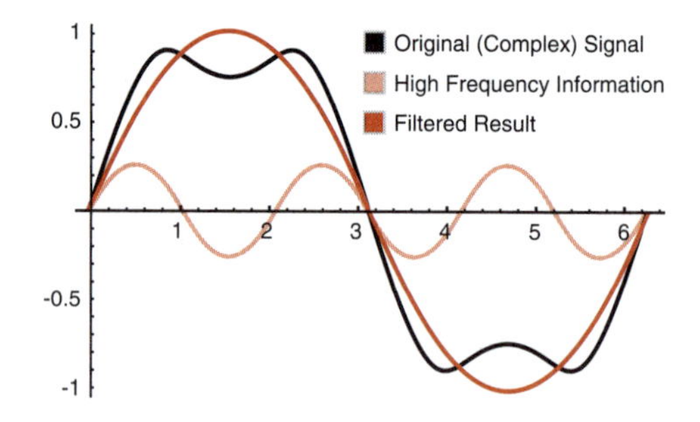


*In black is a complex wave. When the high frequency information (light orange) is filtered out, the result is a signal (orange) that is higher in amplitude than the original!*

**Equipment designers have known for years that because of filtering, the analog output level of**

**complex audio from a DAC can exceed the sinewave value of 0 dBFS** but very few have taken this into account in the design. TC Electronic has performed tests on typical consumer DACs,[10] showing that many of them distort severely since their digital filters and analog output stages do not have the headroom to accommodate output levels which exceed 0 dBFS! While typical 0 dBFS+ peaks do not exceed +0.3 dBFS, some very rare 0 dBFS+ peaks may exceed full scale by as much as 4 or 5 dB with certain types of signals—especially mastered material which has been highly processed, **clipped** (turned into a square wave on top and bottom), and/or brightly equalized.[11] By oversampling the signal, we can measure peaks that would occur after filtering. An oversampling meter (or reconstruction meter) calculates what these peaks would be, but these meters are still rare. Products from TC Electronic (System 6000) and Sony (Oxford) have an oversampling limiter and reconstruction peak meter. RME's Digicheck software includes an oversampling meter.[12]

*"Signals which cross domains can exceed 0 dBFS"*

Reconstruction meters tell us not only how our DAC will react, but what may happen to the signal after it is converted to mp3 or sent to broadcast, both of which employ many filters and post-processes. Many DSP-based consumer players cannot handle the high levels at all and exhibit severe distortion with 0 dBFS+ signals. Armed with this knowledge, no mastering engineer should produce a master that may sound acceptable in the control room but which she knows will likely produce severe distortion when post-processed or auditioned in the real world. If the reconstruction meter is not enough to convince the client, she should also demonstrate that this "loud" signal becomes distorted, ugly, and soft when it is converted to low bit rate mp3. All the harmonics which made the signal seem loud in the control room have been converted to additional distortion. For example, ***Figure C5-1*** in the color plates is a spectragram of the remnant distortion from an mp3 conversion of two different CDs. Time moves from left to right, levels vary from highest in red to lowest in blue, and high frequencies are at the top. At left, a clipped CD master; at right, a "loud" master made without clipping, then brickwall limiting. Note how the high frequency distortion in CD #1 cuts off sharply below about 12 kHz, indicating some kind of aliasing distortion in the mp3 converter.[13]

### Practice Safe Levels

What this means is that if you are mixing with a standard digital meter, keep peaks below -3 dBFS, especially if you are using aggressive bus processing.[14] The more severely processed, equalized or compressed a master, the more problems it can cause when it leaves the mastering studio. We didn't start hearing about this problem, or at least the severity of it, before the loudness race and the invention of digital processing which could be egregiously abused. Maximizing engineers should try to use a

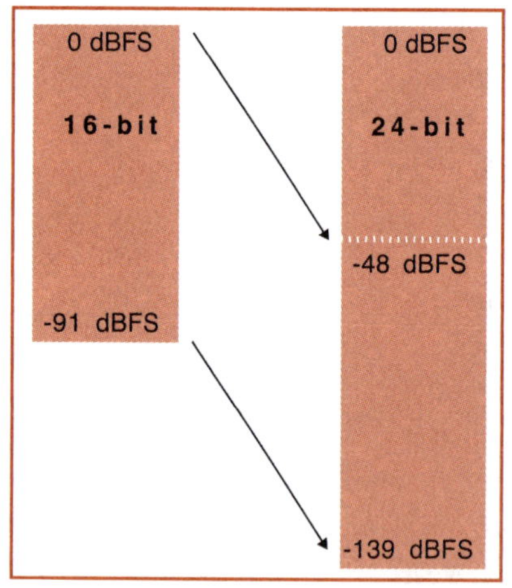


*A 24-bit recording would have to be lowered in level by 48 dB in order to reduce it to the SNR of 16-bit. The noise floors shown are with flat dither.*

reconstruction meter and/or an oversampled brickwall limiter. If these are not available, use a standard peak limiter whose ceiling is set to -0.3 dB (see Chapter 10) and exercise caution. But even the oversampled brickwall limiter is not foolproof; I've discovered that such limiters do not protect from very severe processing and can still make a consumer DAC overload unpleasantly. The best solution is to be conservative on levels. Clipping of any type is to be avoided, as demonstrated in Appendix 1.

### The Myth of the Magic Clip Removal

If the level is turned down by as little as 0.1 dB, then a recording which may be full of OVERs will no longer measure any overs. But this does not get rid of the clipping or the distortion, it merely prevents it from triggering the meter. Some mastering engineers deliberately clip the signal severely, and then drop the level slightly, so that the meters will not show any OVERs. This practice, known as **SHRED**, produces very fatiguing (and potentially boringly similar) recordings.[15]

*"You would have to lower the peak level of a 24-bit recording by 48 dB to yield an effective 16-bit recording!"*

### Peak Level Practice for Good 24-bit Recording

Even though 24-bit recording is now the norm, some engineers retain the habit of trying to hit the top of the meters, which is totally unnecessary as illustrated at left. Note that a 16-bit recording fits entirely in the bottom 91 dB of the 24-bit. You would have to lower the peak level of a 24-bit recording by 48 dB to yield an effective 16-bit recording! There is a lot of room at the bottom, so you won't lose any dynamic range if you peak to -3 dBFS or even as low as -10 dBFS, and you'll end up with a cleaner recording. Since distortion accumulates, if a "hot" recording arrives for mastering, the mastering engineer doing analog processing may have to attenuate the level to prevent the processing DAC from overloading. A digital mix that peaks to -3 dBFS or lower makes it easier to equalize and otherwise process without needing an extra stage of attenuation in the mastering.

A number of 24-bit ADCs are advertised as having *additional headroom*, achieved by employing a built-in compressor at the top of the scale, claiming that the compressor can also protect the ADC from accidental overloads. But this is specious advertising. Level accidents don't occur in a mix studio; engineers have control over their levels and when tracking live musicians, it is better to turn off the ADC's compressor, drop the level and leave plenty of headroom for peaks. The only possible use of this function of a compressor is if you like its *sonic qualities* and are trying to emulate the sound of tracking to analog tape. But since tracking decisions are not reversible, I suggest postponing "analog simulation" to the mixing stage. It's easier to add warmth later than try to take away some mushiness due to an overdriven compressor. As we have just seen, there is no audible improvement in SNR by maximizing a 24-bit recording and no SNR advantage to compressing levels with a good 24-bit ADC.

#### How Loud is It?

Contrary to popular belief, the levels on a digital peak meter have (almost) nothing to do with loudness. Here is an illustration. Suppose you are doing a direct to two-track recording (some engineers do still work that way!) and you've found the perfect mix. Leaving the faders alone, you let the musicians do a couple of great takes. During take one, the performance reached -4 dB on the meter; and in take two, it reached 0 dB for a brief moment during a snare drum hit. Does that mean that take two is louder? No: **because in general, the ear responds to average levels, not peak levels when judging loudness**. If you raise the master gain of take one by 4 dB so that it too reaches 0 dBFS peak, it will sound 4 dB louder than take two, even though they both now measure **the same** on the peak meter.

An analog tape and digital recording of the same source peaked to full scale sound very different in terms of loudness. If we make an analog tape recording and a digital recording of the same music, and then dub the analog recording to digital, peaking at the same peak level as the digital recording, the analog dub will have about 6 dB more intrinsic loudness than the all-digital recording. Quite a difference! This is because the peak-to-average ratio of an analog recording can be as much as 12-14 dB, compared with as much as 20 dB for an uncompressed digital recording. Analog tape's built-in compressor is a means of getting recordings to sound louder (oops, did I just reveal a secret?).[16] That's why pop producers who record digitally may have to compress or limit to compete with the loudness of their analog counterparts.

#### The Myths of Normalization

**The Esthetic Myth.** Digital audio editing programs have a feature called Peak Normalization, a semi-automatic method of adjusting levels. The engineer selects all the songs on the album, and the computer grinds away, searching for the highest peak level on the album and then automatically adjusts the level of all the material until the highest peak reaches 0 dBFS. If all the material is group-normalized at once, this is not a serious esthetic problem, as long as all the songs have been raised or lowered by the same amount. But it is also possible to select each song and normalize it individually, but this is a big mistake; since the ear responds to average levels, and normalization measures peak levels, the result can totally distort musical values. A ballad with little crest factor will be disproportionately increased and so will end up louder than a rock piece with lots of percussion!

**The Technical Myth.** It's also a myth that normalization improves the sound quality of a recording; it can only degrade it. Technically speaking, normalization adds one more degrading calculation and level of quantization distortion. And since the material has already been mixed, it has already been quantized, which predetermines its signal-to-noise ratio—which cannot then be further improved by raising it. Let me repeat: raising the level of the material will not alter its inherent signal-to-noise ratio but will add more quantization distortion. **Of course material to be mastered does not need normalizing** since the mastering engineer will be performing further

MYTH:
Peak Normalization Makes the Song Levels Correct

**MYTH OF THE MAGIC CLIP REMOVAL:** *Turn it down after clipping and the clip will go away.*

processing anyway.[17] Clients often ask: "do you normalize?" I reply that I never use the computer's automatic method, but rather songs are leveled by ear.

### Average Normalization

This is another form of normalization, an attempt to create an intelligent loudness algorithm based on the average level of the music, as opposed to the peak. But when making an album, neither peak nor average normalization nor any intelligent loudness algorithm can do the right job, because the computer does not know that the ballad is supposed to sound soft. There's no substitute for the human ear. However, average normalization or better, a true intelligent loudness algorithm can help in situations where every program needs the same loudness, even if that doesn't sound natural, such as radio broadcast, ceiling loudspeakers in a store, a party or background listening.

### Judging Loudness the Right Way

Since the ear is the only judge of loudness, is there any objective way to determine how loud your CD will sound? The first key is to use a single DAC to reproduce all your digital sources and maintain a fixed setting on your monitor gain. That way you can compare your *CD in the making* against other CDs, in the digital domain. Judge DVDs, CDs, workstations, and digital processors through this single converter.

## III. Analog Studio Levels, Headroom and Cushion

**Protecting the Mix from Clipping the ADC.** Professional mixing studios with analog consoles are still using VU meters to measure average program level and feed the console output to an ADC. I use the term *nominal* to mean the voltage level with a sine wave that corresponds with 0 VU, typically 20 dB below full scale digital (0 dBFS). To set up the system, feed a sinewave through the analog system at 0 VU and adjust the gain of the ADC to produce -20 dBFS, measured with an accurate digital meter. This protects the mix from clipping the ADC since the peak to average ratio of typical music is no more than 20 dB, more typically 12 to 14 dB.

**Headroom of the analog gear.** Protecting your ADC and mix from clipping does no good if your analog console is distorting in front of the ADC! In the mastering suite we usually chain multiple pieces of analog gear, so it's important to learn about the analog levels, distortion and noise in the analog signal chain in front of your ADC.

Not all analog gear is created equal, and the standard nominal +4 dBu[18] may be too high for two reasons:

The first reason is the clipping point of cheaper analog gear has gone down over the years, to save money on parts. Before the advent of inexpensive 8-buss consoles, most professional consoles' clipping points were +24 dBu or higher. But frequently, low-priced console design uses circuits that clip at a lower level, around +20 dBu (7.75 volts). This can be a big impediment to clean audio, especially when cascading amplifiers.

The second reason is that in my opinion, some solid state circuits exhibit an extreme distortion increase **long before they reach the actual clipping point.**[19] So the music peak level must stay below the distortion region, not just the clipping point. To avoid the *solid-state edginess* that plagues a lot of solid state equipment, we should use amplifiers that clip at least 6 dB above the potential peak level of the music, or else lower the level of the signal to suit the amplifier. This means that for a 0 VU level of +4 dBu, the clipping point should be at least +30 dBu (24.5 volts RMS)! That's why more and more high-end professional equipment have clipping points as high as +37 dBu (55 volts!). To be that robust, an amplifier must use very high output devices and high-voltage power supplies. The effects of this are higher cost due to the need for more robust parts but, all other things being equal, the amplifier with the higher clipping point will sound better. Perhaps that's why tube equipment (with its 300 volt B+ supplies and headroom 30 dB or greater) often has a *good* name and solid state equipment with inadequate power supplies or headroom has a *bad* name (see Figure at right).

**Cushion.** Traditionally, the difference between **average level** and clip point has been called the *headroom*, but in order to deal with the increased distortion as the amplifier approaches clipping, I'll call the part of the headroom between the **peak level** of the music and the amplifier clip point a *cushion*. If an active balanced output feeds an unbalanced input, the clipping point reduces by 6 dB, so the situation becomes proportionally worse.[20] Dual-output consoles that are designed to work at either professional or semi-pro levels can be particularly problematic. Ironically, the lower output level in semi-pro mode may sound cleaner.

We should raise the question of whether the professional standard of +4 dBu is still appropriate because not every mastering or mixing studio has gear with extremely high clip points. Short of replacing all the gear, the easiest solution is to lower the analog level that represents nominal level or 0 VU. **I recommend a studio standard nominal analog level of 0 dBu, or 0.775 volts.** Just that simple 4 dB decrease can help produce a cleaner analog chain. Many European studios have been using a 0 dBu standard for decades for this very reason.

**Internal clipping point in DAC.** One of the most common mistakes made by digital equipment manufacturers is to assume that, if the digital signal *clips* at 0 dBFS, then it's OK to install a (cheap) analog output stage that would clip at a voltage equivalent to, say, 1 dB higher. This almost guarantees a nasty-sounding converter or recorder, because of the lack of cushion in its analog output section and the potential for 0 dBFS+ levels.

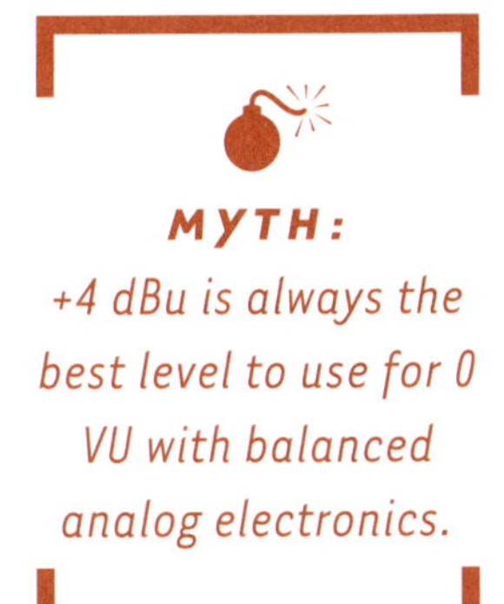


## IV. Gain Staging—Analog Chains

Now that we know how to choose an analog level, it's time to chain our equipment together. To really get a handle on our equipment, we should determine its internal structure. The figures on the next page represent two possible internal structures. All complex equipment structures are variations on these themes.

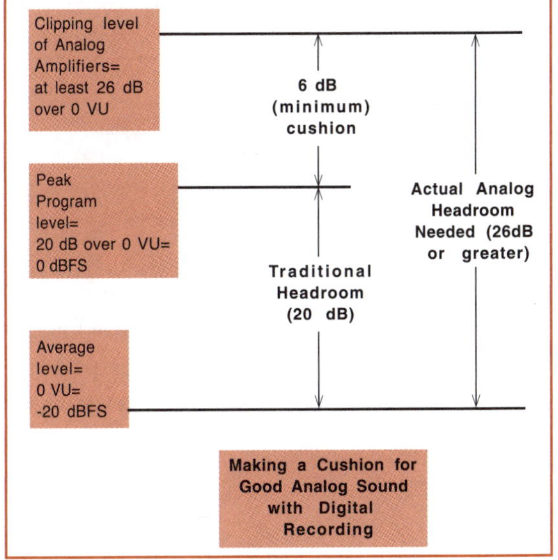

To properly test analog devices and determine their internal makeup, use a good clean monitor system, an oscilloscope, a digital voltmeter and a sine wave generator that can deliver a clean +24 dBu or higher (a tough requirement in itself). There are two different types of devices. The first type has a passive attenuator on its input, which means that we can feed it any reasonable source signal without fear of overload. We can ascertain whether there is a passive attenuator on the input side by turning the generator up and the attenuator down and observing whether or not the output clips. We can also disconnect the generator and listen to the output of the device as we raise and lower the attenuator. There should be no change in noise or hiss, and the output noise should be well below −70 dBu unweighted, preferably below −90 dBu A-Weighted. This is another indication that the device has a passive attenuator on its input. If the output noise changes significantly at intermediate positions of the attenuator, then the internal impedances of the circuit may not be optimal, or there may be some DC offset. The output noise of this device will be limited by the noise floor of its output amplifier. We determine the best **nominal operating level** of this device by taking the output clip point and subtract at least 26 dB for headroom and cushion.

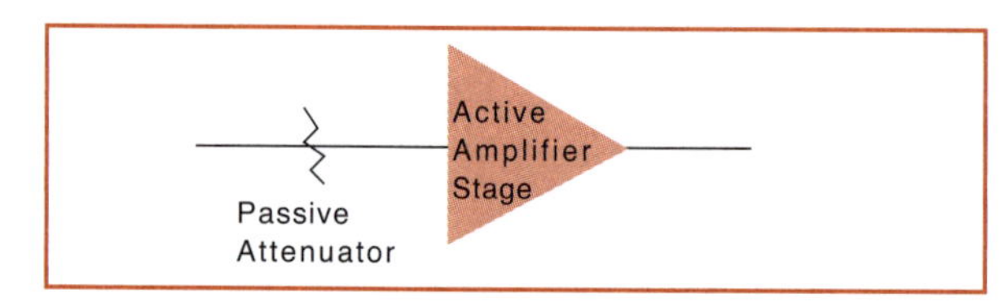


*In the top device, signal enters a passive attenuator and exits through an active amplifier stage. This circuit effectively has infinite input headroom. The bottom device's input headroom is determined by the headroom of the input amplifier.*

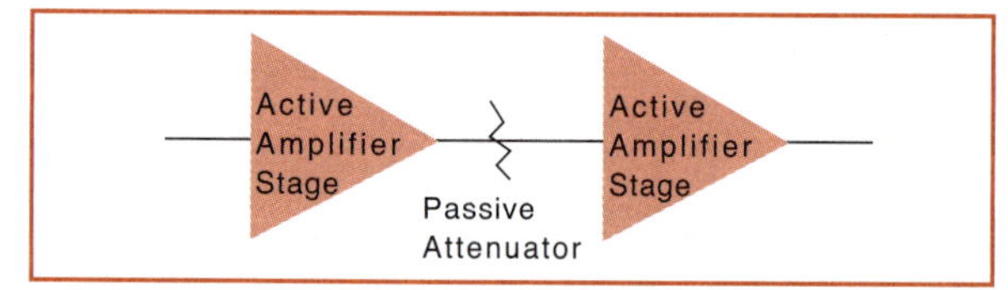


The second type of device has an active amplifier stage on its input, the design of which is much more critical. It is very rare to find a solid state device built this way which won't clip with >+24 dBu input. While raising the signal generator, turn down the attenuator to keep the output from overloading. If we hear clipping prior to the generator reaching +24 dBu, then the device has a weak internal signal path. The clip point determines the nominal analog input level, which should be at least 26 dB below this clip point. Then, we should test to see that the output stage clips at a level no lower than the input stage.

**System noise**. When cascading analog gear, the noise of the system is determined by the weakest link. Set your monitor gain with typical music, then listen closely to the noise floor at the last device in the chain; if the output of the chain sounds good and reasonably quiet, then don't worry about tweaking the chain. In an analog signal chain, raising the music signal level—as high as practical—as early as possible in the chain will improve the signal-to-noise ratio of the entire chain. In general, tube gear has a higher noise floor, so if gain has to be turned up, it should be in front of noisy tube gear, not after it.

## V. Gain Staging—Digital Chains

There is no loss or gain in a digital interconnection such as AES/EBU or S/PDIF but we still have to be concerned about overloads. As we mentioned, equalizers can increase level even when dipping. Many outboard digital processors do not have accurate metering so I recommend patching an external digital meter to their output. If the processor overloads, try attenuating at either the input or the output.

**Headroom of the Processor.** We can test digital systems for headroom, clipping, and noise using digitally-generated test tones and an FFT analyzer. Suppose we have a digital equalizer with several gain controls and equalization; we feed it a 1 kHz sine wave test tone at about -6 dBFS and turn up the 1 kHz equalization by 10 dB, observing that the output clips. Then we turn down the output gain control until the output is below 0 dBFS and verify by listening or FFT measurements that the internal clipping stops. If the clipping does not stop, this indicates that the internal gain structure of the equalizer does not have enough headroom to handle wide range inputs. We may be able to get away with turning down the signal in front of the equalizer, or the EQ's input attenuator if it has one, but the early clipping indicates that this equalizer is not state-of-the-art. Modern-day digital processors should have enough internal headroom to sustain considerable boost in early stages without needing an input attenuator, and clipping can be removed solely by turning down the output attenuator.

**Noise of the Digital Chain.** With a digital chain, we no longer have to maximize the audio signal level in each piece of gear; a low level signal in a 24-bit digital signal chain does not hurt the SNR, considering the inaudible (approximately -139 dBFS) noise of the chain.[21] Instead of getting hung up on the signal level, we should consider every calculation stage as a source of quantization distortion. What matters most in a digital processing chain is to reduce the number of total calculations and use high-quality calculations, e.g. give the job of gain changes and other processing to the components with the highest internal resolution (those which would introduce the least quantization distortion or *grunge*). In fact, we should avoid raising the signal until it reaches a device which has the cleanest-sounding gain control, even if the source audio level is very low. For example, if the workstation has lower resolution than the outboard gear (shorter internal wordlength), we try to hold everything at unity gain in the DAW and reserve the gain changes or EQ for higher-precision devices later in the signal chain. Regardless of the level, pass a perfect clone (bit-transparent copy) of the source from the DAW onto the next device in line to do processing.

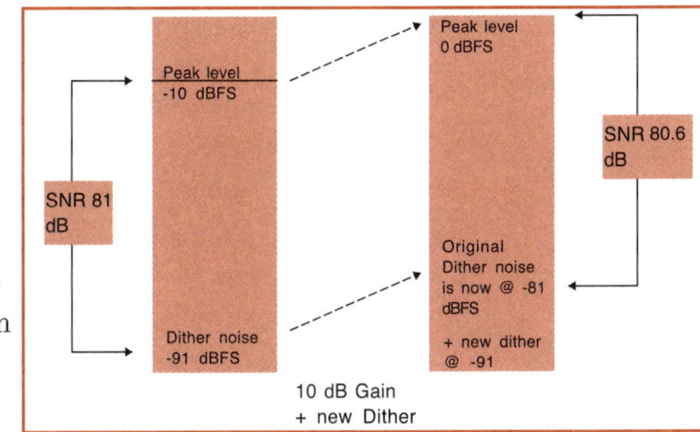


*A 16-bit recording with peak level low at −10 dBFS. When gain is raised 10 dB and redither is added, the original 81 dB peak signal to RMS noise ratio is reduced by about 0.4 dB.*

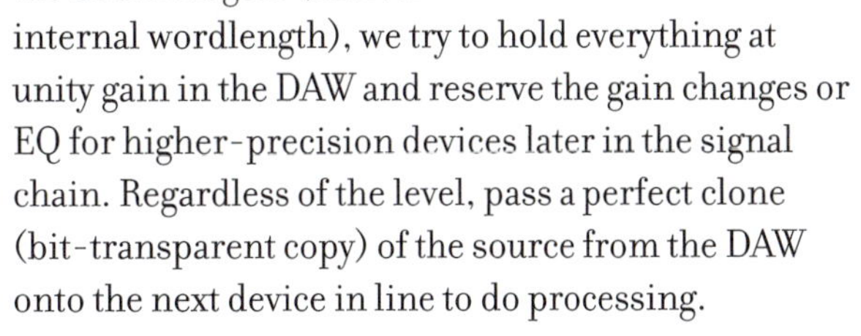

The significant noise floors in a 24-bit chain are not from the digital chain but from the original sources, including mike preamp noise, and our primary concern is with the impact of these higher level noises. We should be aware that noise floors sum in digital in the same way they do in analog, including dither noise. Let's take an example of a 16-bit recording whose peak level is 10 dB low, as in the above figure. In mastering we may choose to raise its level by 10 dB and so must add 16-bit dither before turning it into a 16-bit master. Disregarding the mike preamp and room noise, the original 16-bit recording's dither noise is at -91 dBFS and thus it has a peak signal to RMS noise ratio of 81 dB (-10 - -91).[22] When we raise the signal by 10 dB, both the original signal and the original noise are raised equally, so the original signal to noise ratio is unchanged. However, the new noise floor is the RMS sum of the original dither which has simply become noise at -81 dBFS and the added (new) dither which is at -91 dBFS. We ignore the insignificant noise of the gain processing, well below -130 dBFS, so the new dither raises the noise to -80.6 dBFS, less than 0.5 dB worse, so by raising the gain of the source, the new dither is an insignificant contributor to the total noise. However, if we add two 16-bit dithers without changing the gain of the source, the noise floor goes up by 3 dB—and this may cause a sonic veiling. Despite the potential for degradation, many times we still receive 16-bit sources; and we are forced to adjust the level according to the esthetics of the album. Fortunately I've had considerable success reducing cumulative sonic veiling by using noise-shaped dither.[23]

The manufacturers of the Waves L2 are concerned about losing resolution when converting from 24 to 16 bit. They feel that peak limiting allows raising level enough for it to be significantly above the dither noise, thus increasing the resolution. But exercise caution, because to my ears the apparent noise improvement is more than offset by the loss of transient clarity when peak limiting excessively. Is it worth sacrificing transient clarity just to gain a couple of dB increased signal-to-dither ratio?

If we could avoid 16-bit dither, by producing an output at 24-bit that the consumer could use, then mastering processing and gain-changing could be performed with no significant penalty, with a noise floor 48 dB below the noise of 16-bit. This is the promise of delivering higher wordlengths to the consumer and another reason to record in 24-bit in the first place.

### The Imaginary World of Floating Point

A **fixed point processor** has a fixed maximum peak level of 0 dBFS and a fixed noise floor as per its wordlength, which for dithered 24 bit is approximately -139 dBFS. But a **floating point processor** is capable of doing tricks that do not relate to the real world. It is practically impossible to clip a floating point processor, you can raise the gain by hundreds of dB without clipping. Probably 95% of current native (CPU-based) plugins use floating point processing. Probably 80% of current outboard digital processors use floating point processing. However, **all converters use fixed point**, so wherever a floating point processor meets "the real fixed-point world", the signal must be regulated to a normal level. In other words, you can construct a floating point signal chain only internally within a DAW.[24]

In a floating point system, you can break all the rules: floating point can literally ignore the individual levels in the chain. It's possible to drop the signal level 100 dB, store the signal as a floating point file, then open the file, raise the gain 100 dB and get back the original signal, with little or no deterioration. Or vice versa, you can raise the signal 100 dB and then lower it and get the same result, providing that intermediate products are stored in floating point format. Most floating point processors indicate when signal is above 0 dBFS. Some warn you with a red light that this signal level should not be fed to the real (fixed-point) world. You can test your DAW's internal signal chain for floating point integrity by running the level of the first processor above 0 dBFS. Listening to that processor, it should sound distorted, since your DAC only accepts fixed point numbers but if you then drop the level below 0 dBFS in the last processor of the chain, you should no longer hear any distortion if the chain is working properly.

In Chapter 10 we'll learn more about compressors and limiters, which reduce dynamic range. This figure illustrates the power of floating point by connecting a Waves C1 compressor plugin to a Waves L2 digital limiter. Notice that the output gain of the compressor has been set to +6 dB, and this has brought the highest peaks of the music into overload. But because the compressor is a floating point model, it shows output levels greater than 0 dBFS on its meter, +3.5/+1.3 dBFS (left/right channel). Moreover, these values have been passed on correctly to the L2, as you can see under its threshold slider. With a threshold of -6 dBFS, an input signal of +3.5 dBFS causes a -9.5 dB gain reduction, as you can see in the L2's attenuation meter. The limiter keeps the output level to -0.3 dBFS or below by use of the output gain (ceiling), and no distortion will be heard (other than the dynamic artifacts of an extreme amount of peak limiting!). While this may work, be very careful before abusing a digital chain. Floating point may be capable of doing this without harm, but you never know when you will encounter a fixed point processor, or a chain that converts to fixed point midway. Or you may forget and feed a distorted signal to a real world Aux send.

There is one further advantage to floating point over fixed. 0 dBFS+ signals produced by processes such as filtering can be reduced later in the floating

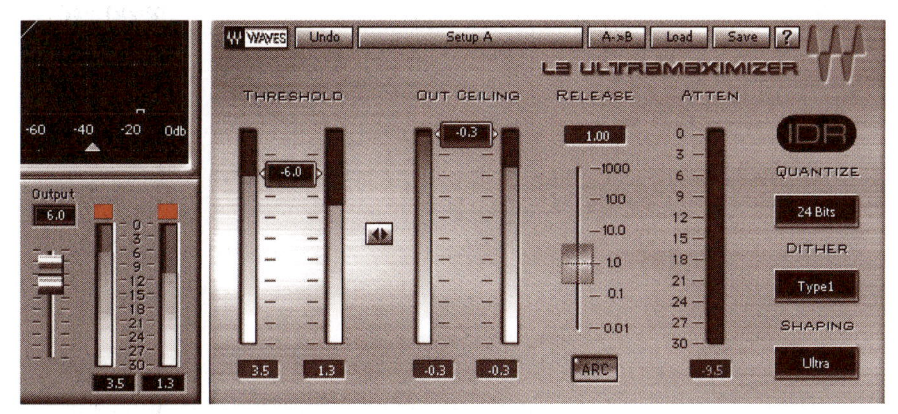


*The Power of Floating Point Processing*

point chain without penalty. In the case of fixed point processors, the majority of which do not contain reconstruction meters, it might be advisable to run maximum levels at -3 dBFS if you are doing severe dynamics processing or equalization. There's certainly no harm in running max levels to -3 dBFS in a 24 bit fixed point chain. How can you tell if a processor or DAW is fixed point besides calling the manufacturer? A digital equalizer with an input attenuator is likely designed in fixed point and should have an overload indicator or meter. The only fixed-point DAW currently produced is Pro Tools HD, which cannot accept or import floating point files.

## VI. Connecting the Analog and Digital Worlds Together

### Dubbing and Monitoring—Translating between analog and digital points in the system

It's very important to **standardize nominal levels** in a studio; each piece of analog gear should have the same nominal voltage level, so when they are monitored or chained the result will be unity gain. All sources, consoles, CD players, music servers, DVD players, and DACs should be adjusted to this level. As mentioned before, I recommend this level to be 0 dBu for cleanest sound, or possibly +4 dBu if the analog gear has the headroom. ADCs and DACs should be aligned so that a sine wave at -20 dBFS produces this standard analog voltage.

### Averaging meters

**The VU and the PPM.** While it's not necessary to have a VU meter to make a good recording, a VU-centric or average-reading studio tends to produce more consistent sonic results, so I recommend mixing studios use an averaging meter like a VU, calibrated so that its 0 point is the same as the standardized nominal level (e.g. 0 dBu or +4 dBu). The whole idea of a VU meter is that it must be used in a system with **headroom** (above 0 VU). So if the operators are using only VU meters, the logical standard is -20 dBFS (sine wave level) for 0 VU. This is commonly called a 20 dB **meter lead**. Other de facto standards for meter lead are 18 dB and 14 dB, but I do not recommend them for a mix room, especially in a VU-centric system, as programs will be encountered with greater crest factors and the digital system will overload. In any case, there is no good reason for a tracking/mixing studio to produce recordings with wildly different nominal levels. Even if a source is compressed and never exceeds 14 dB crest factor, it is more practical to standardize on a -20 nominal level throughout the studio environment and leave the top 6 dB of the digital system "unfilled" most of the time. Use the top of the peak scale for headroom. In Europe, the faster PPM (quasi-peak meter) is a very common analog meter, and a 5 dB meter lead is usually sufficient to protect the digital system from overloading.

**Analog tape** is a special case. It has a headroom of approximately 14 dB and a fragile signal-to-noise ratio, so it is desirable to normalize the tape level to 0 VU—but not if the music has a high crest factor. For example, with percussion, generally forte passages should appear low on the VU or they overload analog tape. Knowledgeable engineers sometimes push analog tape above 0 VU, but they use their experience to know how far to push. If the studio was VU-centric from the beginning and nominal levels were standardized, then a dub to analog would automatically have the correct level in a **standardized level system**. However, if the digital source was not made in a VU-centric studio it may be a *hot* digital master or be peak-normalized to 0 dBFS. In that case, the analog tape transfer will overload. The tape machine's record gain should be turned down, or the output level of the digital source should be reduced.

Vice-versa, when transferring from analog tape to digital, as it is not necessary to peak a 24-bit recording to full scale, the studio can standardize on a -20 dBFS nominal level and even a hot analog tape will not overload the transfer.

### Between the Devil and the Deep Blue Sea

**Mix room blues.** One of the biggest problems in the contemporary audio studio is standardizing monitoring and VU metering levels, because there is no nominal level standard for the compact disc or any digital release medium. In a mix room, contemporary mastered product is auditioned on the analog console along with console mixes; *hot* masters (CDs, or DVDs, or in the case of downloaded product, the output of a music server) come in at a much higher nominal level. If the CD player (or server) output level was standardized to a nominal -20 dBFS, a *hot* CD may damage a mechanical VU meter by *pinning* it, in addition to sounding too loud. The possible solutions (not all of them pretty) are:

1. Replace the mechanical VU with an LED VU, or
2. Get a VU meter like a Crookwood which has a constant impedance switched attenuator, or
3. Disconnect the mechanical VU when playing hot CDs, or
4. Turn down the output level of the CD player, but in that case, normal level CDs will sound too low compared to the console output, or
5. Leave the CD player's nominal analog level at the standard -20 dBFS, but install a monitor controller like the ones discussed in Chapter 2, with a subtrim on the CD input that allows the engineer to adjust the monitor and VU level for better level matching.

**Mastering room blues**. The mastering studio has similar problems, but instead of playing other people's masters, we are creating them. Our masters should not be the ones creating other people's problems.

### In Conclusion

Now that we're standing on firm decibels, let's talk about our mastering tools and techniques. Later, in Chapter 14, we'll discuss a more mature solution to the problems brought up in this chapter.

1 Thanks to Jim Johnston (in correspondence) for helping to clarify some of these definitions.

2 IL (Intensity Level) = SPL - 0.16 dB. "For practical purposes, therefore, SPL and IL are numerically the same for progressive waves in air at STP." Blackenstock (2000) *Fundamentals of Physical Acoustics*, Wiley & Sons.

3 SPL measurements must include the weighting curve used, e.g. A, or C, the speed of the meter (slow or fast), and method of spatial averaging (how many mikes were used and how they were placed).

4 It is controversial whether absolute polarity is audible, or whether it affects the ear's perception of loudness. I believe that the sound quality of transients may be perceived differently depending on their polarity, but over time, the loudness of inverted material is identical to non-inverted.

5 Technically, crest factor should be determined using RMS averaging method. However, in practicality, a common averaging program meter (e.g. a VU meter) will be within about a dB of correct for typical meter readers on typical program material.

6 One of my first lessons in the inaccuracy of the VU meter was in 1972, when I heard William Pierce, voice of the Boston Symphony, clearly and distinctly in the noisy control room at Channel 24, yet he hardly moved the needle. The trained operator must use his ears and learn how to interpret this instrument.

7 Ironically, there's still a tiny disagreement as to **which numeric code to read**, depending on the wordlength involved. Fortunately, a gentleman's agreement has been to use only the top 16 bits to determine level. Full scale 16 bits (positive going, 2's complement) is represented by the number 0111 1111 1111 1111. However, this number is infinitesimally smaller than full scale (positive) 24 bits, 0111 1111 1111 1111 1111 1111. To be exact, the difference is an error of (only) 0.0001 dB, and most people have agreed to ignore the discrepancy!

8 The manufacturers of the Benchmark ADC believe that counting contiguous samples is not a good idea, and they apply an even more conservative standard of any sample hitting 0 dBFS being considered an OVER. And there's nothing wrong with being conservative, especially during initial A/D conversion and especially with 24-bit recording! Paul Frindle (in correspondence) reminds us:

> ...bear in mind that ANY full scale sample is almost certain to be indicative of an illegal signal. Think about it - a full scale sample [from an ADC] which is NOT the result of an illegal signal can only happen in the specific case that the absolute peak of a sinewave (that was gathered via a properly band-limited ADC) happens to exactly occupy max sample value at the very same instant that sampling happened to occur in a non-synchronous clocked system!

9 Thanks to B.J. Buchalter from MH Labs for this simple but powerful explanation.

10 Thomas Lund and Soren Nielsen have investigated a number of hypercompressed pop albums with typical players. They observe that most CD players are still in a distorted mode 200-700 ms after being hit by such peaks, as are radio processors because of SRC on their inputs, phase rotators, and other generally applied tricks. Nielsen, Soren & Lund, Thomas (2000) 0 dBFS+ Levels in Digital Mastering. AES 109th Convention, Preprint #5251.

11 Michael Gerzon published the first material about this issue in the early 1990's and his pioneering Waves L1 limiter is designed to prevent intersample overs. Paul Frindle (in correspondence) notes:

> I have found recorded commercial music which at full modulation (with no clipped max samples) generates reconstruction overs of up to +4 or +5dB for short bursts. This is due to quite dense (or heavily EQ'd) programme being shifted up to abnormally higher levels by output buss comp/limiter processing and happens most often on programme you would least expect—i.e. jazz, vocals, etc...

12 AES SC-02-01 and ITU-R BS.1770 define the standard for an oversampled true peak level meter.

13 The remnant distortion was graphed by subtracting the CD from the mp3 result. I had one client who was simply in the monster level department who spoke of "low class CD players" that distort, forcing us to turn down the level of his CD master. I tried to explain that these excessive levels were causing the problem, but he would have nothing of it.

14 Jim Johnston (in correspondence) points out that processors such as MPEG coders (MP3), Dolby Digital encoders (AC3), WMA, Real, etc. **will add correlated noise, which sounds like distortion**. If you get too close to the edge, they will distort badly unless the input level is reduced. The moral of the story: **do not get too close to digital max**! JJ recommends a maximum sample value peak level <=-0.2 dBFS for the benefit of post-processing.

15 Glenn Meadows and others discuss **shred**, on the Mastering Webboard:

**Glenn**: "Here's where I think all this is coming from, and it's kids oriented. Ever pull up to a stop light, and get blasted from the car next to you? (I assume the answer is yes). Well, besides being aggravated, actually listen to what's going on. ALL of the audio is clipped and distorted on the high end. THAT's what people THINK things sound like, and are SUPPOSED to sound like.

So, for the artists and producers, who are used to "cranking it up in their cars," and having the top and transients clipped/distorted, if they DON'T hear that in their offices, then the mastering is just plain wrong. So, it's once again filtering back to the mix engineers, to provide that hash in the mix to satisfy their clients (remember, we ALL have to satisfy our clients first and foremost), so instead of losing the gig to someone else who WILL provide that edge, everyone is doing the same thing.

**[Unknown respondent:]** In other words, you are stating that the music business is currently conducted by people who don't know what a record should sound like.

**Glenn**: "You got it. Clean is OUT, distorted is in. If it's clean, it's not right. Unfortunately, I've had too many sessions go that way in the past few months."

**Chris Johnson**: "There's no future in that... clipping causes ear fatigue. Ear fatigue means listeners listen less before ceasing the listening. These people are only committing commercial suicide by going for stuff with no longterm sales capacity. It's just the same as if you put everything through an Aural Exciter turned up so far it really HURT, only this time around it's distortion."

16 The saturation (compression) of analog tape and its harmonic content are part of the same mechanism. In other words, a harmonic generator reduces peak-to-average ratio and likewise, a compressor is a harmonic generator.

17 While normalizing after mixing is an extra DSP step and should be avoided, there is no extra step if you go back and remix at a higher level. This sort of "normalization" will raise the signal to noise ratio of the material, especially if you are mixing via analog console. With an analog mix, raising the level of the mix increases SNR by raising the level of the mix signal above the noise floor of the mixdown analog electronics and ADC. If you are mixing digitally, raising the signal level increases the signal above the quantization distortion of the digital mixing DSP, but this distortion is extremely low in a state-of-the-art DSP mixer, around -139 dBFS, only consider raising a digital mix level if it was significantly low (let's say, -10 dBFS to be conservative), for the SNR improvement will normally not be audible.

18 The origin of using +4 **dBu** as a reference for analog audio instead of a more convenient number like **0** goes back to the earliest days of the telephone company. The decibel is a relative measurement, but the reference used by the telephone company was based on **power**. And the telephone company's standard reference for 0 dB is one milliwatt, which across their standard impedance of 600 ohms yields 0.775 volts. This reference is commonly abbreviated as **0 dBm**. The VU meter then came along; it is calibrated to produce a level of 0 VU with 0 dBm, but if put across the 600 ohm line directly it would load it down and cause distortion, so the standard circuit included a 3600 ohm resistor in series with the VU meter. The 3600 ohm resistor attenuates the meter by 4 dB, so the circuit level has to be raised to **+4 dBm** in order to make the meter read 0 VU.

Nowadays, modern-day equipment generally has low impedance outputs (sometimes as low as 10 ohms or less), and high impedance inputs (greater than 10 k ohms), so there is no meaningful power transferred from gear to gear. Instead, a voltage reference is the only thing that is meaningful. And to keep using the same decibel levels we used for telephony, we kept the historical reference of 0.775 volts instead of a more convenient number like 1 volt! Now when the dB is referred to a voltage of 0.775 volts, we call that **0 dBu**. And to make a VU meter read 0 in a modern low impedance circuit with the right resistors, we have to feed it +4 dBu, or 1.23 volts. Also see in Appendix 6 a short table of decibels.

The equations are:

If 0 dBu is 0.775 volts, then +4 dBu is 1.23 volts. 20 * log (1.23/.775) = 4.

I thank Mike Collins for reminding me to include this explanation.

19 This is of course dependent on the skill of the designer. Some IC operational amplifiers perform very well up to the clipping point. Power supply design and regulation has a lot to say about sound quality near the clipping point. To avoid the nasties, use conservative levels, measure and listen.

20 To be more exact, headroom is reduced 6 dB if you unbalance a transformerless amplifier's output. Transformer-coupled amplifiers retain their headroom even if unbalanced.

21 Each processor does add its own quiescent or idle noise, which is cumulative, but in a good chain rarely adds more than 3 to 6 dB to the -139 dBFS RMS noise floor.

22 Simplifying the arithmetic, we assume the peak level is at -10 dBFS RMS and the dither noise is wideband and also RMS-measured at -91 dBFS (rounded from 96-4.77=-91.2). The RMS sum calculation is:

Convert dB to power. $10^{\frac{-91}{10}} = 7.94 \times 10^{-10}$

Convert dB to power. $10^{\frac{-81}{10}} = 7.94 \times 10^{-9}$

sum of powers = $8.74 \times 10^{-9}$

10 log of sum = -80.58 dBFS

Chances are the preamp and room noise on the DAT are much higher than this dither noise, but the dither noise is the absolute minimum noise floor to consider. Though some mastering engineers claim we can hear the degradation of dithering, even at as low a noise floor as -91 dB and even under music levels which are much higher! Thanks to Jens Jorgen for finding my error in the first edition, which neglected to perform an RMS sum.

23 You may ask: Other than the esthetic job of matching one song to another, why are we bothering to raise the level of the recording if the SNR of the source is worsened by the added dither? We may have to match other songs on the album; also consider the noise floor of the final output electronics and DAC, by peaking closer to full scale we may overcome some of the weaknesses of the reproduction system's noisy analog outputs. It's a matter of finding the right balance or an acceptable compromise amongst these several factors.

24 Firewire is capable of passing floating point so it may soon be possible to communicate between DAW and external devices without having to switch to fixed point.

Chapter 6

# Monitor Quality

## I. Philosophy of Accurate Monitoring

The major goal of a professional mastering studio is to make subjective judgments as objectively as possible. What enables this to be done most successfully is the intelligent use of an accurate, high resolution[1] monitoring system. A high resolution monitor system is the mastering engineer's audio microscope, the scientific tool which enables the subtle processing decisions required by her art.

### Elements of a High Resolution Monitor System

The guidelines for constructing an ideal high resolution monitor system probably haven't been written, but we can describe some of the general elements:

1. Mastering engineers work with a single high quality monitor system, with which they are intimately familiar, knowing exactly how its performance will translate to the real world, and please the maximum number of listeners.

2. The mastering room is extremely quiet, with all noise-producing equipment banished to the machine room. Noise floor must be better than NC 30[2], preferably NC 20 or less.

3. There are no significant obstacles between the monitors and the listener within the standard equilateral monitoring triangle.

4. The electronic chain is designed for maximum transparency. Often specialized or customized components are

*"The mastering engineer's monitor system is an audio microscope"*

We use the term high resolution to define an acoustic playback system which has few artifacts to mask the signal we are listening to. In other words, the noise and distortion of a high resolution playback system are low enough to permit the ear to resolve those signal components.

built which incorporate a bare minimum of active stages (see Chapter 2).

5. Monitor loudspeakers have wide bandwidth, high headroom[3], and extremely flat frequency response (up to about 2 kHz, where they begin a slow rolloff). An inaccurate monitor system can lead to inappropriate EQ adjustments. An accurate monitor system reveals differences between recordings—
a system which makes everything sound the same, or makes everything sound beautiful, cannot be accurate.

6. Time-domain problems are minimized: Monitors and listener are in a reflection-free zone,[4] which means that reflections from nearby surfaces arrive at the listener at least 20 ms later than the direct sound and at least 15 dB down (preferably >20,[5] see Chapter 15). Sources of diffraction[6] are minimized. Cabinets are solid, non-resonant, and free of sympathetic vibrations and resonances.

7. The walls in the room are solid and non-resonant and the room is large enough to permit even, extended, bass response, with no significant standing waves. Any remaining standing waves are controlled using techniques including traps or specialized diffusers. Ideally, room length should be at least 20 feet long for stereo and 30 feet long for multichannel, so that all speakers can be far enough from the walls to avoid the bass-resonance proximity effect.[7] The room should be wide enough so that first reflections from the side walls are insignificant, and/or the side walls are treated to minimize reflections. Dimensions should be symmetrical from left to right and a sloping ceiling is an advantage as it reduces floor to ceiling reflections.

8. Acoustical design and electrical layout are accomplished by experienced and trained professionals.

**Subwoofers and Bass Response**

Stereo subwoofers, or prime loudspeakers whose response extends to at least 30 Hz are essential for a good mastering studio. Vocal pops, subway rumble, microphone vibrations, and other distortions will be missed without subwoofers, not just the lowest notes of the bass. Proper subwoofer setup requires knowledge and specialized test equipment (see Chapter 15). If subwoofers are inaccurately adjusted (e.g. "too hot", in a misguided attempt to impress the client) then the results won't translate well to other systems.

Apparent bass response is also greatly affected by monitor level. The equal loudness contours (originally studied by Fletcher, and Munson[8]) dictate that a recording which is mastered at too high a monitor level will seem bass-light when auditioned at a lower level in a typical home environment.

**Monitor Equalization—by Ear or by Machine?**

We must use our ear/brain in conjunction with test instruments to ensure monitor accuracy. Test equipment alone is not sufficien —for example, although some degree of measured high-frequency rolloff usually sounds best (due to losses in the air), there is no objective measurement that says, "this rolloff measures correctly", only an approximation. Different size rooms, monitor distances and monitor dispersions change the rolloff required to make the high end sound right.

Thus, for the high frequencies, the ultimate monitor tweak must be done by ear. But this leads to a chicken and egg situation: if we are using recordings to judge our monitors, how do we know that the recording can be trusted? The answer is to use the finest reference recordings (at least 25 to 50) to judge the monitors, and take an average. The highs will vary from a touch dull to a touch bright, but if the monitor system is accurate the majority will sound right. I try to avoid adding monitor correction equalizers; I prefer first to fix the room or replace the loudspeakers; my techniques include tweaks on speaker crossover components until the monitors fall precisely in the middle of the "acceptance curve" of all 50 reference recordings.

Note however that a variety of factors—the number of people in the room, interconnect cable capacitance, power amplifiers, DACs, and preamplifiers—can all affect low and high frequency response, so if there are any changes to these, I immediately reevaluate the monitors' response with the known 25-best recordings!

#### Why Accurate Monitors Are Needed

Because the mastering engineer strives to create a recording which will play well on the maximum number of playback systems, it is best to work to the middle of a bell curve, as illustrated in the figure (below left). It is obvious that tilting a recording in the bright direction means it will not play well on a lot of small systems that already have too much treble, or that skewing it in the direction of the bass means it will not play well on systems that have too much bass.

*"A monitor which makes everything sound beautiful cannot be accurate."*

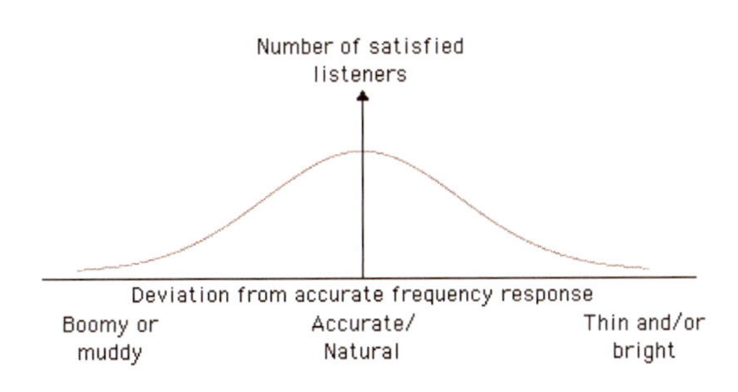


*An accurate monitor system allows us to produce recordings which are in the middle of the curve.*

## II. Debunking Monitor myths

**Myth #1: You must mix (master) with *real-world* monitors to make a recording for the real world.**

Some problems in mixes can be directly traced to the kind of monitor coloration that is endemic in "real-world" speakers: the bass drum is boomy, the vocal is often too low (probably caused by center buildup in the nearfield environment), the reverb is sometimes too low, the midbass of the bass instrument is depressed (caused by resonances or comb filtering artifacts from a console surface), the stereo separation is very small (nearfields exaggerate separation like a big pair of headphones), and the high end is at best, unpredictable.

**Myth #2: Adding high end helps a recording translate to home systems.**

This is an untruth, because adding high end skews a recording away from the mean of the bell curve, resulting in one that's sharp, tinny and fatiguing. Furthermore, radio play will suffer,

**MYTH:**
*Program Compression is required to protect small reproduction systems.*

because the radio limiters will cut back the added highs. But most important, if the midrange is wrong, nothing else is likely to be right, as we will see in Chapter 8.

**Myth #3: Heavy compression is necessary to prevent small monitor systems from overloading.**

I have found the opposite to be true, with few exceptions. When I take my dynamically wide ranging masters to an Aiwa 3-piece system, they sound (comparatively) compressed, with fewer transients and less impact. If I had reduced the transient clarity in the mastering, it would only sound worse on the smaller system, which does its own compressing! Thus I have learned that even if the sound "sticks out a little too much" on a high-headroom mastering system, it's probably going to be fine when played on an inferior system. And vice versa, if you make judgments on a compromised monitor system, you'll never learn if something is over or undercompressed.

*Here's what we're up against.*

## III. Refinements

### Alternate Monitoring Systems

Mastering engineers use alternate loudspeakers as a double-check, not as a benchmark. With the proliferation of iPods, headphones also make a good double-check, although they tend to exaggerate inner details of a mix. We have another room whose system has large, "loose-sounding" woofers, representing an extreme of the bell curve (what may happen to the bottom end in a club, or car). Cars are likely to have very uneven bass response, aggravated by user-set equalizers (as in this photo).

I've learned to watch out for recordings where the client is looking for very hot bass or bass drum. If we boost the bass on the master in the neutral mastering room to get the distorted bass sound they're used to, it will overdrive a typical car system. The boomy alternate listening room demonstrates to them what can happen.

One mastering studio has a radio station transmitter and processor, and invites the client out to their car to hear what it will sound like on the radio. This is a great idea, as long as the client is realistic about the limitations of the car system, for if you try to hear "sparklies" in the car, it will screech at home.

### Narrowcasting

There are boombox, club and car systems whose bass response/resonance is so extreme, they should not be included in the bell curve. It is almost impossible to make a single master that plays well on the club system but doesn't sound thin and lifeless on all others. The best solution is to make a separate (dedicated) master for club playback.

## IV. In Summary

Even the best master will sound different on different systems, but it will sound most correct on an accurate monitor system. Which leads us to this comment from a good client:

> I listened to the master on half a dozen systems and took copious notes. All the notes cancelled out, so the master must be just right!

1 Our use of the term *resolution* applied to monitoring is very similar to its meaning in Chapter 4. In order to measure and hear low level signals cleanly, the monitor system, which includes the room, must be quiet. Time domain interference, such as *diffraction* in a loudspeaker can also be considered a form of noise.

2 NC 30. Noise criterion 30 decibels, follows an attenuation curve whereby at 2 kHz, noise level is 30 dB SPL, and at lower frequencies is permitted to rise.

3 Headroom is another reason for mastering engineers to reject near-field monitoring, as very few typical near-field monitors pass the "bandwidth and compression test" or can tolerate the instantaneous transients and power levels of music without monitor compression.

4 This term was coined by Dr. Peter D'Antonio of RPG.

5 The advantage of this monitoring environment is that time-domain errors in the musical material will be more audible, since they will not be masked or smeared by the monitoring environment. In the absence of TDS equipment, an objective-subjective listening test called the LEDR test can help determine if nearby reflections are interfering with the monitoring. LEDR (Listening Environment Diagnostic Recording) is available from Chesky Records, on JD37, and from Checkpoint Audio (see Appendix 10).

6 Diffraction is the effect of multiple delayed sources in a loudspeaker, caused by the sound reaching a discontinuity at the edges of the cabinet. Diffraction can be reduced at certain wavelengths by using round instead of sharp cabinet edges, soft materials on the edge and cabinet surface, and flush mounting the drivers. The sonic results of diffraction can be a "smearing" of the sound quality and apparent distortion or even edginess.

7 Unless the speakers are placed in soffits within the wall structure, which requires considerable acoustical expertise. It's much easier to design a room with free-standing loudspeakers.

8 Fletcher, Harvey & Munson, W.A. (Oct. 1933) Loudness, Its Definition, Measurement and Calculation. *Journal Acoustical Society of America*, Vol. V, No. 2.

*"The car system should be a double-check, not a benchmark."*

"We'll
FIX it
in the
MIX."
—Anon

# PART II: MASTERING TECHNIQUES

"IT'S not HOW LOUD you make IT. IT'S HOW you make IT LOUD."

—Bob Katz

CHAPTER 7

# Putting The Album Together

## Introduction

Although we are in an era of digital downloads, an emphasis on singles and a shorter attention span of the listening public, the record album is still an important music medium. Sergeant Pepper is often cited as the first rock and roll *concept album*, i.e. an elaborately-designed album organized around a central theme that makes the music more than a simple collection of songs. This started a trend that many assume has more or less died. Is the concept album really dead? Not for me; I treat *every* album that comes for mastering as a *concept album*, even if it doesn't have a fancy theme, artwork or gatefold. Song spacing and leveling contribute greatly to the listener's emotional response and overall enjoyment. It is possible to turn a good album into a *great album* by choosing the right song order. The converse is also true.

## I. How to Put an Album in Order

Sequencing an album is something of an art. Sometimes, the musicians making an album have a good idea of the song order they'd like to use, but many people need help with this task. Traditionally, the label's A&R person would help put the album in order, but with independent productions, that service is not always available and so it falls to the producer, or someone experienced, and politically "neutral".[1] An experienced mastering engineer is well placed to provide useful guidance during this process.

My advice is to avoid over-intellectualizing. One musician decided to order his album by the themes of the lyrics; he started with all the songs about love, followed by those about hate, and finally the songs about reconciliation. It was a musical disaster. The beginning of his album sounded musically repetitive, because all his love songs tended to use the same style, and the progression of intellectual ideas was not obvious to the average listener, who primarily reacted to the musical changes. Even when the listener got the underlying point, it didn't contribute much to the enjoyment of the album. Listening to music is first and foremost an emotional experience. If we were dealing with lyrics without music (poetry), perhaps the intellectual order would be best, but the intellectual point of the album will still come through even if the songs are organized mainly for musical reasons.

Before ordering the album, it's important to have its gestalt in mind: its sound, its feel, its ups and downs. I like to think of an album in terms of a concert. Concerts are usually organized into *sets*, with pauses between the sets when the artist can catch her breath, talk briefly to the audience, and prepare the audience for the mood of the next set. On an album, a *set* typically consists of three or four songs, but can be as short as one. Usually the space between sets is a little greater than the typical space between the songs of a set, in order to establish a breather, or mood change. Sometimes there can be a long segue (crossfade) between the last song of a set and the first of the next. This basic principle applies to all kinds of music, vocal and instrumentals; it is analogous to the spacing in a classical music album, shorter ones between movements of a single composition and longer ones between the compositions themselves.

To make the job of organizing the sets easier, I (or the artist) prepare a rough CD of all the songs, or a DAW playlist to allow instant play of all the candidates—which is a lot easier than it was in the days of analog tape. Then I make a simple list, describing each song's characteristics in one or two words or symbols, e.g. *uptempo*, *midtempo*, *ballad*. Sometimes I'll give letter grades to indicate which songs are the best-performed, most exciting or interesting, trying to place some of the highest grade songs early in the order for a good first impression. I may note the key of the song, although this is usually secondary compared to its mood and how it kicks off. If there's a bothersome clash in keys, sometimes more spacing helps to clear the ear, or else I exchange that song with one that has a similar feel and compatible key.

The opening track is the most important; it sets the tone for the whole album and must favorably prejudice the listener. It doesn't have to be the *hit* or the *single*, but most frequently is up-tempo and establishes the excitement of the album. Even if it is an album of ballads, the first song should be the one that is most likely to engage the listener's emotions.

If the first song was exciting, we usually try to extend the mood, keep things moving like a concert, by a short space, with an up or mid-tempo follow-up. Then, it's a matter of deciding when to take the audience down for a breather. Shall it be a three or four-song set? I examine the other available songs, then decide if it will be a progression of a mid-tempo or fast third song followed by a relaxed fourth, or end with a nice relaxed third song.

At this point, I have provisional track numbers penciled next to the candidates for the first set. I play the beginning of the first song to see how it works as an opener, then skip to the last 30 or 40 seconds, play it out and jump to the start of the second song. If this musical transition doesn't work, then the sequence is faulty regardless of how compatible the two songs seem to be. That's why transitions can let us join different musical feels; an up-tempo song that winds down gently at the end can easily lead to a ballad. If the set doesn't flow, I substitute songs until it does.

Then, I check off the songs already used, and pick candidates for the second set, usually starting with another up-tempo in a similar "concert" pattern. This can be reversed; some sets may begin with a ballad and end with a rip-roaring number, largely depending on the ending mood from the previous set. A set can also be a roller coaster ride, depending on the mood we want to create, but when you consider the album in terms of sets, it all becomes a lot easier to organize. By the way, the ultimate listener doesn't usually realize there are sets; our work should be subliminal.[2] As the list gets filled up, it becomes a jigsaw puzzle to make the remaining pieces fit. Perhaps the third or fourth set doesn't work quite as well as the first, or one of the transitions is clashing, even if we increase the spacing. At that point I may try a one-song set, or try to place this problem song into an earlier set, either replacing a song, or adding to the earlier set.

#### The Odd Man Out

One song may just not fit well musically with the rest. For a Brazilian samba album, the artist also recorded a semi-rock blues number. She said everyone loved this song in Brazil, so we couldn't excise it from the album, but stylistically it did not gel as a part of any set. At first I suggested putting it last as a "bonus track," but this ruined the original album ending, which was a beautiful, introspective song that really did belong at the end. Eventually, we found a place for the offender near the middle of the sequence, as a one-song-set, with a long-enough pause before and after. It served as a bridge between the two halves of the album.

#### The Right Kind of Ending

So, how to end the album? What is the final encore in a concert? It's almost never a big, uptempo number, because the audience always cries "more, more, more." You've got to leave them in a relaxed, comfortable "goodbye mood", otherwise you'll be playing encores forever. That's why the last encore is usually an intimate number, or a solo, with fewer members of the band. The same principle applies with the record album. I usually try to create a climax, followed by a dénouement. The climax is obviously an exciting song that ends with a nice peak. This, followed by one or two *easy-going* songs to close out the album. Finding the perfect sequence is a real treat!

## II. Spacing The Album

The next thing to remember is never to count the seconds between songs. Experienced producers know that the old "4 second" "3 second" or "2 second" rule really does not apply, although it is clear that album track spacing has gotten shorter over the past 50 years, along with the increased pace of daily life. The correct space between songs can never be accurately measured, for different people start counting at different times depending on when they think the decay is over. Counting from the beginning of true silence, the computer may objectively say that a space is only 1 second, but the ear may think it's closer to 2.5. So don't count—just listen. As a general rule, the space between two fast songs is usually short, between

"Never count the seconds between songs"

a fast and a slow song is medium length, and between a slow and a fast song is usually long. After a fadeout, the space is usually very short, because the listener in a noisy room or car doesn't notice the tail of a fadeout. Often we have to shorten fadeouts and make segues[3] or the space will seem overextended, especially in the car. Perception of appropriate spacing can depend on the mood of the producer and the time of day. If you space an album in the morning when you're relaxed, it almost always sounds more leisurely than one which has been paced in the afternoon, when our hearts generally beat faster. To avoid being unduly influenced by these external factors try not to make spaces too short when you're in an energetic mood, or too long when you're very relaxed.

The overall *pace* of an album is also affected by intertrack spacing. We probably want the first set to be exciting, and so control the pace using shorter spaces within the first set and slightly longer spaces thereafter. An interesting observation is that our sense of timing is relative; if we begin with very tight spaces and then revert to "normal" spaces, these seem too long. Manipulate spaces to produce special effects—surprises, super-quick and super-long pauses make great effects. One client wanted to have a long space in the middle of his CD, about 8-10 seconds, to simulate the change of sides of an LP. Respecting the input of a creative individual I tried the super-long space, and it worked! This was largely due to his choices of songs and the order. The set which began side two had a significantly different feel, and the long space helped to set it off, like a concert intermission.

Some people think that it's sufficient to play the last 30 seconds of a tune in order to judge the space before the next. But if you play an entire exciting song, you will most assuredly need more space to catch your breath before the next one can start. To avoid playing a whole song, we try to anticipate this effect by using a slightly longer space; when we play the album through we'll know if we were successful. One technique for judging a space is to cut it shorter and shorter until it is obviously too short and then add just the scintilla necessary to make it sound "just right", especially knowing that it will seem longer on a domestic hi-fi. Another type of space is to make the downbeat of the next song be in time with the rhythm of the previous. This can sound very nice, but not if overused.

We didn't have this luxury in the days of analog tape, and it's interesting to note that when an LP master comes in for conversion to CD, the spaces always seem too long. One reason, as mentioned before, is the current quicker pace of life, but the other is that vinyl and tape noise act as a filler. When there's dead silence between tracks, spaces always seem longer. Remove 2 or more seconds out of an LP space and it will feel just fine on CD.

## III. PQ Coding

### PQ Offsets

Most authorities recommend placing a track start mark (called **Index 1**) at least 5 SMPTE frames before the downbeat to accommodate slow-cuing CD players. This is approximately 12 CD frames, 160 ms (one CD frame = 1/75 second). The DAW can automatically apply these offsets, and show the PQ codes as they will appear on the disc. Sophisticated DAWs let us rehearse the effect of cuing with or without the offsets, critical when the cue has to be very tight.[4] For example, when the previous song is crossfading into the next, if we do not place the track mark extremely close to the downbeat of the next song the CD player may play a piece of the previous sound. I may accept as little as 2 (occasionally 1) SMPTE frames, which risks that a slow-cuing player will miss the downbeat. Pictured at top right is an example of a live album with the track mark located nearly on top of the downbeat to avoid the spoken introduction. Some players clip the downbeat, but on this CD it was less of a problem than hearing the previous sound.

## Spaces and PQ (Track) Coding

**Index 0** is an optional mark between the tracks which defines the end of the previous track; the CD player's time display begins to count backward up to the Index 1. This is called the *pause time*, a misleading term, for there is no requirement for silence and in fact, index 0 can be 0 seconds. I recommend normalizing Index Zeros shorter than 2 seconds to 0 to keep the player's time display from glitching.[5] This doesn't mean that musical spaces cannot be as short as you want them to be, it only means there will be no *official* pause between tracks. When Index 0 is 0 seconds, the player interprets Index 1 as the end mark of the previous track and the start of the next.

## Hiding in the Gap

When a cut from a concert album is played on the radio, it's often desirable to cue the tune on the downbeat, but the listener at home wants to hear the atmosphere between cuts and maybe the artists' introductions. To accomplish this dual feat, the creative mastering engineer can place Index 0 and Index 1 times as in the figure below.

*Track mark placed very tight to the downbeat with no offset to avoid hearing talking which comes before the mark.*

In this example, the song for track 9 ends with applause; the official end of song 9 is at Index 0. There is sound in the pause time between Index 0 and Index 1; this permits consumer choice or the CD player's random play function to ignore the boring or irrelevant parts. Similarly, the introductions, count offs, sticks, and so on, for songs on any kind of album can be placed in the gap so they will not be heard on the radio or in random play. The pause time does not count as part of the official length of either track (which keeps royalty costs down!). Unfortunately, this functionality of the CD standard is eroding, hindering artistry that we have enjoyed for over 20 years. iTunes and some primitive CD players do not read Index 0; they treat the introduction or the countoffs as the end of the previous track, producing some incongruous results in random play. Furthermore, iPods (until the 5th generation and iTunes ver. 7) and some other players briefly mute at each track start, breaking up the continuity of a continuous album. After version 7, iTunes/iPods can play gapless albums without glitches, but

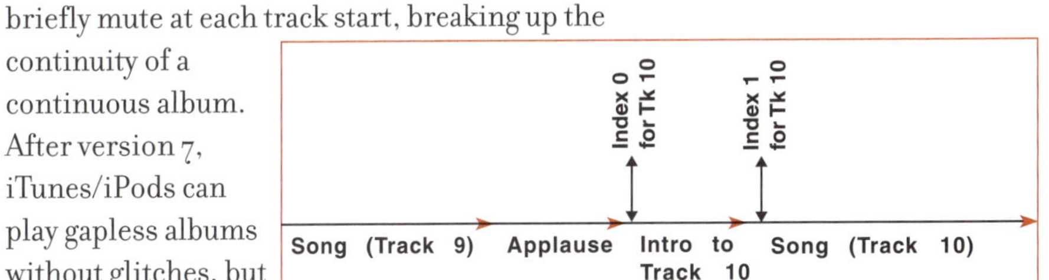

only if the user explicitly marks the tune as part of a gapless album.[6] Regardless, I code masters assuming they will be played on CD players that respect the standard; there is little other choice.

It pays to be vigilant because some CDR duplicators will mute the pause audio, sometimes even taking many seconds OUT and putting just 2 blank seconds IN. These copiers were found to be copying in Track At Once Mode, rather than Disc at Once, instead of simply cloning the disc.[7] Imagine the classic Pink Floyd *The Wall*, which has continuous sound, being gapped by accident at the plant.

#### Redbook Limits

The Sony/Philips Redbook specifies all the parameters for an audio compact disc. A CD may have up to 99 tracks, whose minimum length is 4 seconds, and each of these tracks may have up to 99 indexes (or subindices). Rarely do we code CDs with indexes since many players do not support them and most people don't know how to use them. Classical engineers who used to code movements with indexes are now using a track mark for each succeeding piece. There is no standard CD length; maximum length can be stretched to 80 minutes if the plant tightens the disc spirals to the minimum Redbook tolerance—but not all players can play the outer tracks of these discs. Individual plants specify shorter cutting limits on the order of 78:00 to 79:38 (check with the plant).

#### Standalone CD Recorders

Standalone CD recorders should not be used to make CDRs for replication. There is no provision for Index 0, and the location of Index 1 (the track mark) can only be as accurate as a manual button push. Also, when recording one track at a time, these standalone recorders work in **Track-At-Once** mode, which puts an E32 error onto the disc wherever the laser stops recording. Computer-based machines should be set to work in **Disc-At-Once** mode, which means that the CD must be written in one continuous pass.

#### PQs and Processor Latency

Since the Sonic and SADiE DAWs can cut a physical CDR master in real time with all processors in line, we have to consider the latency (delay) of the external processors. The realignment is achieved by measuring the delay and sliding the PQ marks by this amount.

#### Hidden Tracks

**Find the hidden track** is a little game the producer plays with the record-buyer. To make this possible, the mastering engineer can hide a track by inserting many short, blank 4-second "dummy tracks" at the end of the CD prior to the "hidden" one, which forces the listener to cue many times before he can reach it. Another method is to put several track marks within the "hidden" song, which causes ripping programs to break it up into pieces. Yet another way to hide a track is to have a track mark with no music for a minute or more.

Some CD players have the ability to rewind in front of track one; this is called the pregap or first Index 0. One company claimed the rights to putting hidden tracks in that position, but it's not permitted by the Redbook standard, and many plants will not press CDs with a hidden track in the pregap. To the best of my knowledge, there is no way to produce a DDP with this feature, so only CDR masters can be produced if the DAW allows hidden pregap tracks.

## IV. Editing

I love editing because it can generate a hundred smiles in a day! I think a whole book should be written on digital audio editing techniques, but ultimately the skill of fine editing can only be learned through guided experience: the school of hard knocks, and an apprenticeship. The purpose of this short section is to discuss some of what is possible in editing. Using sophisticated workstations, we can perform edits that were impossible in the days of analog tape and the razor blade. I once spent 30 hours painstakingly editing a spoken-word version of a novel, a task which can now be accomplished in a single day. SADiE's playlist-editing mode, which allows us to "spill" virtual tape, makes this extremely easy.

### The Tale of the Head and Tail

Editing heads and tails is an important skill based on musical knowledge but developed through experience. Because mechanical artifacts can easily distract the listener's attention from the emotional aspects of the music, a mastered work should generally be consistent and *smooth*. Consider the fade-up at the beginning of a song. When the music requires it, this fade-up can be made quite sharp (equivalent to a 90 degree cut). But a fast fade-up often sounds wrong with soft music, especially pieces that begin with solo vocal or acoustic instruments. A delicate acoustic guitar solo can sound abrupt if the noise of the room and the preamp is not brought up from silence at the right speed and timing.

**Natural Anticipation.** We also have to be aware of the important role played by moments of *natural anticipation:* the human breath before the vocal; the movement of the guitarist's hand before a strum; or the movement of the fingers and keys prior to hearing a piano downbeat. Often it sounds unnatural to cut these off, making the opening appear choked. If the breath is better included, but sounds a bit loud, then a gentle fade-up can produce just the right result. I advise mixing engineers not to cut off the tops when sending songs for mastering, for the mastering engineer probably has better tools to fix these, and a quiet, meditative environment to make these artistic decisions properly.

**Tail Noise Cleanup.** Sometimes the tail end of a song contains noise from musicians or equipment, which draws attention to itself by the transition from noise to the silence of the gap. Tools such as Cedar **Retouch** and Algorithmix **Renovator** (see Chapter 12) are very useful in cleaning such noises and creating clean decays. Another common solution is called a *follow fade*, which is usually a cosine or S-shaped fade to silence. A good mastering engineer may spend a minute or more on such a fade to ensure that the tail ambience or reverberation is not cut off, as the hiss or noise is brought invisibly to silence. We can take advantage of the fact that noise is masked by signal of the right amplitude, so the follow fade can and should be slightly slower than the natural decay. The delicate decay of a piano chord at the end of a tune should sound natural, even while we manipulate the fadeout to avoid or soften the thump of the release of the pedal. Fine editing can allow us to raise the gain at the tail, after having previously lowered it, in order to hear some inner detail.

**Fadeouts.** A good-sounding musical fadeout is one that makes us think the music is still playing; we're tapping our feet even after the sound has ceased. Although we can apply the same cosine shape we use for tails, fadeouts are a distinct art in themselves. Typically, a fadeout will start slowly, and then taper off rapidly, mimicking the natural hand movement on a fader—there's nothing more annoying than a fadeout that lingers beyond its artistic optimum. On the other hand, a fadeout should not sound like it fell off a cliff, and often in mastering we get material that has to be repaired because the mix engineer dropped the tail of the fade too fast. Since editing is like whittling soap, I recommend that mix engineers send unfaded material so it can be refined in the mastering. It is difficult to satisfactorily repair a fade that was too fast at the end; sometimes an S-shape helps, and sometimes we can apply a taper on top of the original slope.

*Adding a tail via a crossfade to artificial reverb.*

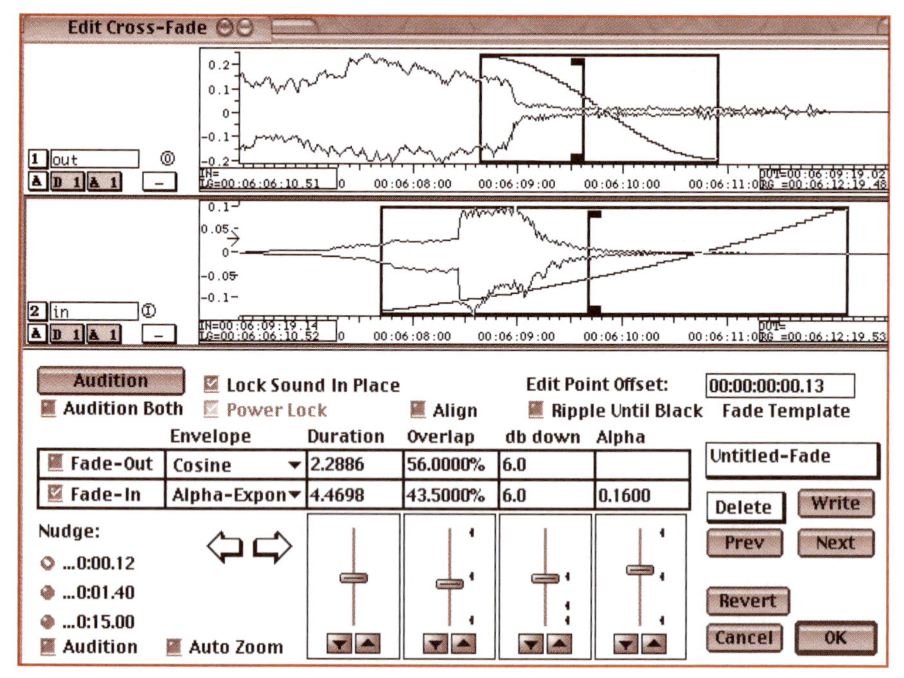


**Adding tails.** Although editing fades is like whittling soap, sometimes we're called upon to make more soap. If the musicians or instruments make a distracting noise during the ambient decay, the ambience will sound cheated or cut off if we perform a follow fade to remove the noises. In the figure at left is a fadeout; at the tail you can see the noise made by the musicians. Unfortunately, these noises occurred during the reverberation, so the ambience sounds cut off. The trick is to feed just the tail of the music into a high-quality artificial reverb and capture that in the workstation, which you can see in the bottom panel. Since the predelay of the reverb postpones its onset, its position can be adjusted in the DAW's crossfade window which allows us to carefully shape, time, and level the transition to this artificial reverb in a manner that can sound completely seamless. Thus we have performed the impossible: putting the soap back on the sculpture!

Sometimes an analog tape may have a lot of echoey print through or hiss noticeable at the tail of the tune. Adding a new artificial tail can help, or editing to a digital safety version of the mix if it exists (so I advise clients to send both versions).

### Adding Room Tone

Room tone is essential between tracks of much natural acoustic and classical music. Recording engineers should bring samples of room tone to an editing session. Room tone is usually not necessary for pop productions, but if a recording gets very soft and you can hear the noise of the room, going sharply to *audio black* can be disconcerting. The object is not to draw the listener's attention to the removal of noise, as illustrated on the next page.

Always record room tone in advance as a separate "silent take" with no musicians in the room. If this is not supplied (10 seconds or more), it is almost impossible for us to manufacture a convincing transition and we have to be satisfied with a fade to/from silence. **Retouch** or **Renovator** are handy at editing out glitches in room tone.

**Repairing Bad Edits.** One type of bad edit is where the reverberation of one section has been cut off by the insertion of a new one. This used to happen in analog tape editing because we could not do intricate crossfades with a razor blade. But the error still occurs in classical music if a highly inexperienced producer instructed the musicians to begin the retake exactly at the intended edit point, instead of a few bars earlier which not only gives the musicians a running start and allows for a better music flow, but also generates the reverberant decay of the preceding note for the editor to work with. If the producer did not record the reverberation, the ear notices the cutoff of the reverb, which is not masked by the transient attack of the next downbeat. Luckily, when it comes to mastering, we can repair some of these bad edits even if the original takes are not available. The trick is to take apart the original edit, then feed an artificial reverb chamber to re-create the missing tail, then join it all back together.

**Editing and assembling concert albums** can be a great pleasure because they are the perfect example of suspension of disbelief. To edit applause requires a familiarity with natural applause, but real-life applause is almost never as short as 15 or 20 seconds, and real-life artists have to stop to tune their instruments. The object is to capture just the essence of the concert so that the home listener is never bored on replay. Cutting applause and ambience between different performances exercises the power of the workstation's crossfades. There must always be some degree of room tone (audience ambience) between numbers but the audience at the end of a loud performance is more enthusiastic than before a quiet one. The transition will not sound realistic by simply dropping the ambience level; it must be done with a crossfade from loud to quiet ambience. My approach is to do the major cutting on one pair of tracks (for stereo), and wherever it needs transitional help, mix in a bed of compensating ambience on another track pair. I once put an audience loop under the only studio cut on a live album, and to this day no one has been able to figure out which track is the ringer!

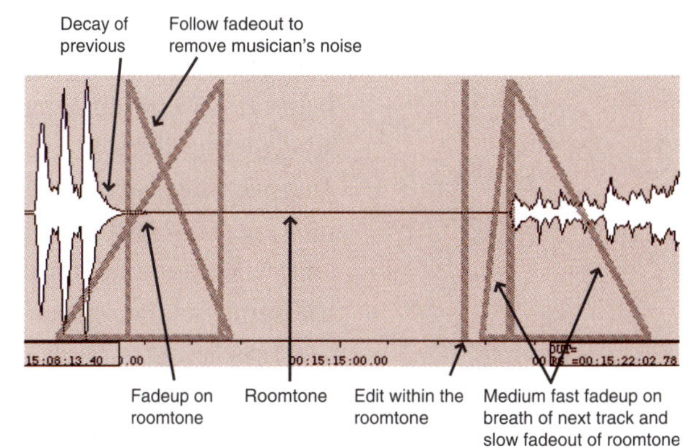


*Editing room tone in an acoustic work requires considerable artistry. An edit must not call attention to itself.*

## V. Leveling The Album

**Context-based leveling.** A piece of music which begins softly but follows one that ends loudly creates a potential problem. We may have to raise these beginnings, but the same soft level could be perfectly acceptable in the middle of a different piece. Similarly, a loud attack is amplified by the ear if it is preceded by silence. This is why albums and singles have to be leveled differently; the ballads often must be raised for single release.[8]

The greater a recording's dynamic range, the harder it is to judge "average level" and you have to listen in several spots. I usually start with the loudest song on the album and find its highest point. I then engineer the processing to create the impact I'm looking for, set the monitor at the gain which works, and make the rest of the songs work together at that monitor gain. This practice also helps prevent overprocessing or overcompression[9] (See Ch. 9-11). The rest of the album falls in line once the loudest song has its proper level and impact. During the processing of this loudest song, it's important to ensure an optimal gain structure in the chain of processors; this is the test for the rest of the album.

Leveling and dynamics processing are inseparable, for the output (makeup gain) of the processors also determines the song's loudness compared to the others. A more compressed song may sound louder than another even if its peaks don't hit full scale. If you change the processing, you have also changed its level, it has to be done by ear. After working on the loudest song and saving the settings, I usually go to the first song and work in sequence. Then the second song, and the transition between the first and second. This transition will usually work without any fine-tuning because we've been monitoring at a consistent gain. If one song appears too loud or soft in context, I make a slight adjustment in level until they work together, or sometimes increase the spacing to "clear the ear". So you can see why it's important to have the album in proper order before mastering!

Consider the size of the ensemble. A song with solo vocalist and acoustic guitar should be naturally softer than the full electric band.

### Everything Louder than Everything Else

After leveling and processing the last song, I review songs one and two, to make sure they still fit well into the context or if there is a tweak that can further optimize them. Or, I might find that the album has grown in amplitude due to ear fatigue and the latter songs may need to be lowered.

Overzealous leveling practice can produce a *Domino Effect*. Suddenly, the song which used to be the loudest, doesn't sound as loud. Not every song can be the loudest! If it was loud enough before, the problem may be unintentional escalation. Instead of trying to push the loudest song further, thereby squashing it with the limiter, it may be wise to lower the previous song by even a few tenths of a dB, which will restore the next song's impact by use of contrast.

1 A neutral producer helps prevent the more me syndrome, and achieve the goal of a cohesive album rather than a collection of different band member's tunes.

2 Though everyone has their favorite album transition, like Sergeant Pepper's between the rooster crow and **Good Morning, Good Morning**!

3 Segue (pronounced seg-way)—a crossfade or overlap of two elements. Webster's: proceed without interruption. Italian: seguire, to follow.

4 The plant only needs PQ lists with the offset times, which appear on the CD master. The other recommended offsets are: first track mark 2 seconds before the beginning of the music; end mark a few frames after music end, to prevent premature muting; and last end mark 2 seconds after music end, so the player can stop spinning without losing the last sound.

5 In the first edition I incorrectly stated that Index 0 times shorter than 2 seconds are not permitted in the Redbook. The confusion is that Redbook requires P flag to be active at least 2 seconds before Index 1. But Redbook permits a short Index 0 to begin after the P flag becomes active.

6 A limitation of coded formats like mp3, which cannot be easily edited or crossfaded. iTunes must have engineered a workaround for gapless albums.

7 Thanks to Dan Stout, as viewed on the Mastering Webboard.

8 But ballads do not have to be raised as much as you think. Read about the acoustic advantage in Chapter 14.

9 Mix engineers follow similar practice, beginning the mix from the loudest point of the song.

Chapter 8

# Equalization Techniques

## I. Introduction

### The First Principle of Mastering

The first principle of mastering is this: *changing anything affects everything.* This principle means that mastering becomes the art of compromise, the art of knowing what is sonically possible, and then making informed decisions about what is most important for the music.

Equalization practice is an especially clear case of where a technique used in mastering is crucially different from an apparently similar technique used in mixing. For example, when mastering, adjusting the low bass of a mix will affect the perception of the extreme highs. Similarly, if a snare drum sounds dull but the vocal sounds good, then many times, the voice will suffer when you try to equalize for the snare.[1] These problems occur even between elements in the same frequency range. During mixing, bass-range instruments that exhibit problems in their harmonic range can be treated individually, but in mastering their harmonic range overlaps with the range of other instruments. For example, although a mix engineer can significantly boost a bass instrument somewhere between say, 700 and 2 kHz, in mastering even a small boost in this range can have detrimental effects. Or when we need to fix a bass drum problem, to minimize affecting the bass guitar it may be necessary to try careful, selective equalization to "get under the bass" at the fundamental of the drum, somewhere under 60 Hz. Sometimes we can't tell if a problem can be solved until we try,

{ *"Mastering is the art of compromise..."* }

"*Practice is the best of all instructions*"
— Chinese Fortune Cookie

so let's not promise a client miracles—but then they're delighted when we deliver them!

## II. What is a Good Tonal Balance?

Perhaps the major reason clients come to mastering houses is to verify and obtain a good tonal balance. But what, exactly, is a "good" tonal balance? The human ear responds positively to the tonality of a symphony orchestra which on a spectrum analyzer, always shows a gradual high frequency rolloff—as will most good pop music masters. The amount of this rolloff varies considerably depending on the musical style of course, so we use our ears, not the spectrum analyzer, as the basis for our EQ judgments.

Everything starts with the midrange: the fundamentals of the vocal, guitar, piano and other instruments must be correct, or nothing else can be made right. The message in the music—and more literally in radio, internet and low-cost home systems—comes from the midrange. Listen to a great recording in the next room. The information still comes through despite the filtering of the doorway, carpets and obstacles. Then try filtering the recording severely below 200 Hz and above 5 kHz (like the sound of an old, bad cinema loudspeaker). A good recording will still translate.

The mastering engineer tries to make the sound pleasant, warm, and clear, if that is appropriate for the genre. While a master can deviate from this to provide a deliberately different color—for example, a brighter, thinner sound,[2] the mastering engineer controls excessive deviation from neutral, ensuring that the sound will translate to the widest variety of playback systems and on the air.[3]

### Specialized Music Genres

The symphonic tonal balance is generally a good guide for rock, pop, jazz, world, and folk music, especially in the mid to high frequencies. But some specialized music genres deliberately utilize very different frequency balances. We could think of Reggae as the symphony spectrum with lots more bass instruments, whereas punk rock is often extremely aggressive, thin, loud and bright. Punk voices can be thin and tinny over a fat musical background. If this straining of the natural fundamental-harmonic relationships is excessive and done for a whole record, most people find it fatiguing, but it can be interesting when it's part of the artistic variety of the record.

### Be aware of the intentions of the mix

Equalization affects more than just tonality—it can also affect the internal balance of a mix. So a good mastering engineer must make sure she understands the intentions and needs of the production team. In fact, mastering equalization may help the producer's balance if his judgment was inadvertently affected by a monitoring problem in the mix environment.

## III. Equalization Techniques

### Two Basic Types: Parametric and Shelving

There are two basic types of equalizers—**parametric** and **shelving**—named after the shape of their characteristic curve. Parametric EQ, invented by George Massenburg in 1967,[4] is the most flexible curve, providing three controls: center frequency, bandwidth, and level of boost or cut. Mix engineers like to use parametrics on individual instruments, either boosting to bring out their clarity or salient characteristic, or selectively dipping to eliminate problems, or by virtue of the dip, to exaggerate the other ranges. The parametric is also the most popular equalizer in mastering since it can be used surgically to remove certain defects, such as overly-resonant bass instruments or enhance narrow ranges of frequencies. On the other hand, shelving equalizers are more popular in mastering than in mixing, since they provide boosts or cuts to the entire spectrum below or above a selected frequency, and can alter the tonality of the entire mix.

#### Parametric: Q and Bandwidth

The parameter Q is defined mathematically as the product of the center frequency divided by the bandwidth in Hertz at the 3 dB down (up) points measured from the peak (dip) of the curve. A low Q means a high bandwidth, and vice versa. The figure (above right) shows two parametric bands with extreme levels for purposes of illustration: On the left, a 17 dB cut at 50 Hz with a very narrow Q of 4, which is 0.36 octaves or a bandwidth of 12.5 Hz. On the right, a 17 dB boost centered at 2 kHz, with a fairly wide (gentle) Q of 0.86, which is 1.6 octaves. The bandwidth is 2325 Hz, represented by the dashed white line.[5]

Gentle equalizer slopes almost always sound more natural than sharp ones, so Q's of 0.6 and 0.7 are therefore very popular. Higher (sharper) Q's (greater than 2) are used surgically, to deal with narrow-band resonances or discrete-frequency noises, though we must listen for artifacts of high Q such as ringing. It is possible to work on just one note with a sufficiently narrow-band equalizer, or we may overturn the first principle of mastering by using a higher Q to attempt to isolate and emphasize a single instrument. For example, a poorly-mixed program may have a weak bass instrument; but boosting the bass around 80 Hz may help the bass but muddy the vocal, so we narrow the bandwidth of the bass boost. Rarely is this totally effective, so if the bass boost is not good for the vocal it's probably not good for the song. But if the vocal is made only slightly bassier, we may try a slight compensatory boost around 5 kHz, as long as that doesn't interact poorly with yet another instrument!

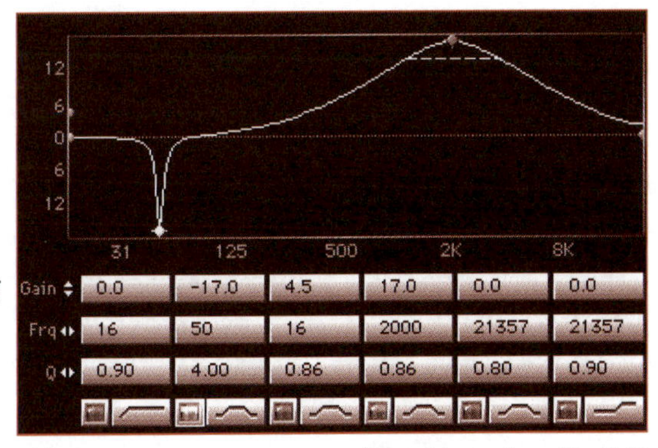


*Parametric equalizer with +17 dB boost centered at 2 kHz with a fairly wide bandwidth of 1.60 oct (Q = 0.86), indicated by the dashed white line at the 3 dB down points. A cut of -17 dB at 50 Hz with a very narrow bandwidth of 0.36 octaves (Q = 4).*

#### Focusing the Equalizer

There are three techniques for finding a problem frequency. The **classic approach** is to **focus the equalizer directly:** starting with a large boost and fairly wide (low value) Q, sweep through the frequencies until the resonance is most exaggerated, then narrow the Q to be surgical, and finally, dip the EQ the amount desired. This technique works well with analog equalizers, but some digital equalizers present ergonomic obstacles: the inefficient mouse, and latency. A **second technique** is for engineers who have a musical background—keep a keyboard handy to determine the key of the song and use your sense of relative pitch to determine the problem

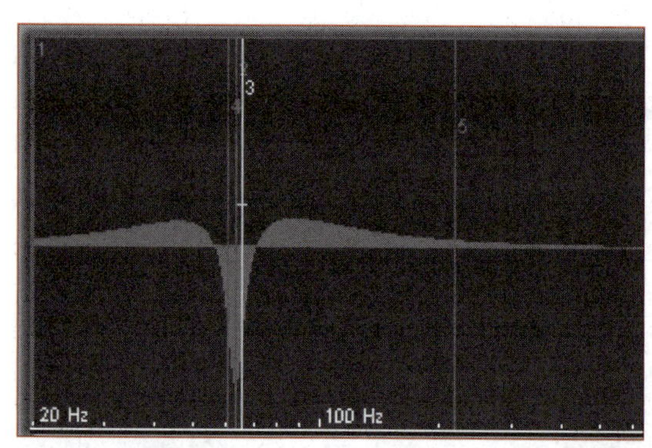

*"One note bass" resonance fixed by a combination of a shelving boost (which was useful to help the rest of the notes that were weak) and a narrow band dip at the resonance frequency.*

note. Then translate that note to frequency with the Carnegie Chart (attached to the front of this book) and dip that frequency. Pictured (at left) is an example of a bass EQ found in seconds using this method. It combines a shelving boost with a single dip at the problem frequency. Engineers with a Cranesong Ibis EQ can skip the chart, because this equalizer is marked directly in musical notes. A **third technique** works great with fast bass runs where several individual notes stick out unacceptably: run a high resolution FFT, watch the notes as they're played—each frequency can be pinpointed to within a couple of Hz.

### Shelving Equalizers

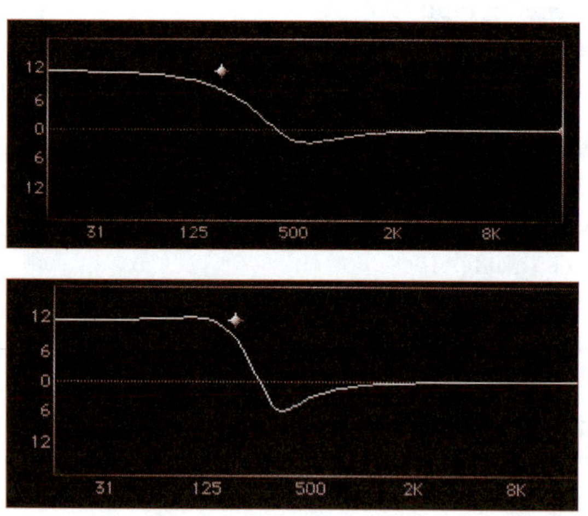

**Top:** *Gerzon resonant shelf with a low Q.*
**Bottom:** *The same with a high Q. The dip just past the shelving boost frequency is characteristic of the Gerzon resonant shelf.*

As mentioned, a shelving equalizer affects the level of the entire low frequency or high frequency range below or above a specified frequency. Some have Q controls, defined as the slope of the shelf at its 3 dB up or down point. One interesting variant on the standard shelf shape can be found in the Weiss EQ-1, Waves Renaissance EQ and Manley Massive Passive. This shape, called a *resonant shelf* (pictured at left), was proposed by psychoacoustician Michael Gerzon. I like to think of it as a combination of a shelving boost and a parametric dip (or vice versa). In the top image, a low Q (0.71) bass shelf of 11.7 dB below 178 Hz is mollified by a gentle parametric dip above 178 Hz; all of which is controlled by a single band of the equalizer. This type of curve can help keep a vocal from sounding thick while implementing a bass boost. The bottom image shows the same boost with a high Q of 1.41.

### EQ Yin and Yang

Remember the yin and the yang: Contrasting ranges have an interactive effect. For example...

- Adding low frequencies makes the sound seem duller and reducing them makes it seem brighter.
- Adding extreme highs between 15-20 kHz makes the sound seem thinner in the bass/lower midrange and vice versa.
- A slight dip in the lower midrange (around 250 Hz) reduces warmth, and has a similar effect to a boost in the presence range (around 5 kHz).
- A harsh-sounding trumpet section can be improved by dipping around 6-8 kHz, and/or by boosting around 250 Hz.

Yin and yang considerations allow us to work in either or both contrasting ranges, whichever is most effective. Thus when the overall level is too high—we pick the range which we would be reducing. When an instrument(s) exhibits upper midrange harshness, we pick the frequency range that would have the least effect on other instruments playing at the same time.

### Using Baxandall for air

In Chapter 3, we described the *air band* as the range of frequencies between about 15-20 kHz, the highest frequencies we can hear. An accurate monitoring system will indicate whether these frequencies need help. An *air boost* is contraindicated if it makes the sound harsh or unintentionally brings instruments like the cymbals forward (closer or louder). A third and important shape that's extremely useful in mastering is the Baxandall curve (at right), named after Peter Baxandall. Hi-Fi tone controls are usually modeled around this curve. Like shelving equalizers, a Baxandall curve is applied to low or high frequency boost/cuts. Instead of reaching a plateau (shelf), the Baxandall continues to rise (or dip, if cutting instead of boosting). Think of the spread wings of a butterfly with a gentle curve applied. This shape often corresponds better to the ear's desires than any standard shelf so a Baxandall high frequency boost makes a great *air EQ*. Shelving equalizers with slope or Q controls can often approximate a Baxandall shape. We can also simulate a Baxandall high frequency boost by placing a parametric equalizer (Q= ~1) at the high-frequency limit (~20 kHz). The portion of the bell curve above 20 kHz is ignored, and the result is a gradual rise starting at about 10 k and reaching its extreme at 20 k.

Be careful when making high frequency boosts (adding *sparklies*). They are initially seductive, but can easily become fatiguing. The principle of yin and yang reminds us that the ear interprets a high frequency boost as a thinning of the lower midrange. In addition, when the highs come up, the cymbals, triangle and tambourine become louder, which changes the balance of rhythm to melody, for better or worse.

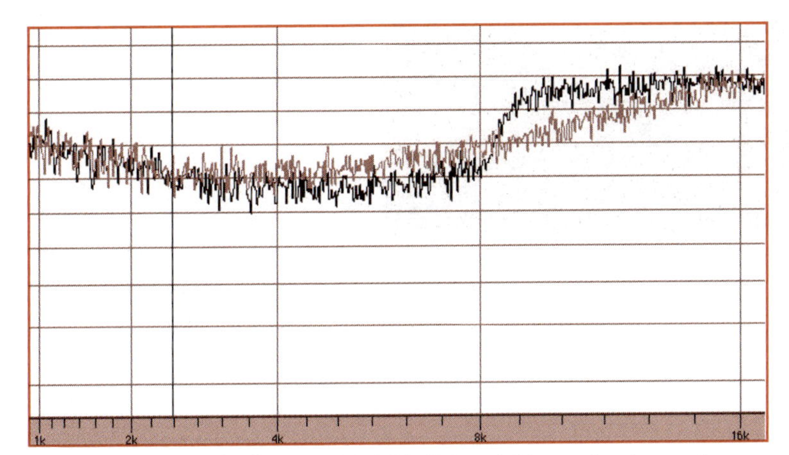


*Gentle Baxandall curve (pink) vs. sharp Q shelf (black). Many shelving equalizers have gentler curves and may approach the shape of the Baxandall. Try a shelf with 3 dB per octave slope for this purpose.*

### High-Pass and Low-Pass Filters

On the left of the figure on the next page is a sharp high-pass (low cut) filter at 61 Hz, and on the right, a gentle low-pass (high cut) filter at 3364 Hz. The frequencies are defined as the points where the filter is 3 dB down. Although pass filters can be used to solve noise problems in mastering, they can also introduce problems of their own, because they affect everything above or below a certain frequency. High-pass filters can reduce rumble, thumps, p-pops and similar noises. Low-pass filters are sometimes used to reduce hiss, though since the ear is most sensitive to hiss in the 3 kHz range, a parametric dip around that frequency is more effective than the radical low-pass filter. For hiss removal, we usually prefer specialized noise-reduction solutions over static filters (Chapter 12).

*"Remember the yin and the yang: Contrasting ranges have an interactive effect"*

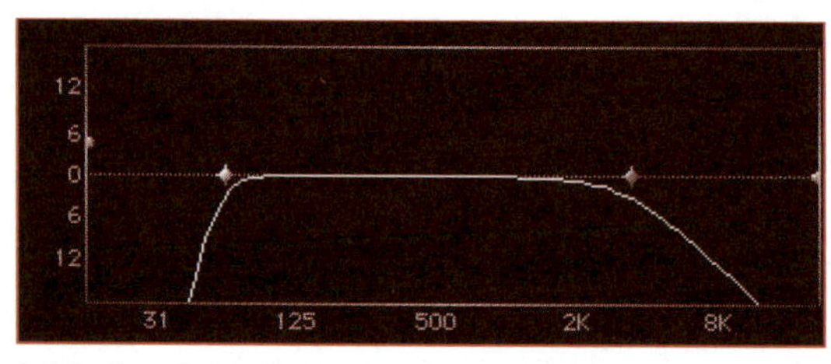


*At left: Sharp high-pass filter at 61 Hz. At right: Gentle low pass filter at 3364 Hz.*

### One channel or both (all)?

Most times making the same EQ adjustment in both (all) channels is the best way to proceed as it maintains the stereo (surround) balance and the relative phase between channels. But sometimes it is essential to be able to alter only one channel's EQ. For example, with a too-bright high-hat on the right side, a good vocal in the middle and proper crash cymbal on the left, the best solution is to work on the right channel's high frequencies.

### Start subtly first

Sometimes important instruments need help, though ideally, they should have been fixed in the mix. The best repair approach is to start subtly and advance to severity only if subtlety doesn't work. For example, if the piano solo is weak, we try to make the changes surgically:

- only during the solo
- only on the channel where the piano is primarily located, if that sounds less obtrusive
- only in the frequency range(s) that help: fundamental, harmonic, or both
- as a last resort, by raising the entire level, because it would affect the entire mix, though the ear focuses on the primary instrument

### The Limitations and the Potential of the Recording

Waiting until the mastering stage to fix certain problems is inviting a compromise since there is only so much that can be accomplished in mastering. There is little we can do to fix a recording where one instrument or voice requires one type of equalization and the rest requires another. In many cases I recommend a remix, however if a remix is not possible, then we might be able to use the specialized techniques to be discussed in Chapter 16.

Comb filtering is a similar issue. Equalization can do little or nothing to fix a comb filtering problem since EQ affects the entire mix. First we discuss the problem with the mix engineer to see if he can address the offending track and remix. If that is not possible, then we may try an overall mastering EQ, for example, a lower midrange EQ boost to help a vocal that sounds thin due to comb filtering.

The better the mix we get, the better we can make the master, which implies, of course, that a perfect mix needs no mastering processing at all! And this is true: we should not automatically equalize; we should always listen and evaluate first. Many pieces leave mastering with no equalization or processing.

### Instant A/B's?

With a good monitoring setup, equalization changes of less than 1/2 dB are audible. Take an equalizer in and out to confirm initial settings, but then avoid making instant EQ judgments because quickly switching back and forth distracts us from the long-term effect of an EQ change. Music is fluid so changes in the music can easily be confused with EQ changes. I usually play a passage for a reasonably long time with setting "A", then play it again with setting "B".

#### Loud and Soft Passages

I usually begin mastering on the loudest part of a song. Why? Because psychoacousticians note that EQ peaks affect partial loudness more than dips, and loud passages accentuate these peaks more than soft ones. Equalization choices which are pleasing during a mezzo-forte passage may well displease during a loud one.

#### Fundamental or Harmonic?

The extreme treble range mostly contains instrumental harmonics. Since the fundamental of a crash cymbal can be lower than 1.5 kHz, boosting the harmonics too much makes a cymbal sound tinny or thin. When equalizing or processing bass frequencies, it is easy to confuse the fundamental with the second harmonic. The detail shot of a SpectraFoo™ Spectragram in ***Color Plate Figure C8-1*** illustrates the importance of the harmonics of a bass instrument. High amplitudes are indicated in red, descending levels in orange, yellow, green, then blue.

Notice the parallel run of the bass instrument's fundamental from 62-125 Hz and its second and third harmonics from 125-250 and up. Should we equalize the bass instrument's fundamental or the harmonic? It's easy to be fooled by the octave relationship; the answer has to be determined by ear—sometimes one, the other or both.

#### Bass boosts can create serious problems

Since the ear is significantly less sensitive to bass energy, bass information requires a lot more power for equal sonic impact: around 6 to 10 dB more below about 50 Hz, and about 3-5 dB more between 50 and 100 Hz.[6] This explains why bass instruments often have to be compressed to sound even. It also means that a low frequency boost introduces so much energy it can reduce the highest clean intrinsic level we can give to a song (in cases where the client is demanding a "loud" master). Fortunately, the ear's tendency to supply missing fundamentals (see Chapter 3) works in our favor, allowing us to save "energy" by cutting with a fairly sharp high pass filter, but only if it does not hurt the quality of the bass drum or the low notes of the bass. The high pass filter must be extremely transparent and have low distortion. Sometimes a gentle filter is a better choice than a steep one, as when dealing with a boomy bass drum or bass. But subsonic rumble or thumps benefit from a steep filter to have minimal effect on the instruments.[7]

Mix engineers working with limited bandwidth monitors run the risk of producing an inferior product. Utilizing accurate subwoofers permits hearing low frequency leakage problems that tend to muddy up the mix, for example, bass drum leakage in vocal and piano mikes. It's much better to apply selective high pass filtering during the mixing process because mastering filters will affect all the instruments in a frequency range. For example, mix engineers may get away with a steep 80 Hz filter on an isolated vocal, but that's generally too high a frequency for mastering a whole mix. A mixing engineer should form an alliance with a mastering engineer, who can review her first mix and alert her to potential problems before they get to the mastering stage.

*{ "The perfect mix may need no mastering processing at all!" }*

## IV. Other refinements

### Linear phase Equalizers: The Theory

All current analog equalizer designs and nearly all current digital equalizers produce phase shift when boosted or cut; that is, signal delay varies with frequency and the length of the delay changes with the amount of boost or cut. The higher the Q, the more the phase shift. This kind of filter will always alter the musical timing and wave shape, also known as *phase distortion*. Daniel Weiss introduces…

> a particular type of digital filter, called the **Symmetric FIR Filter**, [which] is inherently linear phase.[8] This means that the delay induced by processing is constant across the whole spectrum, unconstrained by EQ settings.[9]

In the figure (above right) we graph the amplitude and phase response of two equalizers, a minimum phase and a linear phase. The linear phase EQ's phase response is a straight line.

Jim Johnston outlines the fundamental differences:

> Whenever you have to equalize, you will alter the signal in both the time and frequency domains (as mathematics requires); there will always be a **time artifact**. In the *analog* style equalizer, which is usually mathematically termed *minimum phase*, the alteration will be primarily to spread the signal **downstream**, i.e. does not lead the original signal by much, if any. A **downstream modification** translates into different delays at different frequencies dispersing the original signal. In some cases this effect is quite audible. If one uses a *digital approach*, one can either mimic the analog behavior, or use a *linear phase*, a.k.a. *constant delay* filter. This filter will **equally precede and follow the signal**; part of the filter may create a pre-echo effect, modifying the leading edge of transients and signal changes. A high Q linear phase filter can introduce audible pre-echo in the short millisecond range; it's exactly like a floor bounce but without the comb filtering. Anytime that a high Q filter is used, careful listening with both types of equalization may be necessary to decide which choice is best.[10]

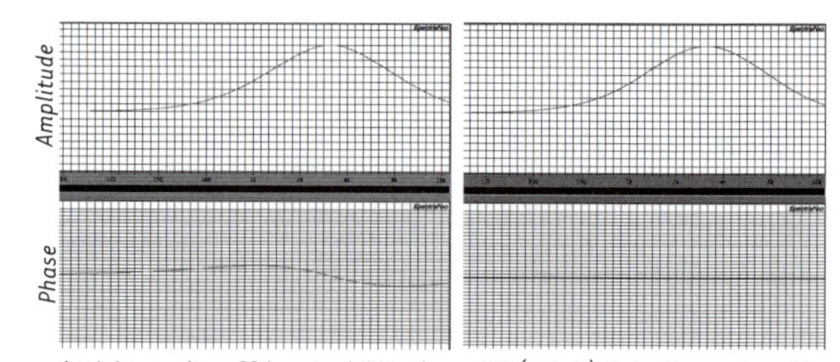

*A minimum phase EQ boost exhibits phase shift(at left), but a linear phase EQ boost has none (at right).*

Since symmetric FIR filters are expensive to implement in real time, linear phase equalizers can be made with different techniques. The Weiss uses a complementary IIR (a.k.a. dual-pass) technique, avoiding one of the down sides of FIR, which can produce ripple in the frequency response unless the designers use proper windowing.

### Linear phase EQ: The Sound

To my ears, the linear phase (abbreviated LP) sounds more analog-like than even analog, but ironically, while mastering a punk rock recording, my EQ proved too *sweet* in LP mode so I had to return to normal mode to give the sound some *grunge*. Clearly much of the qualities we've grown accustomed to in standard equalizers (also known as **minimum phase equalizers**, abbreviated MP) must be due to their phase shift, in fact John Watkinson believes that much of the audible difference between EQs comes down to their different phase response.[11] The Weiss has a very pure tone quality and seems to boost and cut frequencies without introducing obvious artifacts.

Neither approach is fundamentally better. It's easier to be aggressive with MP. Because of the pre-echo alluded to by Johnston, the LP subtly reduces transient response, which might contribute to its "sweeter", softer sound. The MP's tendency to smear depth can yield a pleasant artificial depth where none existed before. However, the LP's ability to preserve the depth of the mix is frequently an advantage when mastering, because altering a band's level does not move instruments forward or backward in the soundstage, as would happen with a minimum phase equalizer. Narrow-band peaks and dips can be accomplished with no artifacts in linear phase, avoiding the smeary quality that occurs in minimum phase. I especially love the ability of LP to keep high hats from moving forward when boosting in the cymbal range. Alan Silverman (in correspondence) says,

> EQ'ing frequency ranges [in LP] is more like raising or lowering a fader in a mix than EQ'ing.

### Dual-pass vs. Frequency Domain LP

Algorithmix's Orange LP EQ uses the dual-pass technique, while the Red accomplishes linear phase with a frequency-domain technique. To my ears, the Red combines the depth-preservation of LP with the liveliness and transparency normally associated with MP. To support my hypothesis, I compared the time-domain (impulse) responses of the Weiss in MP and LP modes, and the Red (the Orange was nearly identical to the Weiss LP). An impulse should look like a narrow vertical line if there is no phase shift or time dispersion. Pictured below is a graph of an impulse played through the three equalizers, set to the same high frequency shelving boost. Amplitude is on the vertical scale, time on the horizontal. Vertical scale is linear, so twice the amplitude is 6 dB. At left, MP; in the middle, dual-pass LP; at right, frequency-domain LP.

Notice in the MP, the time dispersion mentioned by Jim Johnston; in the dual-pass LP, the

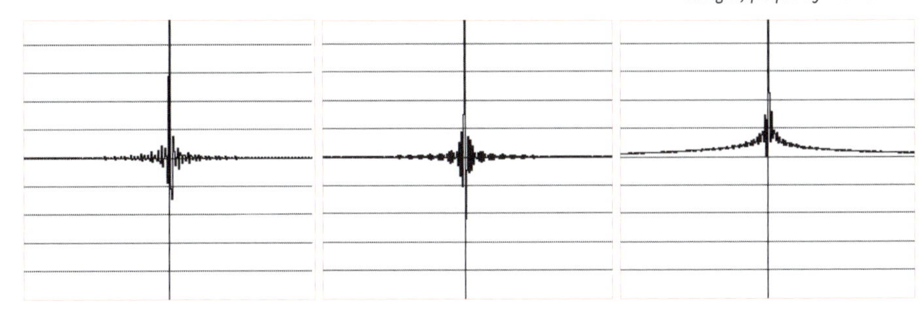

*Impulse responses. At left, MP; in the middle, dual-pass LP; at right, frequency-domain LP.*

distinct pre and post echos; while the frequency domain LP has significantly less echo and noise, it has a clean, sharp trace with at least 12 dB lower level echos (though there appears to be a DC bias in the residual).

### Dynamic Equalization

Dynamic equalizers (like the Weiss EQ1-Dyn) emphasize or cut frequency ranges *dynamically*, as opposed to *static*, or *fixed* EQ. Thresholds set a level above or below which a band is dynamically boosted or cut. This extra amount is added to a *static* setting. For example, above the threshold we can lower the high frequency response; we could start a static band at, say + 1 dB, but to prevent harshness at high levels, slowly cut the band's level when the signal exceeds the threshold. Dynamic EQs can be used as noise or hiss gates, rumble filters that function at low levels (especially useful for traffic control in a location classical recording), sibilance controllers, or ambience enhancers. They can enhance inner details or clarity of high frequencies at low levels, where details are often masked by noise. Or enhance warmth by raising the respective band at low levels but prevent the sound from getting muddy at high levels. Multiband dynamics processors (see Chapters 10 and 11) can also perform dynamic equalization.

1 Chapter 16 will discuss ways to overcome the first principle, and reduce the compromise.

2 We may believe we have "the absolute sound" in our heads, but are surprised to learn how much the ear/brain accommodates. If we play a bright album immediately after a dark one, at first there is ear shock, but we quickly adapt, though the new sound continues to affect subliminally. The same thing happens in photography and motion pictures, after an initial shock, the change to "Kodachrome" becomes subliminal.

3 Overly bright records become dull on the air due to high frequency FM broadcast limiters. Thus radio processing makes brightness self-defeating.

4 In 1967, George Massenburg began the search for a circuit which would be able to independently adjust an equalizer's gain, bandwidth and frequency. The key word is *independent*, for most analog circuits fail in this regard and the frequency, Q, and gain controls interact with each other. His circuit, which he calls a **parametric equalizer**, remains proprietary today.

5 Some equalizers define bandwidth in octaves instead of Q. Appendix 6 contains a convenient table for converting between Q and bandwidth.

6 This is dictated by the psychoacoustic *Equal Loudness Curves*, first researched by Fletcher, Harvey and Munson in the 1930's.

7 Historically, the high pass filter was crucial when making LPs, to prevent excess groove excursion and obtain more time per LP side, but digital media do not have this physical problem.

8 **FIR** mean Finite Impulse Response, and **IIR** Infinite Impulse Response. Read John Watkinson's **The Art of Digital Audio** for an explanation of the differences.

9 Weiss, Daniel, check www.digido.com/links/ for the Weiss website.

10 Jim Johnston, in correspondence.

11 (9/1997) *Studio Sound Magazine*.

CHAPTER 9

# How To Manipulate Dynamic Range for Fun and Profit

## Part One: Macrodynamics

## I. The Art of Dynamic Range

The term **dynamic range** refers to the **difference** between the loudest and softest passages of the body of the music; it should not be confused with loudness or *absolute* level. The dynamic range of popular music is typically only 6 to 10 dB, but for some musical forms it can be as little as a single dB or as great as 15 (very rare). In typical pop music, soft passages 8 to 15 dB below the highest level are effective only for brief periods, but in classical, jazz and many other acoustic forms, low passages can effectively last much longer.

### Microdynamics and Macrodynamics

Dynamics can be divided further into **Microdynamics**—the music's rhythmic expression, integrity or bounce; and **Macrodynamics**—the loudness differences between sections of a song or song-cycle. Usually dynamics processors (such as compressors, expanders) are best for *microdynamic manipulation*, and manual gain riding is best for *macrodynamic manipulation*. The micro- and macro-work hand in hand, and many good compositions incorporate both microdynamic changes (e.g. percussive hits or instantaneous changes) as well as macrodynamic (e.g. crescendos and decrescendos). Think of a music album as a four-course meal: The progression from soup to appetizer to main course and dessert is the macrodynamics. The spicy impact of each morsel, is the microdynamics. In this chapter, we concentrate on macrodynamics.

### The Art of Decreasing Dynamic Range

The dynamics of a song or song cycle are critical to creative musicians and composers. As engineers, our paradigm sound quality reference should be a live performance; we should be able to tell by listening if a recording will be helped or hurt by modifying its dynamics. In a natural performance, the choruses should sound louder than verses, ensembles louder than soloists, and the climax should be the loudest. Many recordings have already gone through several stages of transient degradation, and indiscriminate or further dynamic reduction can easily take the clarity and the quality downhill. However, usually the recording medium and intended listening environment simply cannot keep up with the full dynamic range of real life, so the mastering engineer is often called upon to raise the level of soft passages, and/or to reduce loud passages, which is a form of **manual compression.**[1] We may reduce dynamic range (compress) when the original range is too large for the typical home environment, or to help make the mix sound more exciting, *fatter*, more coherent, to bring out inner details, or to even out dynamic changes within a song if they sound excessive.[2]

Experience tells us when a passage is too soft. The context of the soft passage also determines whether it has to be raised. For example, a soft introduction immediately after a loud song may have to be raised, but a similar soft passage in the middle of a piece may be just fine. This is because the ears self-adjust their sensitivity over a medium time period, and may not be prepared for an instantaneous soft level after a very loud one. Thus, meter readings are fairly useless in this regard. How soft is too soft? The engineers at Lucasfilm discovered that having a calibrated monitor gain and a dubbing stage with noise floor NC-20[3] do not guarantee that a film mix will translate to the theatre. During theatre test screenings, some very delicate dialogue scenes were "eaten up" by the air conditioning rumble and audience noise in a real theatre. So they created a calibrated noise generator, labeled "popcorn noise" which could be turned on and added to the monitor mix whenever they wanted to check a particularly soft passage. For similar purposes, our alternate listening room at Digital Domain has a ceiling fan and other noisemakers. Whenever I have a concern, I start the DAW playing a loud passage just before the soft one, and take a walk to the noisy listening room.

**MYTH:**
*"Of course I've got dynamic range. I'm playing as loudly as I can!"*[7]

### The Art of Increasing Dynamic Range...

...can also make a song sound more exciting, by using contrast or by increasing the intensity of a peak. The key to success here is to recognize when an enhancement has become a defect—musical interest can be enhanced by variety, but too much variety is just as bad as too much similarity. Another reason to increase dynamic range is to restore, or attempt to restore the excitement of dynamics which had been lost due to multiple generations of compression or tape saturation.

### The Four Varieties of Dynamic Range Modification

We always use the term **Compression** for the reduction of dynamic range and **Expansion** for its increase. There are two varieties of each: **upward compression, downward compression, upward expansion, and downward expansion**, as illustrated in the figure at right, next page.

**Downward compression** is the most popular form of dynamic modification, bringing high level passages down. Limiting is a special case—downward compression with a very high ratio (to be explained in Chapter 10). Examples include just about every compressor or limiter you have ever used.

**Upward compression** raises the level of low passages. Examples include the encode side of a Dolby® or other noise reduction system, the AGC which radio stations use to make soft things louder, and the type of compressor frequently used in consumer video cameras.[4] In Chapter 11 we will introduce a more effective upward compression technique that is extremely transparent to the ear. For clarity, we will always use the short term **compressor** to mean **downward compressor** unless we need to distinguish it from **upward compressor.**

**Upward expansion** takes high level passages and brings them up even further. Upward expanders are relatively rare; in skilled hands they can be used to enhance dynamics, increase musical excitement, or restore lost dynamics. Examples include the peak restoration process in the playback side of a Dolby SR, the DBX Quantum Processor, the various Waves brand dynamics processors, and the Weiss DS1-MK2 when used with ratios less than 1:1 (to be explained).

**Downward expansion** is the most common type of expansion: it brings low level passages down further. Most downward expanders are used to reduce noise, hiss, or leakage. A dedicated noise gate is a special case—downward expansion with a very high ratio (to be explained). Examples of downward expanders include the classic Kepex and Drawmer gates, Dolby and similar noise reduction

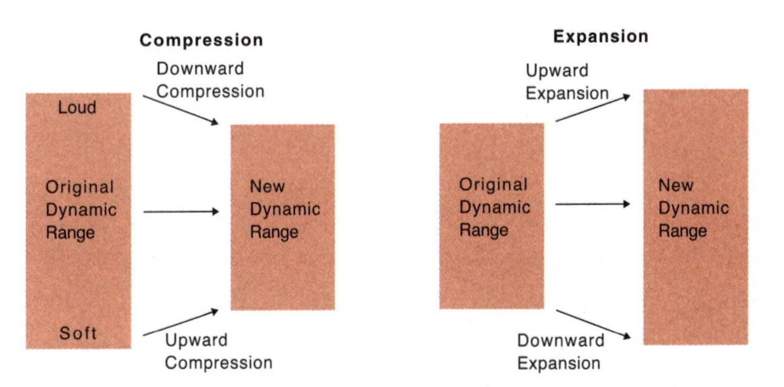


Any combination of these four processes may be employed in a mastering session

systems in playback mode, expander functions in multi-function boxes (e.g. TC Finalizer), and the gates on recording consoles. Again for clarity, we will use the simple term **expander** to mean the downward type unless we need to distinguish it from the upward type.

## II. The Art of Manual Gain-Riding: Macrodynamic Manipulation

### In General

During mixing it is difficult to simultaneously pay attention to the internal balances as well as the dynamic movement of the music between, for example, verse and chorus. Sometimes engineers inadvertently lower the master fader during the mix to keep it from overloading, which strips the climax of its impact. In mastering we can enhance a well-balanced rock or pop mix by taking the dynamic movement of the music where it would like to go. Delicate level changes can make a big difference; it's amazing what a single dB can accomplish. It's also our responsibility to make sure the client's level change was not intentional!

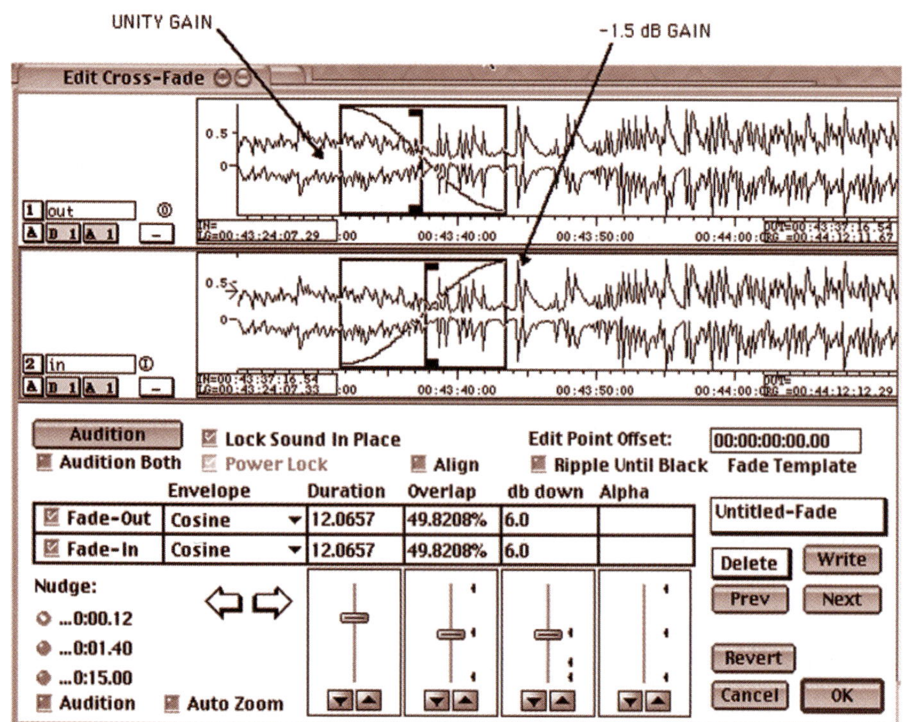


*The modern version of fader-riding. In Sonic Solutions' "classic" edit window the outgoing edit is on top, the incoming on the bottom. Note that the gain drop is performed in the soft passage preceding the loud downbeat, thus preserving the apparent impact of the downbeat.*

## How and When to Move the Fader

Artistic level changes can really improve a production but they need to be made in the most musical way. To this end, internal level changes are least intrusive when performed manually (by raising or lowering the fader), as little as a 1/4 dB at a time, as opposed to using processors such as compressors or expanders, which tend to be more aggressive.

When riding the gain, aim just to augment the natural dynamic flow: if the musicians are trying for upward impact, pulling the fader back during a crescendo can be detrimental since it will diminish the intended impact. Extra-soft passages require special attention. If the highest point in the song sounds "just right" after processing, but the intro sounds too soft, it's best to simply raise the intro, finding just the right editing method to restore the gain using one or more of these approaches:

- a long, gradual lowering of the gain, which might occur at the end of the intro, or slowly during the first verse of the body.
- a series of 1/4 or 1/2 dB edits, taking the sound down step by step at critical moments. This is useful when we don't want the listener to notice that we're cheating the gain back down and we may be forced to work against the natural dynamics.
- a quick edit and level change at the transition between the raised-level intro and the normal-level body. This can have a nice effect and be the least intrusive.

## The Art of Changing Internal Levels of a Song

Some soft passages must be raised, but if the musicians are trying to play something delicately, pushing the fader too far can ruin the effect. The art is to know how far to raise it without losing the feeling of being soft, and the ideal speed to move the fader without being noticed. In a DAW, physical fader moves are replaced by crossfades, or by drawing level changes on an automation curve. The mastering engineer's aim is to be invisible; if the sound is being audibly manipulated, the job has not been done properly.[5] Here's a technique for decreasing dynamic range in the least damaging way which I learned many years ago from Alec Nesbitt's book **The Technique of the Sound Studio** (see Appendix). If we have to take a loud passage down, the best place to lower the gain is at the **end of the preceding soft passage before the loud part**

**begins**. Look for a natural dip or decrease in energy, and apply the gain drop during the end of the soft passage before the crescendo into the loud part begins. That way, the loud passage will not lose its comparative impact, for the ear judges loud passages in the context of the soft ones.

The figure at left on the previous page, from a Sonic Solutions "classic" workstation, illustrates the technique. The gain change is accomplished through a crossfade from one gain to another.

The producer and I decided that the *shout chorus* of this jazz piece was a bit overplayed and had to be brought down from triple to double forte (which amounted to one dB or so).[6] To retain impact of the chorus, we slowly dropped the level during the soft passage just before the drum hit announcing the chorus. In the trim window, we constructed a 12 second crossfade from unity gain (top panel) to -1.5 dB gain (bottom panel); the drum hit is just to the right of the crossfade box. This drum hit retains its impact by contrast, because the musicians' prior delicate decrescendo has been enhanced during mastering.

## Where to Begin

As mentioned in Chapter 7, start mastering by going directly to the loudest part of the loudest song, which reduces the later temptation to raise the loud part so much that it might be squashed by excessive processing. After getting a great sound with the necessary processing, return to the beginning of the song to see how it sounds in that context.

Having heard that, we may decide to **reduce** the level of a song's introduction. In the figure above, I

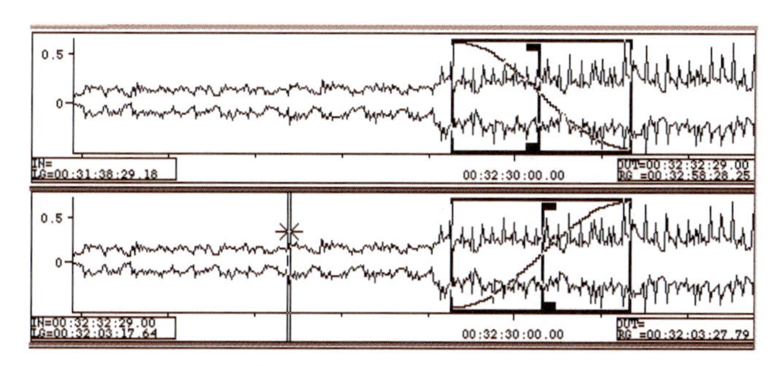

*A soft introduction has been reduced even further, and the impact of the body of the song is enhanced by gradually increasing the gain during the beginning of the main part of the song.*

reduced the intro and slowly introduced a crescendo (20 seconds long) that enhances the natural build as it goes into the first chorus. The top panel is at -1 dB gain, bottom panel is at unity (0 dB) gain, achieved at the end of the crossfade.

Increasing the space between two songs is another way of increasing the dynamic impact as it extends the tension caused by silence.

## In Conclusion

**Macrodynamic** manipulation is a sometimes overlooked but powerful tool in the mastering engineer's arsenal. In the next chapter we move on to the use of compressors, expanders and limiters to manipulate **microdynamics**.

1. Please do not confuse the term dynamic range reduction (compression) with data rate reduction. Digital Coding systems employ data rate reduction, so that the bit rate (measured in kilobits per second) is less. Examples include mp3, or Dolby Digital (AC-3). Since it's not good to refer to two different concepts with the same word, we should encourage people to use the term **Data Reduction System** or **Coding system** when referring to data. Use **Compression** only when referring to the reduction of dynamic range.
2. Excessive is definitely a subjective judgment! It's very important to develop an esthetic which appreciates the benefits of dynamic range, and which also knows when there is too much—or too little. This is a matter of taste, as well as objective knowledge of the requirements of the medium and listening environment.
3. A room with NC-30 rating is quiet. The quietest rooms may have noise floors lower than NC-20.
4. AGC (automatic gain control) has been given a bad name by a poorly-implemented use in camcorders, which often exhibit pumping and breathing artifacts.
5. This is true for most of the "natural" music genres, with some exceptions being hip-hop, psychedelic rock, performance art, etc. where the artists invite the engineer to contribute surprising or rococo dynamic effects.
6. Producers don't always use classical Italian dynamic terms to describe their needs. The mastering engineer should choose the bonding language which is best for the client—"Make it louder, man!"
7. A common misconception. Thanks to Gordon Reid of Cedar for contributing this audio myth.

CHAPTER 10

# How to Manipulate Dynamic Range for Fun and Profit

PART TWO: DOWNWARD PROCESSORS

## I. Compressors and Limiters: Objective Characteristics

Parts two and three of this series discuss micro- (and some macro-) dynamic manipulation, which is achieved primarily through the use of dedicated **dynamics processors**. In this chapter we look in detail at how compressors and limiters work, because we must first study their objective characteristics to learn how to use them effectively.

### Transfer Curves (Compressors and Limiters)

A transfer curve (or plot) displays the input-to-output gain characteristic of an amplifier or processor. Input level is plotted on the X axis, and output on the Y. A linear amplifier shows a straight line (not a curve), hence the name. This figure shows a family of linear plots at 3 different gains. **Unity gain** means the ratio of output to input level is 1, or 0 dB, so a unity-gain amplifier shows a straight diagonal line across the middle at 45°, called the unity gain line. From left to right: unity gain; 10 dB gain; 10 dB attenuation. Notice that the middle plot would yield distortion for any input signals above -10 dBFS.

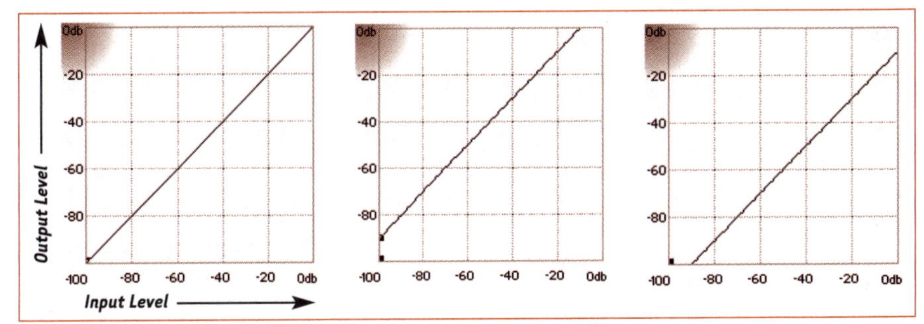


***Three transfer curves.*** *At left, a unity-Gain Amplifier, then an amplifier with 10 dB gain, then with 10 dB loss (attenuation).*

The **threshold** of a compressor is the level at which gain reduction begins, and **compression ratio** describes the relation between input and output above the threshold. At left in the figure (below) is a simple compressor with a fairly gentle 2.5:1 compression ratio, and a threshold at around -40 dBFS (which is quite low and would yield strong compression for loud signals). 2.5:1 means that an increase in the input of 2.5 dB will yield an increase in the output of only 1 dB, or for an input rise of 5 dB, the output will only rise by 2 dB, or as can be seen in the plot, an input change of 20 dB yields an output change of a little less than 10 dB (once the curve has reached its maximum slope). A compressor such as this would actually make loud passages softer, because the output above the threshold is less than the input; this is always the case unless the compression is followed by gain makeup (a simple gain amplifier after the compression section).

At the right-hand side of the figure, using makeup we can restore the gain so that a full level (0 dBFS) signal input will yield a full level signal output. For illustration purposes we show an amplifier with an extreme amount of gain, 20 dB, which would considerably amplify soft passages (below the threshold). In typical use, however, makeup gains are rarely more than 1 to 3 dB. Loud input passages from about -40 to about -15 are still amplified in this figure, but above about -15 dBFS, the curve slopes back to unity gain and resembles that of a linear amplifier. Far below the threshold, it's a fairly linear 20 dB amplifier and can have pretty low distortion because there is no gain reduction action. At full scale, 20 dB of gain makeup is summed with 20 dB of gain reduction, yielding 0 dB total gain. This particular compressor model's curve levels off towards a straight line above a certain amount of compression, so the ratio only holds true for the first 15-20 dB above the threshold. Other compressor models continue their steep slope, thus maintaining their ratio far above the threshold. There are as many varieties of compression shapes as there are brands of compressors, and they all give different sounds. To get the greatest esthetic effect from any compressor, most of the music action must occur around the threshold point, where the curve's shape is changing; thus, a real-world compressor's threshold would likely be -20 to -10 dBFS or higher.

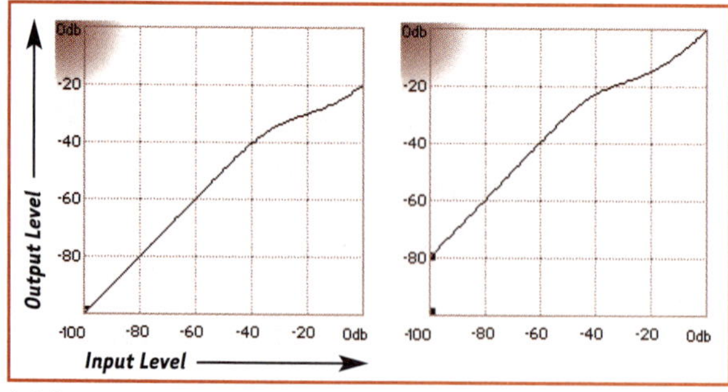


*At left, compressor with 2.5:1 ratio and -40 dBFS threshold and no gain makeup. At right, the same compressor with 20 dB gain makeup.*

### Knee

The figure (top right, next page) shows a very high ratio of 10:1; above the threshold, the output is almost a horizontal line, which is very severe compression, commonly called **limiting**.[1] The portion of the curve near the threshold is called the **knee**, which marks the transition between unity gain and compressed output. The term **soft knee** refers to a rounded knee shape, or gentle transition (at left), and **hard knee** to a sharper shape (at right), where the compression reaches full ratio immediately above threshold. Soft knee can sweeten the sound of a compressor near the threshold. For those models of compressors that only have hard knees, some of the effect of a soft knee can be simulated by lessening the ratio or raising the threshold, which will result in less action.

### Attack and Release Times

**Attack time** is the time it takes for the compressor to implement full gain reduction after the signal has crossed the threshold. Typical attack times used in music mastering range from 50 ms to 300 ms (or longer on occasion), with the average used probably 100 ms. Because digital compressors react with textbook speed, a digital compressor set to 100 ms may sound similar to an analog compressor set to, say, 40 ms; so it's probably better to remove all the labels on the knob (except *slow* and *fast*) and just listen! **Release time**, or **recovery time**, is how long it takes for the signal level to return to unity gain after it has dropped below the threshold. Typical release times used in music range from 50 ms to 500 ms or as much as a second or two, with the average around 150-250 ms.[2] The terms **short** or **fast** with attack or release time are used interchangeably, as are **slow** and **long** attack and release times. Oddly enough there is no standard for defining of these times; manufacturers may measure times to 90% of gain reduction or another empirical approach.

**The preview**, or **look-ahead function**, allows very fast, or even instantaneous (zero) attack time, which is especially useful in a peak limiter. In effect the unit has to react to the transient before it has occurred! This requires a delay line, so there is no look-ahead in an analog processor. Every compressor has a sidechain, which is the control path (as opposed to the audio path), as illustrated in this figure (at right).

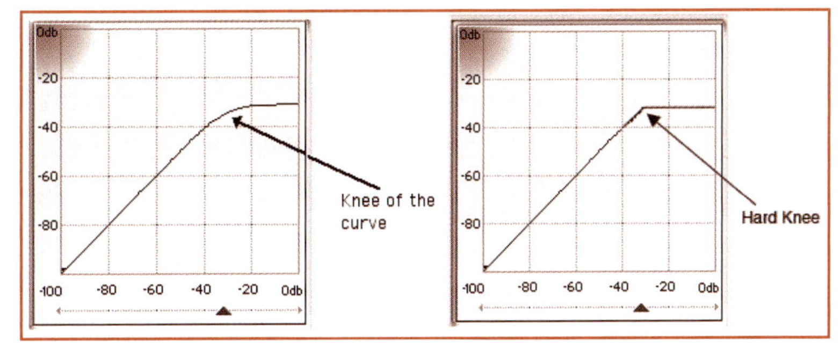


*At left, compressor with soft knee, at right, hard knee.*

The compressor places a time delay in the audio signal chain, but not in the sidechain, which allows the sidechain time to "anticipate" the leading edge of a transient and nip it in the bud. The delay time only has to be as long as the shortest transient we want to control in addition to the reaction time of the sidechain circuit. Look-ahead is only relevant when desiring short attack times since if we want a long attack then we probably also want to let initial transients pass through. Certainly analog compressors have gotten along splendidly without preview delay; in fact much of their sonic virtues come from their **inability** to stop initial transients. Because they contribute to the life and impact of a recording, I only remove sharp transients when they are audibly objectionable. The exceptions to this might be transients shorter than the ear can hear, which often occur in digital recording, and would prevent the program from having a higher

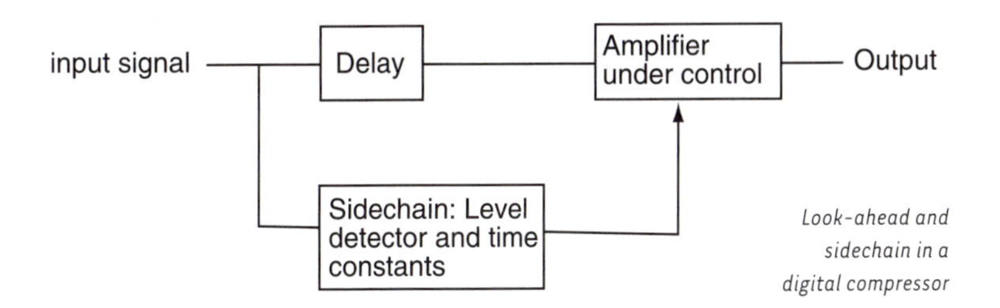


*Look-ahead and sidechain in a digital compressor*

*At left, a simple tone burst from high to low level and back. At right, the same tone burst passed through a compressor with very fast attack, high ratio, and fast release time.*

intrinsic level. Removing these was the initial purpose of the brickwall limiter, designed to be short, quick, and invisible.

At the left side of the above figure is the envelope shape of a simple tone burst, from a high level to a low one and back again. At the right side is the same tone burst passed through a compressor with a very fast attack, high ratio, and fast release, and whose threshold is midway between the loud and soft signals. Note that the loud passages are instantly brought down, the soft passages are instantly brought up and there is less total dynamic range, as shown by the relative vertical heights (amplitudes).

At left in the figure below is the envelope of a compressor with a low ratio, slow attack time and a slow release time. Notice how the slow attack time of the compressor permits some of the original transient energy of the source to remain until the compressor kicks in, at which point the level is brought down. Then, when the signal drops below threshold, it takes a moment (the release time) during which the gain slowly comes back up. A lot of the compression effect (the "sound" of the compressor) occurs during the critical release period, since except for the attack phase, the compressor has actually reduced gain of the high level signal.

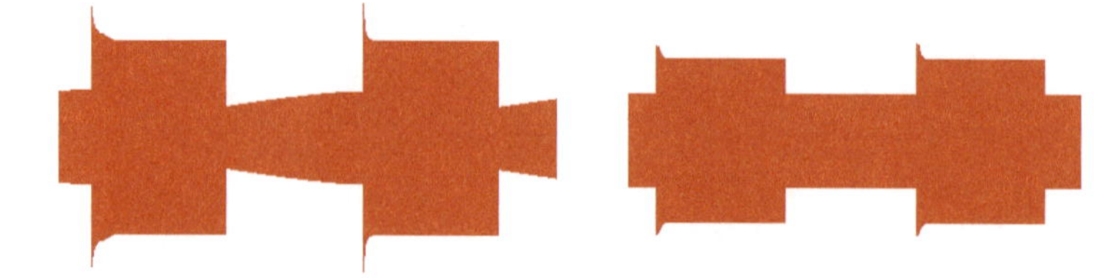

*At left, a compressor with a low ratio, slow attack time and slow release time. At right, higher ratio, faster attack and very fast release.*

Contrast this with the compressor at right, below, which has a much higher ratio, faster attack, and very fast release time. The higher ratio clamps the high signal down farther, and with the fast release, as soon as the signal drops below threshold, the release time aggressively brings the level up. This type of fast action can make music sound *squashed* because it quickly brings down the loud and raises the soft passages.

The essential fact here is that (downward) compressors take the loudest passages **down**. Gain makeup allows the **average** level to be raised, but the loudest passages end up proportionally lower. In mixing, this allows the engineer to mix an instrument at a lower level without losing its low passages, or at a louder level without having intermittent loud passages interfere with hearing other instruments. Although a compressor can add or enhance **punch** in mastering, since its essential mechanism is to reduce the partial loudness of peaks, if overused or if too many transient attacks are softened, this can actually produce the opposite effect.[3] So a compressor should ideally only be applied when loud passages in a recording sound problematic and/or when some parts sound too loud or too soft. Remember that **manually** raising the soft passages (avoiding the processor entirely) can leave the loud passages alone and yield a more impacting production. In the next chapter, we will investigate how upward compressors can increase soft and mid-level passages with little effect on the important loud ones, a technique which can produce a recording that is dynamic, loud and has impact at low levels.

### Release Delay

A release delay control allows more flexibility in painting the sound character. Very few compressors provide this facility but it is useful when we want to retain more of the natural sound of the instrument(s), and not exaggerate its sustain when the signal instantly goes soft, or reduce "breathing" or hissing effects when the source is noisy (illustrated below).

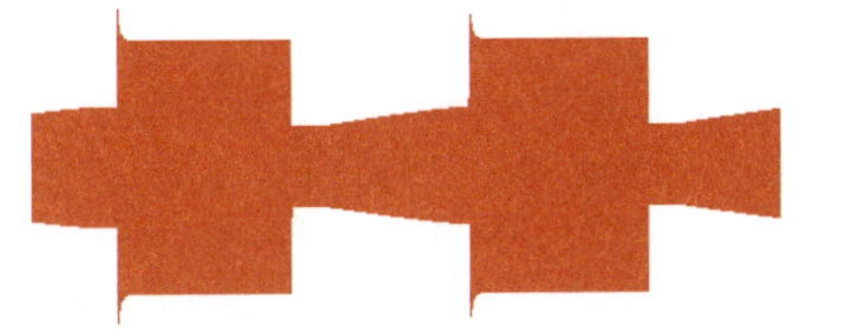

*Output of a compressor with a low ratio, slow attack time, slow release time plus release delay.*

### Attack/Release Distortion

This figure (below) illustrates what happens when the attack and release times are much too fast.

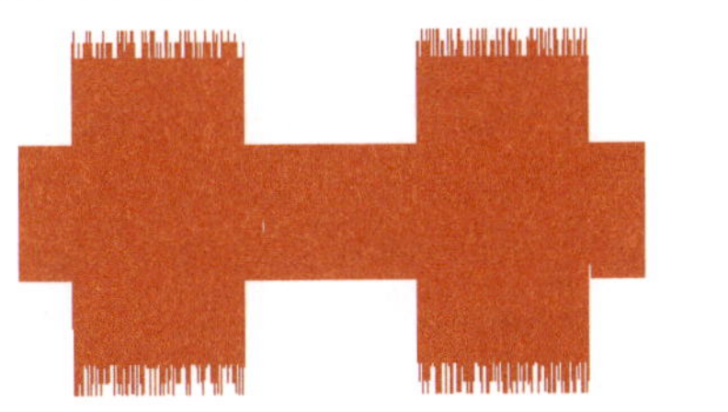

*With attack/release too fast, a compressor can produce severe distortion.*

The distortion shown here is caused by the compressor's action being so fast that it follows the shape of the low frequency waveform rather than the overall envelope of the music. This problem can occur with release times shorter than about 50 ms and correspondingly short attack times.

## II. Microdynamic Manipulation: Adjusting the Impact of Music with a (downward) Compressor

### The Engineer as Artist

Compressors, expanders and limiters form the foundation of modern-day recording, mixing and mastering. With the right device we can make a recording sound more or less percussive, more or less punchy, more or less bouncy, or simply good or bad, mediocre or excellent.

Used in skilled hands, compression produces wonderful recordings. A skilled engineer may intentionally use creative compression to paint a sound and form new special effects; and a lot of contemporary music genres are based on the sound of compression, both in mixing and mastering, from Disco to Rap to Heavy Metal. The key words here though are *intent and skill*. Surprisingly, however, some engineer/artists don't know what uncompressed, natural-sounding audio sounds like. While more and more music is created in the studio control room, it's good to learn how to capture natural sound before moving into the abstract. Picasso was a creative genius, but he approached his art systematically, first mastering the natural plastic arts before moving into his cubist period. Similarly, it's good practice to know the real sound of instruments. Recording a well-balanced group in a good acoustic space with just two mikes is a lot of work, but also a lot of fun! Before multitracking was invented, there was much less need for compression, because close miking exaggerates the natural dynamics of instruments and vocals. At first, compressors were used to control those instruments whose dynamics were severely altered by close miking, e.g. vocals and acoustic bass. When modern music began to emphasize rhythm, many instruments began to get masked by the rhythmic energy, inspiring the creative possibilities of compressors and a totally new style of recording and mixing. The advent of the SSL console, with a compressor on every individual channel, changed the sound of recorded music forever.

### Compression and Limiting In Mastering

Mastering requires us to develop new skills since it is concerned with overall mixes rather than individual instruments. Compression is a tool which can change the inner dynamics of music, e.g. beef up low and mid-level passages, enhance rhythmic movement or make a stronger musical message. But limiters generally change sound very little, simply enable it to be louder.[4] That's why they are used more often in mastering than in mixing. Even the best limiter is not perfectly inaudible, softening the transients and even fattening the sound slightly.

BBC research in the 1940s demonstrated that distortion shorter than about 6-10 ms is fairly inaudible, hence the 6 ms integration time of the BBC PPM meter. But this result reflected also the then-current state of technology: with digital recording and solid-state equipment, some transient distortion as short as 1 ms will audibly change the sound of the initial transient, particularly for acoustic piano. With good equipment and mastering technique, program material with a peak-to-average ratio of 18 to 20 dB can often be reduced to about 14 dB with little effect on the clarity of the sound but some reduction in transient impact. That's one of the reasons 30 IPS analog tape is so desirable: it has this limiting function built-in. A rule of thumb is that short duration (a few ms) transients of **unprocessed digital sources** can often be transparently reduced by 2 dB, in rare cases as much as 6 dB with little effect on the sound; **however, this cannot be done with analog tape sources,** which have already lost the short duration transients.[5] Any further transient reduction by compression or limiting, whatever its desirability, will not be transparent. Limiter distortion is especially audible on material which already has little peak information because a limiter is not designed to work on the RMS portion of the music and it can sound harsh when pushed. So the less limiting, the more *snappy* the sound. In an ideal mastering session, if a limiter is used at all, it should only be acting on occasional inaudible peaks, or perhaps a bit more if we like the slight fattening effect. A manual for a certain digital limiter reads "For best results, start out with a threshold of -6 dBFS." This is like saying "always put a teaspoon of salt and pepper on your food before tasting it." One modern R&B album is so overlimited that the bass drum punches a hole in the vocal on every attack; I doubt this is artistically desirable.

It is a common misconception that a limiter is a *peak protection device* for mastering. It may be used as such in radio broadcasting (to protect the transmitter) or sound reinforcement (to protect loudspeakers), but in mastering (or mixing), the engineer has total control over his levels and makes the choice whether to turn them down or raise them and use a peak limiter.

### The World's Most Transparent Digital Limiter

**The most transparent limiter is to use no limiter at all!** If there is a very short peak (transient) overload, for example, a drumbeat within a section which needs to be made louder, a skilled mastering engineer can use the DAW's editor to perform a short-duration gain drop that can be quite inaudible. This **manual limiting** technique allows us to raise a song's apparent loudness without inducing distortion from a digital limiter, so it is the first process to consider when working with open-sounding music that can be ruined by too much processing. We can often get away with 1 to 3 dB manual limiting typically for a duration of less than 3 ms. But, longer duration manual gain drops will affect the sound as much as or more than a good digital limiter.

### Equal-Loudness Comparisons

Since loudness has such an effect on judgment, it is very important to make comparisons at equal apparent loudness. If played louder, during an instant A/B comparison the processed version may initially seem to sound better, but long-term listeners prefer a less fatiguing sound which "breathes." When we compare at matched loudness, we may be surprised to discover that the processing is making the sound worse, and the "improvement" was an illusion. When making an album at "competitive loudness level", it's a relief if the mastering has not degraded the sound of the program, and ecstasy if it has improved it.

### The Nitty-Gritty: Compression in Music Mastering

Consider this rhythmic passage, representing a piece of modern pop music:

shooby dooby doo **WOP**...
shooby dooby doo **WOP**...
shooby dooby doo **WOP**

The accent point in this rhythm comes on the backbeat (**WOP**), often a snare drum hit. If we strongly compress this music piece, it might change to:

**SHOOBY DOOBY DOO WOP...**
**SHOOBY DOOBY DOO WOP...**
**SHOOBY DOOBY DOO WOP**

This completely removes the accented feel from the music, which is probably counterproductive.

A light amount of compression might accomplish this...

shooby dooby doo **WOP...**
shooby dooby doo **WOP...**
shooby dooby doo **WOP**

...which could be just what the doctor ordered for this music as strengthening the sub accents may give the music more interest. Unless we're trying for a special effect, or purposely creating an unnatural sound, it's counterproductive to go against the natural dynamics of music (like the TV

**The Real Recipe for Radio-Ready**

The real recipe for Radio-Ready includes:

1) Write a great original song, use fabulous singers and wonderful arrangements.
2) Be innovative, not imitative.
3) Make sure the music sounds good at home. Keep the dynamics lively, interesting and unsquashed, and some of that virtue will make it through the radio processing.

weatherperson who puts an accent on the wrong syllable because they've been taught to "punch" every sentence: "The weather **FOR** tomorrow will be cloudy"). Much hip hop music, for example, is intentionally unnatural—anything goes, including the eradication of any resemblance to the attacks and decays of real musical instruments.

One way to start compressing in order to help obtain **punch** or **attitude** is first to find the threshold, using a very high ratio (say 4:1) and very fast release time (say 100 ms) then to adjust the threshold until the gain reduction meter bounces as the "syllables" you want to affect pass by and you hear this bounce. This ensures that the threshold is optimally placed around the musical accents you want to manipulate, the "action point" of the music. Then reduce the ratio to very low (say 1.2:1) and raise the release time to about 250 ms to start. From then on, it's a matter of fine tuning attack, release and ratio, with possibly a readjustment of the threshold. The object is to put the threshold in between the lower and higher dynamics, creating a constant alternation between high and low (or no) compression within the music. Don't get fooled by your eyes, since many compressors' meters are too slow, they go into gain reduction before the meters move, so 1 dB of metered gain reduction can mean a lot. Note that too low a threshold will defeat the purpose, which is to differentiate the "syllables" of the music; with too low a threshold, **everything will be brought up to a constant level**.

### Typical Ratios and Thresholds

When working on microdynamics in the above fashion, compression ratios most commonly used in mastering are from about 1.5:1 through about 3:1, and typical thresholds in the -20 to -10 dBFS range. But there is no rule; some engineers get great results with ratios of 5:1, whereas a delicate *painting* might require a ratio as small as 1.01:1 or a threshold of -3 dBFS. One trick to compress as inaudibly as possible is to use an extremely light ratio, say 1.01 to 1.1 and a very low threshold, perhaps as low as -30 or -40 dBFS, starting well below where the action is. In this case the compressor is not bouncing on the syllables but rather giving a gentle, continuous form of macrodynamic reduction. We may choose a low ratio to lightly control a recording that's too *jumpy* or to give a recording some needed *body*. It's unusual to see such low ratios in tracking and mixing but very common in mastering, partly because with full program material, larger ratios may create breathing, pumping or other artifacts.

### Compressors With Unique Characteristics

Part of the fun is discovering the specialities of different compressors. Even with the same settings, some are *smooth*, some *punchy*, some nicely fatten the sound and others make it brighter, harder or more percussive. This is often due to differences in the curve or acceleration of the time constants, how the device recovers from gain reduction, whether the gain returns to unity on a linear, logarithmic, or even an irregular curve.

Analog compressor designers choose from several styles of gain manipulation. The most common are **optical** (abbreviated **opto**), **VCA** (voltage controlled amplifier), **Vari-Mu**, **PWM** (pulse width modulation) and their various subcategories. Digital designers may emulate their characteristics, as in the Waves Renaissance series of digital compressors which have both *opto* and *electro* modes. In opto position, the release time slows down for the last portion of the release, while in electro it accelerates. Electro can yield a more aggressive sound, while opto is good for gentle, easy-going purposes. Analog optical compressors are great on vocals in tracking or mixing, not as good for aggressive mastering of overall program material because they are generally too slow. However, digital opto models can be faster than their analog counterparts.

Another feature is to add supplementary low frequency harmonics to change tonality, as in Waves "warm" setting of the Renaissance Compressor.[6] Note that the addition of harmonics slightly compresses sound by reducing the peak-to-average ratio.

While generally analog optical models are more suitable for "gentle" mastering, one model, the Pendulum OCL-2, has a proprietary optical sensor whose reaction time is much faster than others on the market as well as a very transparent tube circuit which can provide very subtle warming. This makes the OCL-2 perhaps the only optical analog compressor with the gentleness of optical (useful for adding body), but also capable of providing a bit of punch. However, it is not as fast as a VCA- or PWM-based compressor, which may be necessary to help achieve attitude or punch in rock and roll. In my opinion, closest to a Swiss Army Knife is the Cranesong Trakker, a solid-state compressor which can emulate the tonality and speed characteristics of several different types of compressors and embody some of the warming characteristics associated with tubes. The only downside of the Trakker is the learning curve (the best way to learn is to experiment with each of the presets). Another style of compressor is the Manley Vari-Mu whose ratio varies with dynamics, but its action may be a bit slow for music with fairly fast dynamics.

### Sidechain Manipulation

Most of the time the sidechain (control path) is identical to the audio signal, but interesting things can happen when it is not. For example, in a stereo or multichannel compressor, each channel has its own sidechain, but it is possible to feed or *link* all sidechains from one channel's signal. By linking the sidechains, one channel controls the gain reduction of both equally. The linking switch prevents image wandering, for if a drum hits much louder in one channel than the other, with an unlinked sidechain, the image will momentarily move towards the opposite channel. When unlinked, the box operates as two independent mono compressors. In some models, unlinked is labeled **dual** and linked is labeled **stereo**. In multichannel compressors, there may be a separate sidechain for front and surround, or all channels may be linkable.

While it seems desirable to link a compressor or limiter, apparent stereo separation can increase if the linking is removed. Just be careful to check for image wandering, which may even be desirable and add an artificial sense of space. Don't be afraid to use the **dual** position for stereo if it sounds better. Running limiters unlinked can reduce clamping effects, where the sound appears to drop and not recover fast enough (because the channel which did not drop will mask short-duration drops in the other). As with the unlinked compressor, watch out for instantaneous image shifts caused by extreme transients located only in one channel.

Often sidechains are fed an equalized signal. Perhaps the most popular sidechain EQ is a highpass-filtered signal, which helps prevent the bass drum from pushing down (or *modulating*) the rest of the music. In an analog compressor, it is very easy to implement this without needing a separate equalizer, simply by inserting a capacitor (approximate value 0.1 µf) in series with the sidechain inserts; the exact value depends on the source impedance. Measure the amount of gain reduction with a test tone and experiment. It's useful to have a switch to add or remove the filter.

With a sidechain boost in the upper mid or low treble range, the compressor becomes a **de-esser**, or it can be tuned to deal with troublesome cymbals. The only problem with sidechain-based de-essing is the entire range of audio frequencies is brought down whenever an "s" goes by, so generally the gain reduction has to be kept subtle, no more than a dB or two, unless you are looking for a special effect.

### Multiband processing: Advantages and Disadvantages

Splitting a compressor's signal into multiple bands (and multiple sidechains) avoids the problem of modulation with a single sidechain since compression action in one band will not affect another band. For example, the vocal will not pull down the bass drum (or vice versa). This is perhaps the biggest selling point of multiband, because with the same amount of gain reduction it can sound superior to wideband or sidechain equalization. Or a higher amount of compression and average level can be achieved in a multiband with fewer interaction artifacts. Another advantage is that high frequency transients can be left unaffected while compressing the midrange more severely, producing a brighter, snappier sound. However, loud action in one frequency band can dynamically change the tonality, producing an uncohesive sound especially if all the bands are moving in different amounts throughout the song, but even this property of multiband can be put to advantage. When slightly more compression is applied in the high frequency region, the sound gets duller as it gets louder, which is a way to construct an analog tape simulator, to sweeten digitally-recorded material, or to soften distortion that gets harsh when it gets loud.

Multiband units make good de-essers. Sibilance can be controlled by using selective compression in the 3 through 9 kHz range (the actual frequency has to be tuned by listening to the vocalist). Try a very fast attack, medium release, crest factor set to peak (to be explained), and a narrow bandwidth for the active band. The Weiss DS1-Mk2 is hands-down the best-sounding mastering de-esser I've encountered probably due to its peak sensitive compression, linear phase band-split filters, dual-speed release, and very effective presets.

The multiband device's virtues permit louder average levels than were previously achievable—making it the most powerful but also potentially the most deadly audio process that's ever been invented. "Deadly" because multiband compression helps fuel the loudness race (see Chapter 14). The technique has been hyped as a cure for all ills (which it is not) and it can easily produce a very unmusical sound or take a mix where it doesn't want to be. But it can also be used to help improve ("repair") a bad mix when a remix is not possible, and some mastering engineers become experts at this technique. I once received a rap project that was mixed with very low vocal, extremely loud percussion and bass drum. A remix was not possible, but by compressing and then raising the level of the frequencies in the vocal range (circa 250 Hz) I was able to *rebalance the piece* and turn the vocal up. Just don't be fooled into believing that toning down loud instruments through multiband compression is the same as a remix.

Multiband processing was probably first introduced by TC Electronic in their M5000, then in their ubiquitous Finalizer, and brought to great sophistication and versatility in their System 6000 with the MD4, perhaps the first multiband compressor to introduce linear-phase band-split filters.[7] But for most downward compression purposes multiple bands are rarely needed; one or two bands are usually enough. The Weiss DS1-Mk2 has one active band; the compressed signal can be isolated to one frequency range and the rest of the spectrum left unaffected. Most of the compression I perform in mastering is with a full-band compressor, or a full-band compressor with a high-pass sidechain, or the Weiss with the active band above some low frequency. Regardless of this advantage, rarely do even hip-hop recordings need more than two bands to sound punchy and strong. I use a downward compressor with more than two active bands in my mastering a small percentage of the time, when multiple bands have been a lifesaver on bad mixes, but let us not forget that the key to a great master is to start with a great mix!

**Before trying multiband, first**

- See if simply raising the attack time in a one-band compressor permits sufficient transient energy to come through. Or, try upward expansion (see Chapter 11) instead.
- Try using few bands, only two if possible. This reduces potential "phasey" artifacts if the filters in the compressor are not linear-phase.

### Equalization or Multiband Compression?

When multiband processing is used, the line between equalization and dynamics processing becomes nebulous, because the output levels of each band form a basic equalizer. Use standard equalization when instruments at all levels need alteration, or use multiband compression to provide spectral balancing at different levels. This is a form of dynamic equalization, so depending on one's point of view a multiband compressor can be looked upon as a dynamic equalizer.

When already using a multiband unit, we make our first pass at equalization with the outputs (makeup gains) of each band. Multiband compression and equalization work hand-in-hand. Tonal balance will be affected by the crossover frequencies, the amount of compression, and the makeup gain of each band. As we know, the more compression, the duller the sound, so first try to solve this

*"The key to a great master is to start with a great mix."*

"*Never in the history of mankind have humans listened to such compressed music as we listen to now.*" — Bob Ludwig [8]

problem by using less compression, or altering the attack time of the high-frequency compressor, and as a last resort, use the high frequency band's makeup gain or an equalizer to restore the high-frequency balance.

### Emulation vs. Convolution

Digital compressor designers have the choice of **emulating** the transfer characteristics of a source compressor and implementing them in DSP, or sampling the source compressor and **convolving** that sampled characteristic with the incoming audio. Convolution works well with reverbs and equalizers, but I have not personally heard a successful convolution-based digital compressor. These are very difficult to implement, the source device has to be sampled at many different levels to ascertain its dynamic characteristics, which are then very hard to interpolate accurately and with the right timing. The convolution processor has to be fast and operate with great resolution in the oversampled domain to prevent digititis.

### Fancy Compressor Controls

Some compressors provide a **crest factor** control, usually expressed in decibels, or a range from RMS (or full average) to quasi-peak through to full peak. What this means is that the compressor can be set to act on either the average parts of the music, the peak parts, or somewhere in between. Ostensibly, compressors with RMS characteristics sound more natural as they correspond with the ear's sense of loudness, but one of the best-sounding compressors is peak-sensing. When the crest factor control is set to peak, short transients tend to control the action, and at RMS, more continuous sounds control it. The TC MD4 has a continuously-variable crest factor control. In most cases leave it at RMS, but for de-essing and for better control of transients (such as a too-loud snare drum) move it closer to peak.

The Weiss model DS1-Mk2 has **two different release time** constants, *release fast* and *release slow*. The user sets a threshold of average transient duration, such as 80 ms, above which a sound movement is called *slow*, and below which it is called *fast*. Instantaneous transients receive a faster release time, but sustained sounds a slower one, which results in a more natural-sounding, yet louder compression. Indicator lights on the front panel make adjusting this a snap.

The Roger Nichols Dynamizer is a dynamics processor with **multiple thresholds**. Compressors with multi-thresholds can be simulated by running two (or more) compressors in series. For example, the first compressor performs a gentle overall compression with a low threshold and ratio, and the second more aggressively controls some offensive peaks at high levels. It's not uncommon to run several dynamics processors in series in mastering, each doing a small part of the job. To keep a peak limiter from clamping down on the signal too much, try preceding it with some form of *preconditioner*, which may be an analog compressor that can gently soften a transient so that the limiter which follows doesn't have to work as hard.

### Clipping, Soft Clipping and Oversampled Clipping

Clipping is the result of attempting to raise the level higher than 0 dBFS, producing a square wave, a severe form of distortion. Clippers are devices which electronically cut momentary peaks out of the waveform to allow the overall level to be raised. Soft clipping attempts to do this with less distortion. I don't like the quality of distortion produced by clipping or soft clipping, at least at 44.1 kHz (see Chapter 17).The better approach is not to raise the level at all, for many CDs are already too hot for their own good. In this book's first edition, we recommended the use of a brickwall limiter instead of a clipper, but as CD levels have risen to insanity, the tradeoff has changed, as we will discuss in Chapter 16. In Appendix 1, radio gurus Bob Orban and Frank Foti explain why clipping is a severe problem for radio processors.

### Compression, Stereo Image, and Depth

Compressors tend to amplify the mono information in a recording, which affects stereo width. Compression also deteriorates the depth in a recording as it brings up the inner voices in musical material. Instruments that were in the back of the ensemble are brought forward, and the ambience, depth, width, and space are degraded. Not every instrument should be "up front". Pay attention to these effects comparing processed vs. unprocessed and listen for a long enough time to absorb the subtle differences. Variety is the spice of life. As always, make sure the cure isn't worse than the disease.

### The Mastering Engineer's Dilemma

Without compressors in CD changers and in cars, it is extremely difficult for the mastering engineer to fulfill the needs of both casual and critical listeners. It is our duty to satisfy the producer and the needs of the listeners, so we should continue to use the amount of compression necessary to make a recording sound good at home. But try to avoid using more compression than is required for home listening; this will actually help radio play (see Appendix 1). If compromises have to be made for car or casual play, try transparent-sounding techniques such as parallel compression (see Chapter 11), which satisfy even critical listeners and offer their own unique advantages.

*"Not every instrument should be up front."*

1. It is really a matter of degree, but most authorities call a compressor with a ratio of 10:1 or greater a **limiter**. The knee should also be very sharp for most effective limiting. Very few analog compressors have higher ratios, however, some digital limiters have ratios of 1000:1 to prevent the minutest overload.
2. One manufacturer, DBX, measures release time in dB/second, which is probably more accurate, but hard to get used to.
3. Punch comes directly from microdynamic contrast, especially at bass frequencies. The level has to dip just a little in order for it to come back up and "hit you"; without a counterswing there can be no swing. In general **punch** is achieved during the recording and mixing process, and frequently it is lost during mastering if the engineer is not careful.
4. As with compressors, it is the gain makeup process that permits the output of a limiter to be louder. When the peaks have been brought down, there is room to bring the average level up without overloading. For example, snare drum hits that stick out above the average can be softened by the peak limiter (if this change in sound proves desirable) and the average level can then be raised.
5. Limiter release time is important: the faster the release time, the greater the distortion, which is why the only successful limiters which use extra fast release times have **auto-release control**, which slows down the release time if the duration of the limiting is greater than a few milliseconds. The effective release time of an auto-release circuit can be as short as a couple of milliseconds, and as long as 50 to 150 milliseconds. If limiting a very short (invisible) transient, the release time can be made very short.
6. Waves calls the alternative setting (when the extra harmonics are turned off), "smooth", because they do not want to prejudice the user with a term such as "cool". But this has nothing to do with "smooth" so in retrospect they should simply have labeled one position of the switch "warm" with no label in the other position.
7. The Weiss DS1-Mk2 uses a linear-phase band -split filter, but it only has one active band, so it cannot be properly called a multiband compressor.
8. In correspondence. A variation of this quote is in Owsinsky, Bobby (2000). *Mastering Engineer's Handbook.*

CHAPTER 11

# How To Manipulate Dynamic Range for Fun and Profit

PART THREE: THE LOST PROCESSES

## Introduction

This chapter introduces two processes which should be part of every audio engineer's vocabulary. To be successful with them, you have to learn to think like a contrarian, but it is well worth it.

## I. Upward Compression

It is a psychoacoustic fact that the ear is much more forgiving of the upward "cheating" of soft passages than of the awkward "pushing down" of loud passages. The latter—downward compression—sometimes feels like an artificial loss while the former—upward compression—can feel very natural.

Let's introduce an upward compression technique which requires just a single knob—no need to adjust attack, threshold, release or ratio—and produces such a transparent sound quality[1] that careful listening is required to even know the circuit is in operation! New Zealand radio engineer Richard Hulse discussed with me his practice of **parallel compression**,[2] which is a means of creating an upward compressor. Richard was using analog components and got acceptable results, but he thought that a digital implementation could sound even better and suggested I try one. The digital version of this technique was so successful that it is now one of the most common techniques I use in mastering. The principle is quite simple: Take a source, and mix the output of a compressor with it.

In the digital domain, it is possible to sum the source with a compressor without any side effects, by using a precise time delay for the "dry" signal which exactly matches that of the compressor, as shown in this block diagram (below, one channel only of stereo shown).[3]

In principle, the distortion of the parallel compression technique can be much lower than standard (downward) compression, since the main signal has an unaffected linear path with the non-linear path added to it.[4] The amount of compression is controlled by the attenuator or makeup gain. When building a parallel compressor from plugins, a DAW with automatic latency compensation may not need the extra time delay. Test by adjusting the parallel compressor to a 1:1 ratio and unity gain, and invert the polarity (commonly called *phase* in most DAWs) of either half of the chain, which will produce a complete null (no sound) if the time delay is correct to the sample.

#### Transparent Parallel Compression

I've found two ways to approach parallel compression. This first one is the transparent approach taken by Hulse, where the compressor is as invisible as possible, producing no obvious tonal shifts and little or no loss of transients. In many cases this technique is indistinguishable from manual fader riding, and highly suitable for acoustic music. The transparent parallel compressor raises gain at very low levels and contributes less to the total sound as the signal gets louder. Here is a recipe:

*The Parallel Compression technique employs a matched time delay in the "dry" signal path to avoid phase shift or comb filtering. This yields very transparent-sounding upward compression.*

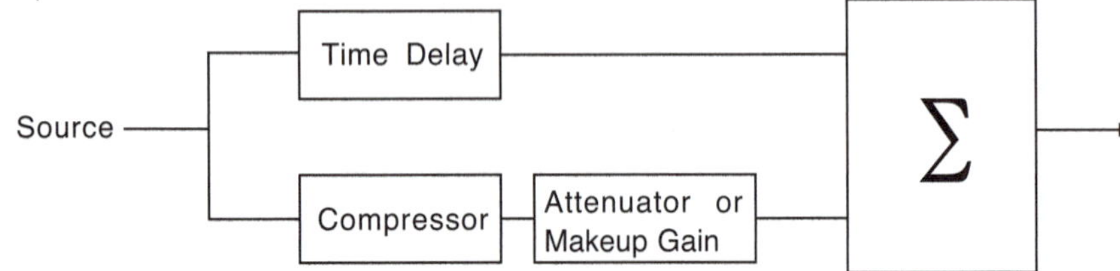


- Threshold -50 dBFS. A very low threshold puts the parallel compressor into heavy gain reduction almost all the time, ensuring that it will be into extreme gain reduction during loud passages. Because the output of the parallel compressor is pushed down during loud passages, it contributes only negligibly to the highest level. In principle, if you add in a second signal that is 20 dB or more down on the main signal, this second element will not perceptibly contribute to the total level.
- Attack time as fast as possible. One millisecond or less if available. This ensures that the transient impact of the original sound will be preserved, for as soon as a loud transient hits, the compressor goes into gain reduction. The faster the attack time, the more invisible the parallel compressor, requiring accurate look-ahead (see Chapter 10).
- Ratio 2:1 or 2.5:1 (I prefer 2.5). This is the root setting of the compressor, but the net ratio of the parallel sum varies depending on the output level of the parallel compressor. Richard has developed a chart by the numbers, but I prefer to go by ear.
- Release time medium length. Experiments show that 250-350 milliseconds works best to avoid breathing or pumping, although in cases where the reverberation is very exposed, particularly *a capella* music, as much as 500 ms may be needed to avoid overemphasizing the reverb tails.
- Crest factor set to Peak. I've found the most transparent parallel compressors are peak-sensing.
- Output level or makeup gain adjusted to taste. With the parallel compressor off (-∞ gain), there is no compression. Above about -5 dB, compression will be very noticeable, with soft or even medium-level passages being raised in level. A nice subtle

compression can be achieved with settings of -15 through -5 dB.

### Parallel Compression for Tonalization or Attitude

This second approach to parallel compression is a way of achieving attitude or punch without damaging the loud peaks, or to warm or clarify the low to mid levels of the music. The attitude parallel compressor effectively brings up the mid levels, where the meat of the music lives, which can help achieve that desirable *epic* quality in rock and roll. Parallel technique can often fatten up sound better than a normal compressor because it concentrates on the mid-levels without harming the highest levels.

- Threshold. Set the threshold in the middle of the musical action, as described in Chapter 10, resulting **in up to** 5-7 dB of gain reduction for this application, often as little as 1-3 dB.
- Attack time, medium (start with 125 ms), because too short an attack time will subdue transients. Look-ahead is unnecessary.
- Ratio to taste, set for the aggressiveness of the desired action, in conjunction with the output level.
- Release time to taste, set to work in tandem with the attack time to obtain maximum rhythm and punch.
- Crest factor set to RMS. I've found the best attitude compressors are RMS-sensing.
- Output level or makeup gain adjusted to taste. Since the gain reduction is much less than in the invisible technique, expect to mix the compressor at a higher level, whatever obtains the desired effect, rarely past -6 dB. Still, it will be far less harmful to loud peaks than a standard downward compressor, because even at 0 dB, the dry signal is mixed higher than the compressed because of the gain reduction.

Tonalization is my term for a form of dynamic equalization which is accomplished by using a multiband compressor in parallel mode. By manipulating the gain of each frequency band that is being mixed in parallel, the tonality of the program can be changed. It allows warming up the mid-levels of the program without losing clarity at high levels, adding presence at low levels, and so on. The nice thing about parallel bass compression is that the body of the bass instrument gets fatter without destroying its transient impact. Or when increasing the presence frequencies at low levels, the sound can be clearer and better defined without becoming harsh at mid or loud levels.

Three digital compressors have integrated parallel compression, the TC Electronic MD4 in the System 6000, Weiss DS1-Mk2, and the plugin PSP Mastercomp. The MD4 excels at the tonalization or attitude style; though it is no slouch at the transparent style, it's not quite as transparent as the Weiss. Correspondents have told me they have successfully created parallel compressors within Pro Tools, Digital Performer, and SADiE DAWs.

As with any process, if upward compression is pushed too far, it will call attention to itself. Like fill flash in a camera—too much fill and the picture becomes overexposed. The first audible artifact will be increased sustains and emphasized reverberation, then, finally, breathing or pumping. These artifacts can sometimes be reduced by raising the release time of the parallel compressor. However, if the music is so open or delicate that the process continues to call attention to itself, the only solution is to abandon the processor and manually raise the passages which are too soft.

## 11. Upward Expansion

Another underused but incredibly useful processing technique is **upward expansion.** Some people think of an upward expander as the **uncompressor**, but it is far more than that (indeed there is a limit to how much a sound can be restored once it has been excessively compressed). Rather, upward expanders can be used to emphasize different parts of the dynamic rhythm from those affected by downward compressors. For example, upward expansion is great for adding liveliness to typically uninteresting musical samples and can also put the snap back into a slightly-squashed snare drum. Upward expansion is definitely a technique worth learning, no more difficult to use than a downward compressor.

Upward expanders were not easy to build until the advent of the VCA,[5] but it is a simple matter to turn any VCA-based device into an upward expander by reversing the sign (polarity) of the sidechain signal. Probably the first commercial dedicated upward expander was in the DBX model 117, from 1971, which was designed to enhance dynamics in a hi-fi system. Another early upward expander was the Phase Linear Peak Unlimiter. The honor for the first digital upward expander goes to the Waves C1 (plug-in), algorithms designed by Michael Gerzon; ever since then, every Waves dynamics processor includes fractional ratios for upward expansion. The first stand-alone digital upward expander was in the DBX Quantum mastering unit, followed shortly by the Weiss DS1-MK2. The Waves C4 (plug-in) is the first single processor to perform all four dynamics processes, though it can only do one process at a time per band. It would be heavenly to be able to do simultaneous upward compression, upward expansion, and limiting in a single box.

*An upward expander with .75:1 ratio, expressed in decimal (1:1.33 expressed as a fraction). Threshold is -32 dBFS, and without attenuation, the output will overload if input exceeds approximately -10 dBFS.*

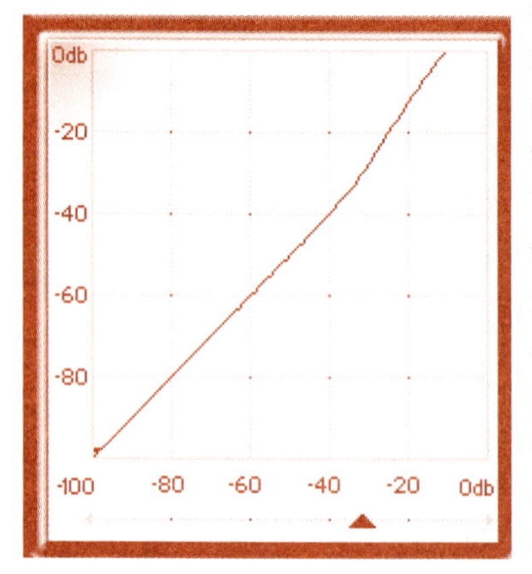


In downward compression, the loudness **increases** when the incoming level **decreases** (below the threshold), which is against the natural movement of the music. In contrast, upward expansion **increases further** the loudness of incoming passages that are already **increasing** in loudness—a process which is in sync with the motion of the music, though it may be necessary to use output attenuation instead of makeup gain to prevent the output from overloading. There is an increase in dynamic range, so if used delicately, the upward expander becomes as valuable a production tool as the downward compressor.

For illustration (pictured left), this transfer function shows an upward expander with a severe .75:1 ratio and threshold at -32 dBFS. Without attenuation it will overload with input levels exceeding about -10 dBFS. Note that the ratio of an upward expander can be expressed in decimal or fraction form, depending on the manufacturer's preference. The Waves and DBX units use decimal form, while the Weiss unit expresses this as a fraction, 1:1.33. Typically, the usable range of ratios for mastering with upward expansion is small, from a very gentle 1:1.01 through about 1:1.2 (fraction); equivalent to 0.99 through .83 (decimal). A common value used for music enhancement is around .95 decimal (1:1.05 fraction).

The figure (top right) shows an upward expander with fast attack and slow release, and one with slow attack and fast release. As you can see, the dynamic characteristics are the opposite of the compressor examples shown in the previous chapter.

The best way to learn how to use an upward expander is to compare it to a downward compressor, described in the chart on the next page. Upward compressors and upward expanders make a nice team which can fatten sound at low levels, raise the average level, and yet keep dynamics lively at high levels. While downward compressors tend to "sweeten" or warm up sound, upward expanders tend to brighten sound because they increase the strength of transients. If the sound becomes too bright or "snappy" consider expanding only the bass through the lower midrange. This warms up the sound **by raising the bass as loudness increases**, which is subtly different from a tape saturator, that **lowers high frequencies as loudness increases**. Both warming approaches have merit depending on the problems in the source.

### Compromises When Making Hot Masters

Neither downward compression nor upward expansion work very well above a certain intrinsic loudness. With the first technique, when we are asked to make a "louder" master, the sound becomes more squashed. With the second technique, to prevent peak overload, we have to use more limiting, which eventually counteracts the expansion and impact is again lost. If we cannot live with the degradation, the only solution is to master at a lower intrinsic level.

### Does Compansion Really Work?

*At left, upward expander with fast attack, slow release. At right, slow attack, fast release.*

Is there such a thing as an uncompressor which can bring material with squashed dynamics back to life? If a dynamically-challenged source has no incoming dynamic variation, the expander can do nothing, or will make the sound worse. But if there is some dynamic movement left in the source, an expander with the right parameters can improve its movement, pace, rhythm and transient impact. **Compansion** means *compression followed by complementary expansion*. The figure (below) shows that compansion can work. A toneburst alternates between -15 and -5 dBFS, followed by a downward compressor, then a complementary upward expander. As illustrated, the average levels are restored, but the initial transient attacks and to some extent the decays are not well preserved.

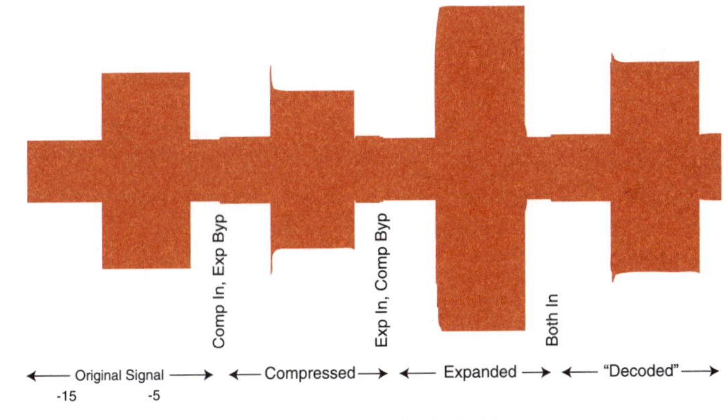


*Does Compansion Really Work?*

| DOWNWARD COMPRESSION | UPWARD EXPANSION |
|---|---|
| makes sound louder during the **descent** of the music (release phase). | makes sound louder during the **rise** of the music (attack phase). |
| tends to make sound fatter and exaggerate low frequencies (subject to time constants and threshold). | tends to exaggerate transients and high frequencies (subject to time constants and threshold). |
| Attack times that are too short (fast) cause transients to be lost. | Attack times as short as a few ms can restore and sharpen lost transients (e.g. from analog tape or overcompressed sources). |
| Typical attacks 100 ms through 300 ms. Less than 100 tends to blur transients. | Typical attacks 1 ms through 300 ms. If a transient still sounds too sharp and trying >150 ms attack, perhaps this is not the right process for this music, or consider a touch of limiting after the expansion. |
| tends to make things sound duller or warmer. | tends to make sounds brighter or sharper. |
| If sounds "jump out" too much, raise the ratio, shorten the attack, and/or speed up the release. | If sounds "jump out" too much, lower the ratio, lengthen the attack, and/or slow down the release. |
| If attacks seem too sharp, shorten the attack time. | If attacks seem too sharp, lengthen the attack time, or consider compression. |
| If sustains seem too long or too prominent, lengthen the release time. | If sustains seem too short, lengthen the release time. |
| If attacks seem too dull, lengthen the attack time. | If attacks need enhancement, shorten the attack time. |
| If you don't like the percussiveness (e.g. snare drum), speed up the attack. To increase the ratio of rhythm to melody, lengthen the attack. Downward compression does not help the impact of percussion instruments. | If you don't like the percussiveness (e.g. snare), slow down the attack. To increase the ratio of rhythm to melody, shorten (speed up) the attack. Upward expansion is very good at helping the impact of percussion instruments, however, sometimes at the expense of the vocal balance because the percussion becomes more prominent. |
| | can work very well with upward compression, which fills in any perceived low level "holes" or lost sustain. |
| Very easy to degrade the liveliness or "bounce" of the music if time constants are not optimized or if overused. | Very easy to enhance the liveliness or "bounce" of the music, but watch out for too much "bounce" or exaggerated dynamics. |
| tends to go **against** the natural movement of the music, especially when the parameters are not optimized. | tends to work **with** the natural movement of the music, especially when the parameters have been optimized. |
| tends to de-emphasize musical accents and emphasize the sub accents and sustains in reverse proportion to their original movement. | tends to emphasize the hottest musical accents and to a lesser degree, the sub accents in increased proportion to their original movement. |
| | Very useful to follow with a limiter, as loud passages are being brought up by the expander. As long as the limiter is used to cheat down very short, momentary transients, it will not significantly diminish the effect of the upward expansion. The limiter's gain reduction meter should be moving very little and on brief occasions, while the expander's gain increase meter should be bouncing with the syllables of the music that's being enhanced. However, if the limiter's gain reduction meter starts to mirror the expander's gain increase meter, then the two processes are canceling each other out and there's too much limiting. |
| can decrease the overall dynamic range of the song (macrodynamics), in addition to affecting the microdynamic bounce of the music. | can increase the overall dynamic range of the song (macrodynamics), making a climax seem even more climactic, which can be very effective. |

## III. Changing Microdynamics Manually

We can change musical microdynamics by doing manual edits and gain changes in a DAW. In this figure, the attack of the first note of a song has been artificially enhanced with very brief manual upward expansion (it's the brevity which makes it microdynamic).

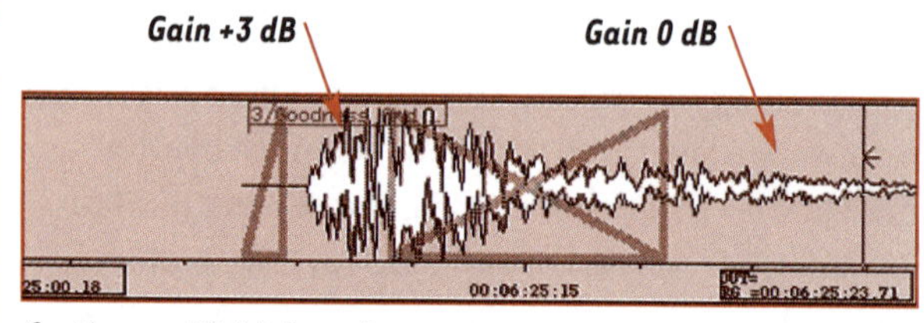


*Creating an artificial sforzando*

At left, the first few milliseconds of the note have a greater gain (in this case, 3 dB), and then there is a crossfade to a gain of 0 dB, resulting in a *sforzando*. Interestingly, the producer was looking for a surprise when this track entered, and I initially had the beginning attack at +5 dB, but when he took the reference CD home, he was really startled, so I took it back a bit for the final master.

*This chapter completes our dynamics trilogy.*

1 For me, **transparent** means the signal path sounds as clean as the source.

2 Which he initially called sidechain compression, but I suggested a name change to avoid confusion with the sidechains of compressors. This technique was publicized by Mike Bevelle in the article Compressors and Limiters, **Studio Sound**, October 1977 (also reprinted June 1988). Engineers have been playing with parallel compression techniques for many years.

3 There are other theoretical methods of achieving upward compression, for example, summing a downward expander with the dry signal in opposite polarity—or upward expansion, summing a downward compressor with the dry signal in opposite polarity. I have not experimented much with either technique. Thanks to Cris Allinson for this hint.

4 This was the principle of the Dolby A/SR systems, which used a direct signal path summed with a compressed one, *doing as little harm to the audio as possible*.

5 Voltage controlled amplifier. When automating a mix in an analog console, the audio passes through a VCA unless it has moving fader automation.

CHAPTER 12

# Noise Reduction

*"No single-ended noise reduction system is perfect; all noise reduction systems take away some degree of signal with the noise."*

## I. Introduction

### Noise and Distortion

Specialists in any developed subject create a vocabulary to mark distinctions which *are not generally noted, or even noticed, by the non-specialist.* Although a layperson would lump **distortion** and **noise** together, an audio engineer characterizes **distortion as a particular form of noise: one that is correlated with the signal**. Distortion can be low level and sound much like what is normally called **noise**, or it can be high level and quite obtrusive, lying on the peaks of the signal.

### Continuous Noise

Noise can be **continuous** (regular in its dynamic character), or **impulsive** (intermittent or periodic). **Impulsive** noise can be characterized as crackle, click(s), tic(s) (very short duration clicks), or pops (primarily low frequency). **Continuous** noise is further divided into two categories: **broadband** or **tonal**. What distinguishes broadband noise from tonal is that although it can have a frequency response character or **color**, it has no obvious identifiable single frequency component(s). Broadband noise is further characterized as *white* (wideband with a rising high frequency response), *pink* (wideband with a flat frequency response), *rumble* (narrowband with a distinctive bassy character) or *hiss* (narrowband with strong components in the 2 to 10 kHz range). In contrast, tonal noise contains distinct components at single (or multiple) frequencies, i.e. feedback, buzz or hum. **Hum** consists of the lower frequency components of the power line (in Europe and Asia the fundamental is 50 Hz, in the U.S. 60 Hz, with its

**MYTH:**
*"I know that you can't hear anything but noise on this tape, but if you get rid of it all, you'll be able to hear my husband having sex with his lover."*
CONTRIBUTED BY GORDON REID (CEDAR)

lower harmonics 120 and 180 Hz). **Buzz** consists of the higher harmonics of the power line frequency, running in a series possibly including 240, 360, right on up to 2400 Hz and higher in severe cases.

### Why Reduce Noise?

The first idea that comes to mind when we mention noise is **tape hiss**, a potential problem when restoring old analog tapes. We are far less aggressive about hiss in analog tape restoration practices than when the CD first came out, only processing extremely noisy analog tapes. Listeners have become used to the idea that a classic analog source master might be a little noisy, and engineers recognize that noise reduction has its tradeoffs.

With modern mixes, a lot of project studio mixing rooms are not as quiet as professional studios; air conditioner rumble, airflow noise, and fans in computers cover up noises in the mix. Regardless, the mix engineer should be concentrating on other things besides whether the singer produced a mouth tic. Consequently, when the mix arrives at the quiet mastering suite, we notice problems that escaped the mix engineer—click track leakage, electrical noises, guitar amp buzz, preamp hiss or noises made by the musicians during the decay of the song. We use our experience to decide if these are tolerable problems or if they need to be fixed. Hiss which can be traced to a single instrument track is more transparently fixable at mix time; I ask the mix engineer to send me the offending track for cleanup, then return it for a remix. Or I may suggest a remix, bringing his attention to vocal noises between syllables, which he can mute. But clients don't always have the luxury or time to remix, and so mastering houses have the most advanced noise reduction tools which affect the surrounding material as little as possible.

The noise reduction methods we describe in this chapter are **single-ended** as opposed to **complementary**. Single-ended systems attempt to separate noise from signal without having a specially-recorded track whereas a complementary, or two-step, noise reduction system, e.g. the Dolby™ system, applies one process during recording and an equal and opposite process during playback.

### Noise Reduction Processors (a selected list)

- Stand-alone (outboard): Cedar Cambridge (every form of noise reduction plus other features such as linear-phase EQ); GML 9550 (broadband denoiser); TC Backdrop (broadband and tonal noise reduction for the System 6000); and Weiss DNA-1.
- DAW-integrated: Cedar Retouch and Algorithmix Renovator are customized to each DAW as is NoNoise, with Sonic Solutions or Pro Tools; Sequoia and Wavelab integrate noise reduction tools.
- A la carte plugins in standard format (VST, Direct-X): There are now many, my personal favorites are products from Algorithmix (broadband, declicker, descratcher); and products available from TC for their Powercore platform. The DAW Cube-Tec's highly regarded plugins are now separately available for certain DAWs.

### The Remedies

Each kind of problematic noise requires its own dedicated technical cure, but the most powerful cure is just to ignore the noise! We engineers tend to forget that the ear has a built-in noise reduction mechanism which gives us the ability to separate signal from noise, and hear information buried within the noise. Thus the key to effective noise reduction is not to attempt to remove all the noise, but to accept a small improvement as a victory. Remember that louder signals mask noise, and that the general public does not zero in on the noise as a problem. They're paying attention to the music, as should we! So before considering the use of any noise reduction technique, we need to judge whether a noise is truly distracting.

Some noise reduction units either require or can optionally use a *fingerprint*, and some are designed to be used without a fingerprint. A **fingerprint** is a sample of noise without signal (even one second will do). Those systems which can optionally function without a fingerprint are usually less effective in that mode, and require far more operator expertise. Systems which do not employ a fingerprint are more rare and usually produce worse results than fingerprint-based systems, one exception being the Weiss DNA-1.

LP transfers and other sources may contain hiss, hum, rumble, crackle, clicks, pops, tics, and peak-level distortion. Each type of noise or distortion requires a dedicated correction algorithm; for example, when noise is continuous we choose either a **broadband** or **tonal** processor. Broadband processors can be further tailored to work in selective frequency ranges such as rumble or hiss or by manipulating any number of band thresholds. Some noises which seem to be continuous, such as air conditioner rumble, have periodic (repetitive) components that may need to be treated as impulsive. On the other hand impulsive noise cannot be fingerprinted, so the algorithm has to focus in on known characteristics of a click, for example, and set a semi-automatic threshold. Impulsive noise reduction systems specialize separately in clicks, pops, crackle, scratches, etc. The dividing line between *clicks* and *crackle* is fluid; crackle consists of many closely spaced clicks. Sometimes crackle can be removed with a declicker, and vice versa; sometimes a continuous denoiser with a fingerprint can perform decrackle or debuzz. Cedar's *Declickle*™ algorithm distinguishes clicks and crackle from signal.

## II. Noise Reduction—Simple to Complex

When it comes to denoising, each processor has its specialties, learning curve and artifacts. Ease-of-use is usually (but not always) inversely proportional to effectiveness; consumer programs with their simple setups provide the least satisfactory results. If you require effective denoising with the least amount of artifacts, expect to dedicate serious time, and probably more than one tool, not always from one manufacturer. Everyone is initially attracted by noise-reduction power, but then the artifacts call attention to themselves. Each practitioner has his priorities, some value power over sonic transparency, and vice versa. For example, preserving depth and space is more important to me than having a perceptibly silent background.

### Simple Equalization

A passage with obtrusive hiss-like noise which contains no high-frequency instruments can be treated with a simple high-frequency equalizer. For example, an electric piano solo introducing a song may be hissy, but that noise will be masked when the rest of the instruments enter. This is a candidate for a selective filter active only during the piano introduction; say 1 to 4 dB dip around 3-5 kHz (this is the range where the ear is most sensitive to hiss).

P-pops are a signal-related noise, so they are a form of distortion, and since they are primarily low frequency, can be treated with a selective high-pass filter, up to about 100 Hz. If the filter is applied briefly, the result can be artifact-free. In my DAW, I capture a short section with the filter, then, using the crossfade editor, narrow the extent of the filter to the p-pop, which edits out just the offending portion. With practice the technique can be extremely fast (but see Retouch below).

### Complex Filtering for Tonal Noise

Tonal noise can be diminished by using narrow-band selective filtering. The practical limit of Q is 40 to about 100 before filters produce ringing artifacts. **Sonic Solutions No-Noise** has a complex filtering option that permits the insertion of many high-resolution narrow-band filters, suitable for removing hum and buzz. Before inserting the filters, it's useful to do an FFT analysis of the noise floor to determine which harmonics are present so as to apply only the filters that are needed. Sequoia's phase-linear FFT filter can do a similar job. SADiE has enough DSP power to insert many narrow-band filters in real time; my dehumming preset has about 25 filters set for a Q of 40 or higher. I selectively bypass each filter to hear if it's needed, and set its reduction just enough to reduce the tonal component below the annoyance level. TC's **Backdrop** has a preset which, with a fingerprint, can be very effective on hum and buzz.

### Narrow-Band Expansion

Compression techniques used in mixing and mastering can bring up noise in original material from tape, preamps, guitar and synth amplifiers, all of which could be perceived as problematic. Since compression aggravated the noise, expanders are its cure. As little as 1 to 4 dB of reduction in a narrow band centered around 3-5 kHz can be very effective. Typically these units have 3 to 4 bands, but we will only use one. Start by finding a threshold, with initially a high expansion ratio, fast attack and release time. Zero in on a threshold that is just above the noise level. You'll hear ugly *chatter* and bouncing of the noise floor because the time constants are so fast. Now, reduce the ratio to very small, below 1:2, perhaps even 1:1.1, and slow the release until there is little or no perceived modulation of the noise floor. Too much expansion, and you will hear artifacts such as pumping or ambience reduction. The attack will usually have to be much faster than the release so that fast impulses will not be affected. Depending on the music, its dynamic characteristics and its original SNR, this subtle approach can yield artifact-free noise reduction. The other expander bands should be bypassed or ratios set to 1:1. An expander's look-ahead delay (see Chapter 10) allows it to open before the signal hits it, thereby conserving transients. If the expander approach does not work, then we will have to apply more sophisticated, dedicated noise reduction processors.

### Broadband Processors

Complex broadband noise reducers are sophisticated multiband downward expanders with many bands. Those which can use fingerprints calculate the expansion threshold of each band, which can then be fine-tuned by the operator. **Algorithmix NoiseFree**, **Cedar Denoise**, and **Sonic Solutions NoNoise** all work best with fingerprints. The task of finding a fingerprint can be made easier when the client sends in samples of the noise with no music playing; thus, when sending material in for noise reduction, the mix engineer should not tightly cut the beginnings of material; the silence just before the downbeat is an excellent candidate for a fingerprint. By manipulating thresholds or the bands of a "noise-reduction equalizer", a skilled operator tailors the frequency response of the noise reduction curve for the best compromise between artifacts and perceived noise reduction.

Those systems which do not use fingerprints separate the signal from the noise through an automatic algorithm or by giving the operator manual control over the threshold of each band or range; Weiss's standalone DNA-1 is one such unit with algorithms for broadband and impulsive noises. Two standalone units are designed for real-time manipulation by the operator, the Cedar DNS1000 and the GML 9550.

#### Declickers

Test for artifacts with the difference button. Aggressive automatic declicking can distort the peaks of high frequency instruments such as trumpets. If there are artifacts, lower the sensitivity and try again, or replace compromised portions of the auto-declicked file with the source and then use surgical manual declicking.

#### Distortion Removal

A **decrackler** or **descratcher** can make an excellent distortion-softener or remover (selectively applied). These algorithms replace the distortion with an interpolation of the original sound. Cedar's **Declip** and Cube-Tec's Declipper remove clipping distortion. A declipper must drop level and leave headroom for the newly-restored peaks. Sonic Solution's **E Type** manual declicker is another excellent fixer for overload distortion. **Retouch** in its filtering mode can soften the level of high-frequency overloads.

#### Specialized Processors

**Cedar's** powerful **Retouch** and **Algorithmix Renovator** can very transparently remove noises that no previous system could handle, such as a baby crying, chair squeaks, even people talking in the middle of a take. Additionally, I've used Retouch to help construct seamless room tone, to soften or reduce distortion and as a manual de-esser. I've used its patching tool to edit music, replacing sections over damaged pieces of music. For p-pops, Retouch is much faster and more selective than the old method of a high-pass filter followed by DAW editing. Both products have become essential mastering tools, superceding and performing better than most classic manual declickers. When a tonal noise is varying in frequency, as from analog tapes with varying speed, a special kind of **tracking filter** is required, usually found in forensic suites. **Dethump** from Cedar is dedicated to long low frequency thumps and scratches. A **Deplop** algorithm (available from Cedar and Cube-Tec) handles the low-mid frequency ringing artifact that may be left after a click is removed. **Phase, time**

**and azimuth** correctors are available from Cedar and Cube-Tec. Cedar's dedicated **debuzzer** simplifies the task of filtering line-related frequency products.

### Artifacts and Perspective

No single-ended noise reduction system is perfect; all noise reduction systems take away some degree of signal with the noise and may add noises of their own. Ironically, a quieter original recording can be more effectively processed, because the more separated the original signal is from the noise, the more easily the noise reduction system can operate without hurting the signal. So a really noisy recording probably cannot be fixed without creating artifacts. Artifacts of overaggressive denoising include: comb-filtering, swishing or phasing noises (known semi-affectionately as *space monkeys*), low level thumps and pops (that can be worse than the disease).

*One of the by-products of noise reduction can be* loss of ambience and stereo separation. On the Mastering Webboard, Gordon Reid of Cedar explains,

> The difficulty lies in the fact that reverberation tends to decay to noise. However, much of the directional information and ambience we perceive is from reverberation. Therefore, remove the reverb with the noise, and—in effect—you remove the walls, floor and ceiling from the room.

To test for problems, use the *difference* button, if provided, to see if signal is being taken away with the noise. Listen for swishing noises (artifacts) in the difference signal. Even if the difference signal is perfect (just contains noise with no signal), be aware that psychoacoustically, the presence of noise increases apparent high frequency response, so only remove the annoying portion of the noise and be prepared to restore high end which seems to be lost.

Another consideration is the client's perspective. I once mastered an album where the opening of a tune had an obvious electrical tic on top of the bass player's note. I removed the tic, restoring the note to its beauty, I thought. But then the producer asked me to restore the tic—demonstrating that many noises are considered to be part of the music. Become familiar with each musical form—sometimes "dirty" is "clean".

What distinguishes good noise reduction work from bad is finding the optimum amount, because as noise is removed, more noise is revealed (noise itself masks other noise below it)! Beneath each layer of the onion is another layer. If you remove hiss, you may then hear crackle which was not previously audible, creating a potentially larger problem since crackle and other impulsive noises are more objectionable than continuous noise. In all cases, careful judgment is required to ensure that the music has been better served.

## III. Production Examples

### Manual Declicking

The top figure on the next page shows a *thunk* from an LP record. The left channel (top panel) has already been *dethunked*, as can be seen by the horizontal red marker above the waveform. When reproduced, the slight DC level shift that remains does not translate to an audible noise. The right channel contains a severe thunk manifested by an instantaneous upward, then downward DC level

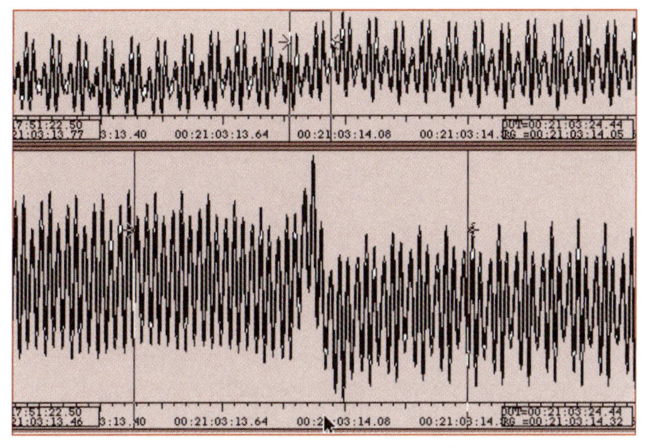

*A: LP Thunk in Sonic solutions. Left channel, area indicated by red bar has been denoised. Right channel has not yet been processed. Different panel heights reflect different visual magnifications, not different amplitudes.*

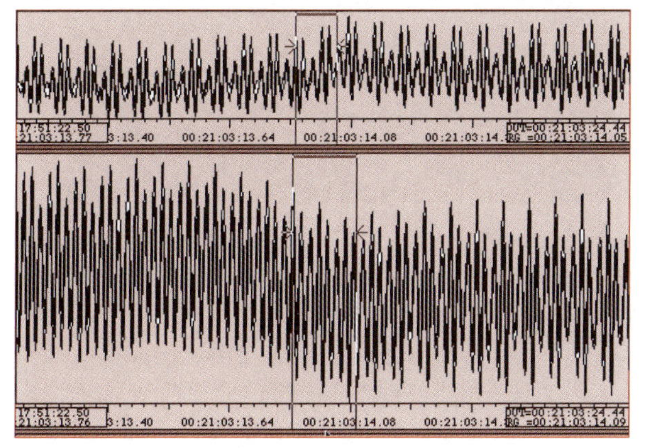

*B: After manual declicking, the right channel thunk has removed.*

shift (which causes woofers to rattle). With Sonic Solutions manual declicking, the correction process is as simple as marking the noise with the gates and selecting **D Type** from the menu. D Type is a powerful interpolator which can stitch together "impossible" waveforms and even remove brief dropouts or holes with no audible artifacts. In the bottom figure, the low frequency thunk and most of the DC discontinuity have been repaired; the ramped DC level shift that remains (probably record warp) does not produce an audible noise.

In the next figure, top right, a severe click is marked manually by the gates, and on the bottom it has been removed. Note that Sonic Solutions' automatic vertical gain conveniently amplifies the display to the highest amplitude in the view.

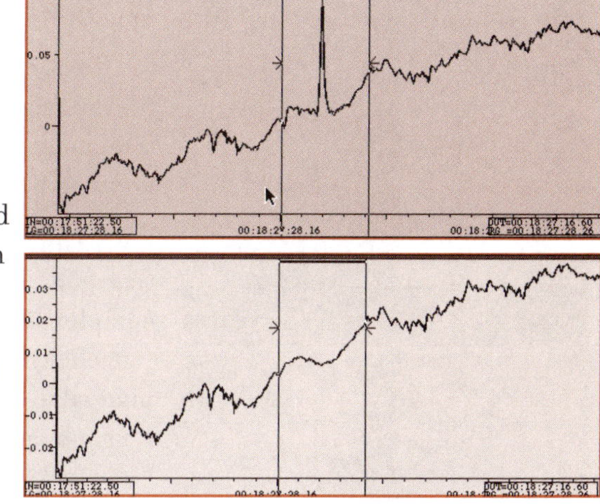

*On top, a click is surrounded by the gates. At bottom, after choosing D Type from the No-noise menu, the click is removed (marked by red bar).*

**Retouch**

The Retouch display is a full color spectragram like Spectrafoo, illustrated in the color plates. But even in black and white its power is dramatic, as seen on the next page. Cleaning up mouth tics is as easy as 1,2,3: **Step 1**, the tic is easily seen within the surrounding music. Notice that it occupies the upper midrange to high frequency portion of the spectrum. **Step 2**, we draw a box around the tic (bold inner rectangle). Retouch then adds a dashed rectangle, indicating the area that will replace the tic. Using surrounding material to replace damaged material is called *interpolation*. Retouch is the first tool that lets us choose the source for interpolation. **Step 3**, we press the **Retouch** button, and the tic instantly disappears. Compared to older audio repair technologies, we can be so surgical there are usually no audible artifacts.

Retouch can deal with the absence of sound just as easily as remove an annoying one. For example, fill in holes caused by analog tape dropouts (next page).

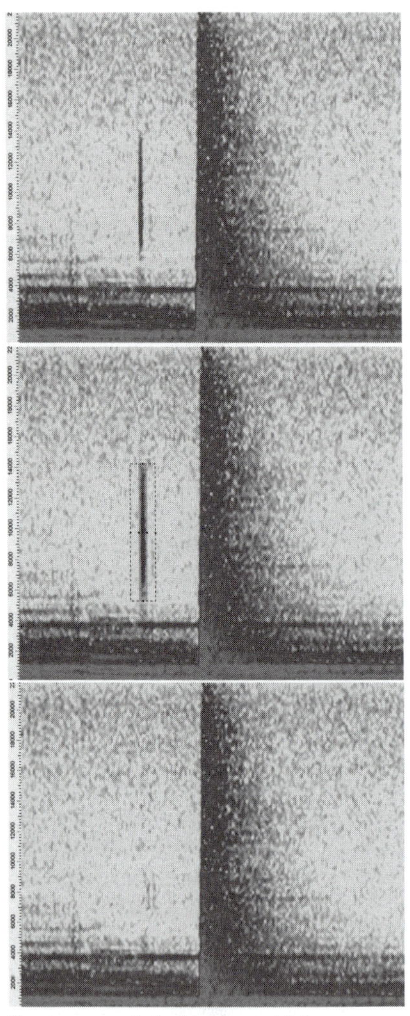

*Cleaning up mouth tics is as easy as 1,2,3*

## IV. Basic Order of Processing

To minimize artifacts and deal with the interaction of processes, it is best to treat noise in a particular order (each step is optional):

- First any tonal artifacts that stand out (e.g. hum, buzz), using a **simple or complex filter**, followed by
- **declicking, first automatic, then manual** to deal with any remnants not caught by automatic declicking.
- **decrackling** (which can also remove some remnant clicks)
- **"de-distortioning"**
- **broadband** denoising.
- Finally, overall program equalization, filtering, other processing if needed.

Each successive process should be saved to a new file including the names of the previous processes used. For example, "The Look of Love fl.wav" is the filtered file. "The Look of Love fl+dc.wav" was first filtered, then declicked. If using floating point plugins, save the intermediate products in 32-bit float format.

Denoising has become far less labor-intensive with inventions such as Retouch and Renovator. It's still work, but very rewarding; like hiring a meticulous gardener to remove each weed in your garden by hand, instead of using harmful chemicals.

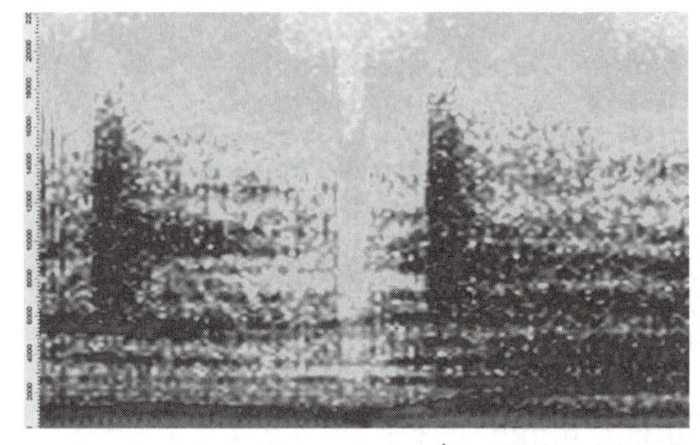

*A dropout in an analog tape (located between two major beats of the music).*

*After Retouch, the hole is gone and the sound completely restored. Whatever visual remnants that remain of the hole are audibly masked.*

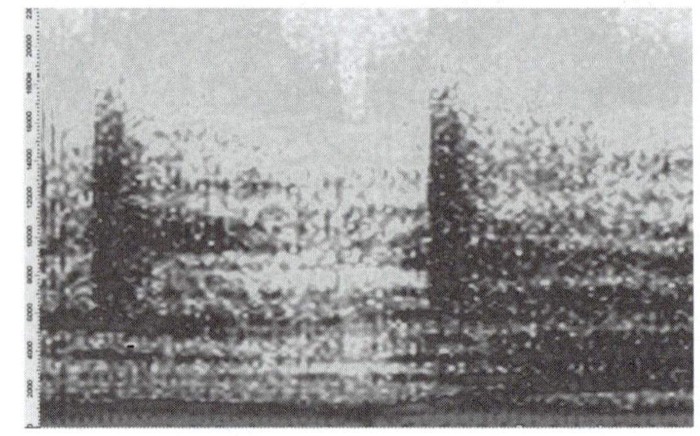

# Chapter 13
# Top Processors

## Introduction

This chapter presents a collection of processors useful for high-quality mastering. The inclusion of a particular unit in this set represents items that either I have used or which have gained a strong reputation among trusted mastering engineers.

## I. Acoustical and Analysis Tools

### GIK Acoustic Panels

GIK creates diffusers and absorbers that can turn a good sounding room into a great one, if properly set up. Trapping is generally used to smooth the bass response, while diffusion prevents specular reflections in the mid to upper ranges, and can help keep a room lively. Schroeder curve measurements determine whether there is sufficient balance of live and dead surfaces.

### Metric Halo Mobile I/O (MIO)

The Metric Halo Mobile I/O is a portable high-resolution recording studio or an analysis tool in conjunction with SpectraFoo. It serves as a multi-channel Firewire interface, portable spectrum analyzer for digital and analog audio problems. Attached to a Macintosh laptop, it's a highly functional measurement and analysis system. The jitter and distortion analyses in this book were made with the MIO and SpectraFoo.

**Real Traps Mondo Traps**

Real Traps Mondo Traps are ideal for low frequency absorption. Careful placement is the key; they can really optimize a room if properly used. We have seven mondo traps in our 22' long Studio A.

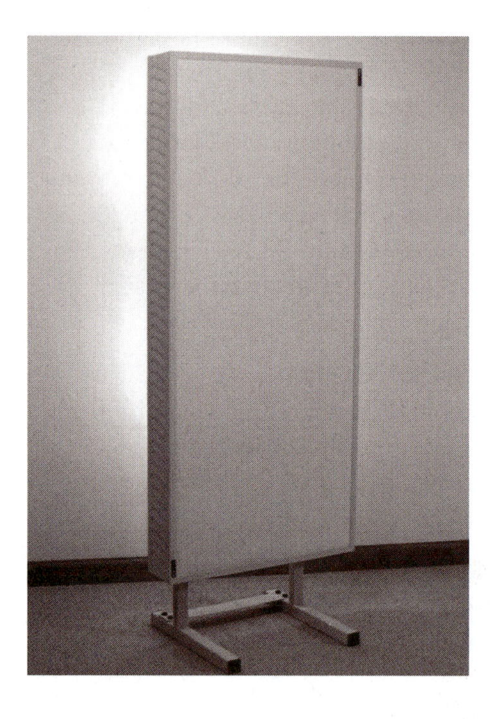

**RTW Surround Meter**

RTW provides surround sound metering, including a unique "Jellyfish" display.

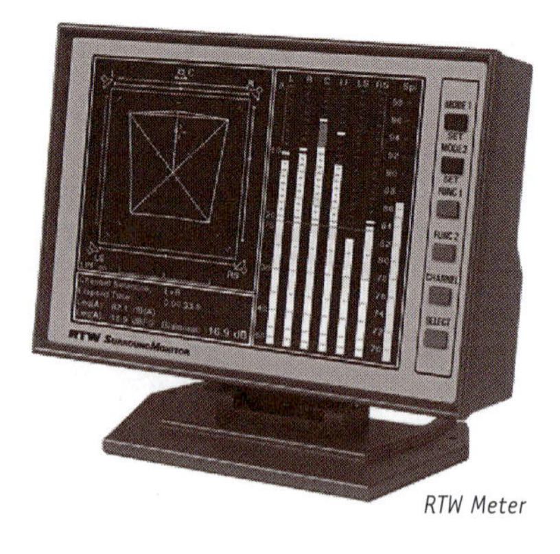


*RTW Meter*

**Terrasone/Sencore Audio Toolbox**

This is another useful portable measurement and setup device. Complete with measurement microphone, it can be used to align a monitor system or align analog system levels.

## II. Converters

All the converters mentioned here are at least A grade. The difference between an A and an A+ is extremely small, perceptible by only the most discriminating listeners, and opinions vary on which is better.

**Benchmark DAC1**

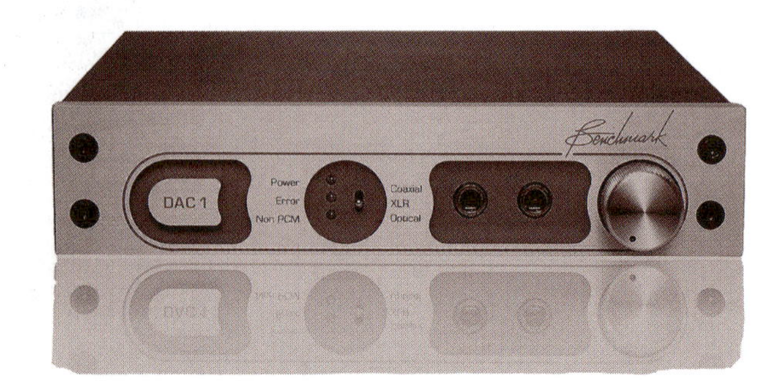

The Benchmark DAC1 is a 2-channel 192 kHz/24-bit digital-to-analog converter, utilizing their UltraLock™ jitter reduction system. The ADC1 has textbook-accurate measurements and is very transparent.

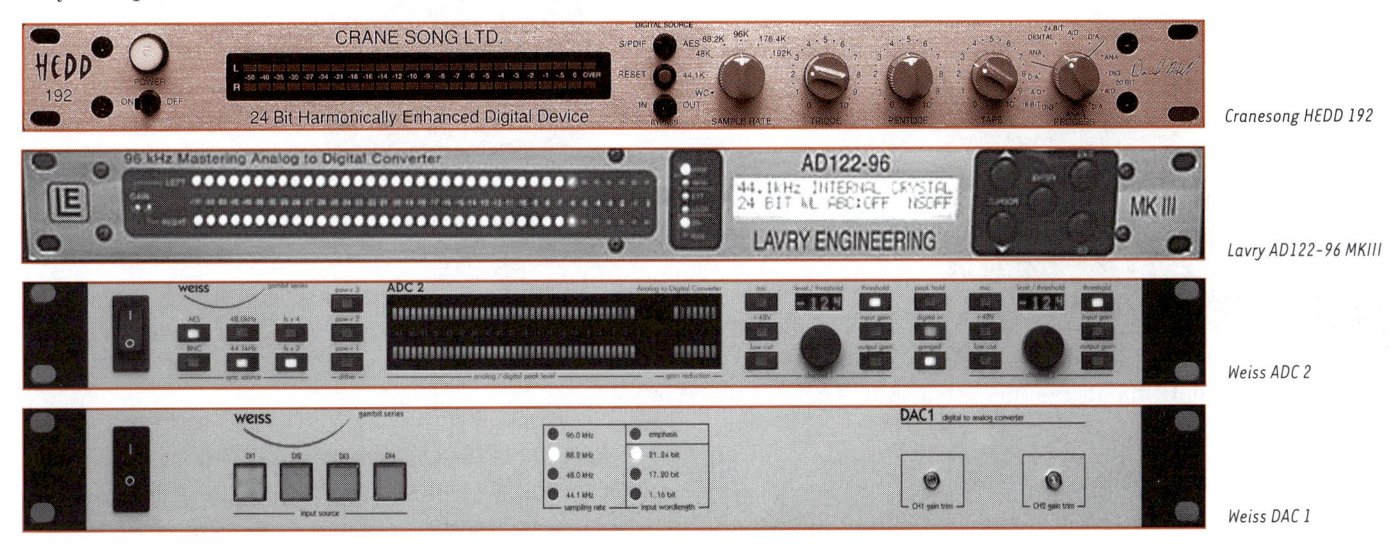

Cranesong HEDD 192

Lavry AD122-96 MKIII

Weiss ADC 2

Weiss DAC 1

**Cranesong HEDD-192**

The processor section of the Cranesong HEDD-192 is described in detail in Chapter 17. Its ADC and DAC have excellent sound, at least A grade.

**Lavry AD122-96 MKIII and DA-924**

The Lavry Gold converters are premium models built with discrete parts, with an extremely quiet noise floor and pristine sound quality. The DAC uses their Crystallock jitter elimination circuitry, whose PLL does not change the data. This is one candidate for best-sounding converter regardless of price.

**Weiss SFC2, ADC2 and DAC1**

The Weiss SFC-2 is a synchronous sample rate convertor that uses a fixed-ratio scheme where the output is directly derived from the input sampling frequency, which contributes to its low distortion. There are two independent stereo converters in a single 1U rackmout. Weiss's excellent ADC 2 and DAC 1 with a data-accurate PLL are two other candidates for "best-sounding".

**Other Converters**

Other highly respected converters that I have used are dCS, Mytek and Prism.

## III. Monitor Controllers

### Cranesong Avocet

We discuss this excellent monitor controller in detail in Chapter 2.

### Crookwood

We discuss this complete patching, routing and monitor controller system in Chapter 2.

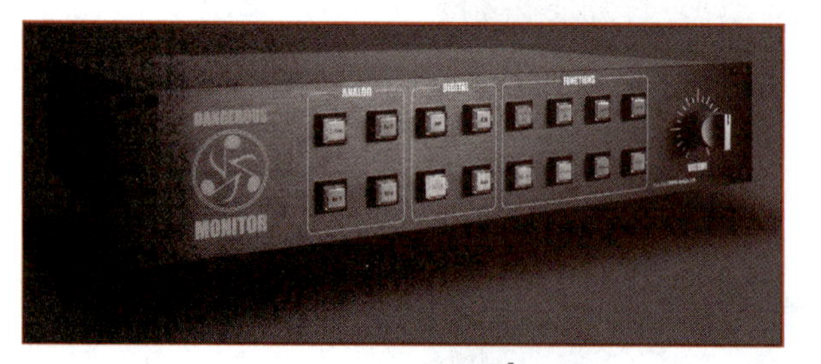


### Dangerous Monitor

Designed for clean sound, with comprehensive monitoring functionality, it allows for both analog and digital switching. But note that its monitor attenuator is not calibrated in 1 dB steps.

### Grace M904

It accepts multiple balanced and unbalanced analog and digital inputs with precision level controls for two pairs of studio monitors and dual headphone outputs. The monitor gain is calibrated in 1 dB steps, and via a software revision could conceivably be changed to meet the K-System requirements.

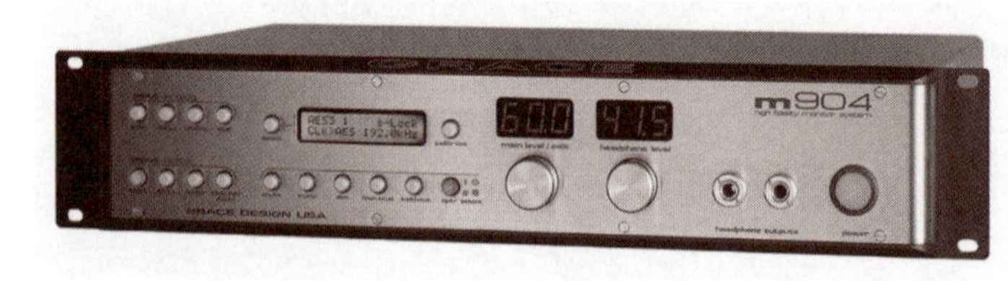

## IV. Loudspeakers

### JL Audio Fathom f112 Subwoofer

We have two of these 1000 watt woofers in Studio A. They have a 4" excursion capability and very low distortion. With every possible control that's necessary for accurate subwoofer adjustment, they are extremely versatile for any room and placement, and we can hear every bass note at the right level. They are also capable of playing a shuttle launch recording at "performance level" without overloading!

### Lipinski L-707 Loudspeakers and Amp

These excellent loudspeakers were designed for accurate impulse response down to about 60 Hz so they must be used with a subwoofer. Lipinski also manufactures a superb Class D power amplifier which we have in Studio B, whose quality equals or exceeds the Class A Pass X-250 which we are using in Studio A.

## V. Outboard: Analog

### Chandler LTD-2

Handmade analog compressor with discrete components, that according to fellow mastering engineers sounds warm and sweet. It uses class A amps and transformers from their 2254 modules.

### Fairchild tube limiter and Pultec EQ

These have not been constructed since the 1960s, but have attained such legendary status for their *fat sound* that I am obliged to mention these unobtainables en passant. There may be some modern-day substitutes which do as well or perhaps better, with cleaner, quieter electronics. If you're looking for the Pultec or Fairchild sound or beyond, consider units from Cranesong, Manley, or Millennia.

*Chandler LTD-2*

*The Cranesong STC-8 is a high quality stereo analog compressor combined with a peak limiter.*

### Cranesong STC-8 and Trakkers

A high quality stereo analog compressor combined with a peak limiter, the STC-8's attack and release times are fairly gentle, optimized for mastering purposes, and it is capable of emulating vintage equipment and creating distinctive new sounds. The Trakker (two units required for stereo) is capable of more "aggressive" compression for "attitude".

### GML-9500

**George Massenburg** is the design engineer for GML and the inventor of the parametric equalizer. The model 9500 is the mastering version of the popular 8200 parametric, which has been an industry standard popular with mastering engineers for over 20 years. GML also manufactures an analog dynamic range controller and a digital noise reduction unit.

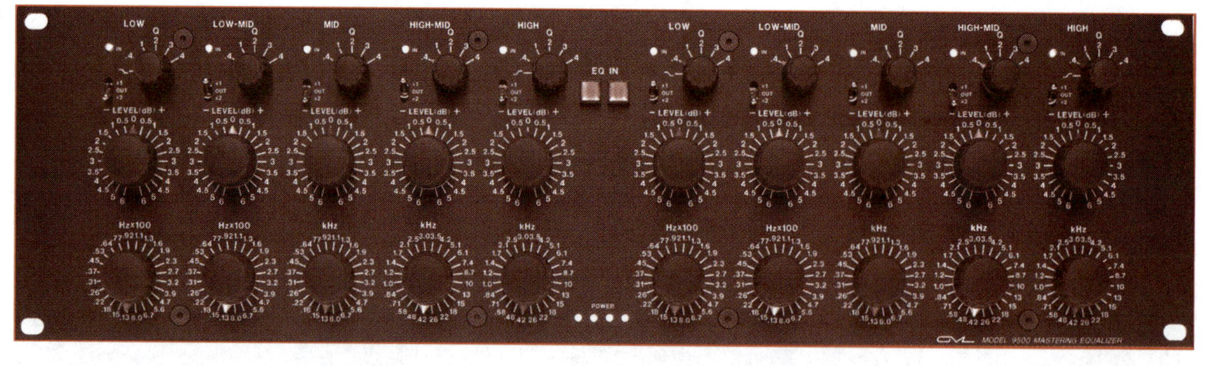


*Massenburg Equalizer Model GML-9500*

**Manley Massive Passive**

This is remarkably transparent and quiet for a tube equalizer. It gains its name by employing a passive equalizer section followed by a quiet, high-gain tube amplifier. To my ears it has just the right amount of tube distortion yet retains clarity without being too "fat". It also has far more versatility than the apparent four bands-per-channel because the Q or shape control affects the shelving curve as well as the bell, giving the effect of a 7 or 8 band equalizer. It's well worth downloading the informative and humorous manual written by Manley's versatile Craig "Hutch" Hutchinson.

**Manley Vari-Mu**

This tube compressor can help provide desirable punch and fatness with modern rhythmic music and is a good replacement for the classic Fairchild, which also employed variable Mu techniques (**Mu** is tube shorthand for gain). Distortion can be varied from very low to screaming by changing the input/output gain ratio.

**Millennia NSEQ-2**

Millennia Media manufactures a Twin Topology line which can be either tube or solid state at the flip of a switch. The NSEQ-2 equalizer probably has the shortest internal signal path of any analog equalizer, with a single DC-coupled solid state or tube opamp performing the duties of input conditioning, equalization, and line driving. In common with other top-of-the-line analog processors, headroom is exceptional, clipping at +37 dBu (solid state) and in solid state mode it is as close to an analog straight wire with equalization as I have ever heard (see Chapter 17).

**Pendulum Audio OCL-2**

A two channel electro-optical compressor/limiter designed for transparency, detail and versatility. Its short signal path uses a custom optical input attenuation network in front of a tube class A transformerless gain stage. The result is an open, detailed sound with an expanded sound stage and clarity that can only be achieved with modern tube circuitry. The designer added a drive control to our unit for additional tonal flexibility. See Chapter 10.

*Manley Massive Passive Stereo Equalizer.*

*Manley Stereo Variable MU Limiter Compressor.*

*Millennia Media NSEQ-2 Tube and Solid State Analog Equalizer.*

*Pendulum OCL-2*

## VI. Outboard: Digital

### DBX Quantum II

This is a powerful multi-function digital processor with up to 96 kHz operation. All DSP is calculated in 48-bit fixed-point notation, accurately dithered to 24-bit on its output for low-distortion. A wide variety of dithering options are provided. It has multiband and M/S options as well as parametric EQ, compression, expansion and limiting. One of the rare dynamics processors which include ratios below 1 (see Chapter 11), it's particularly valuable for *uncompression*. However, ergonomics are daunting, since all the functions are crammed on one LCD screen with multiple menu levels. This is the case with many such multi-function units; examine and test the menu structure before you buy—in the best units, critical functions will be no more than one or two menu levels below the top.[1]

### Drawmer DC2476

An extremely sophisticated, all-digital stereo mastering processor. Both analogue (balanced XLR) and digital (AES/EBU and SPDIF) I/O are provided. Noise shaped dithering is included.

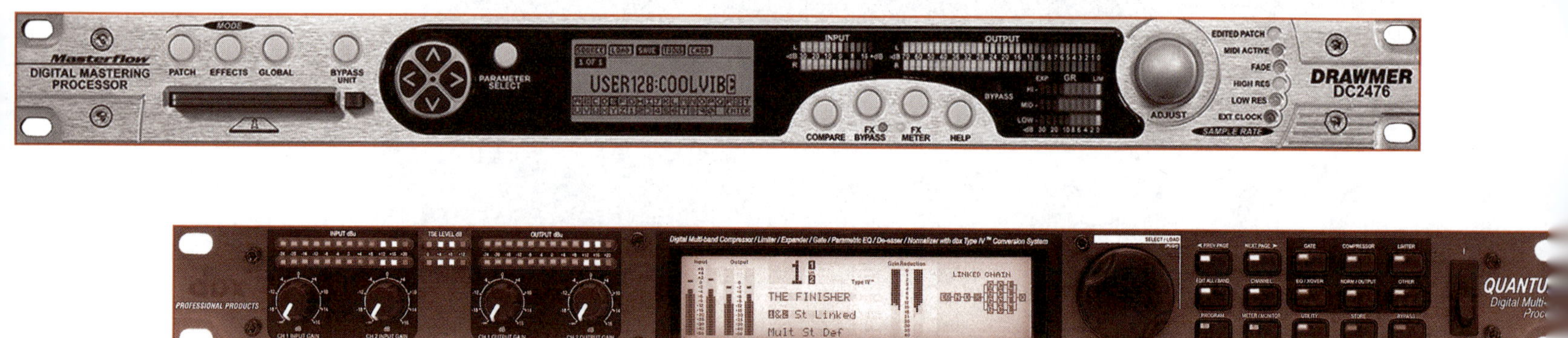

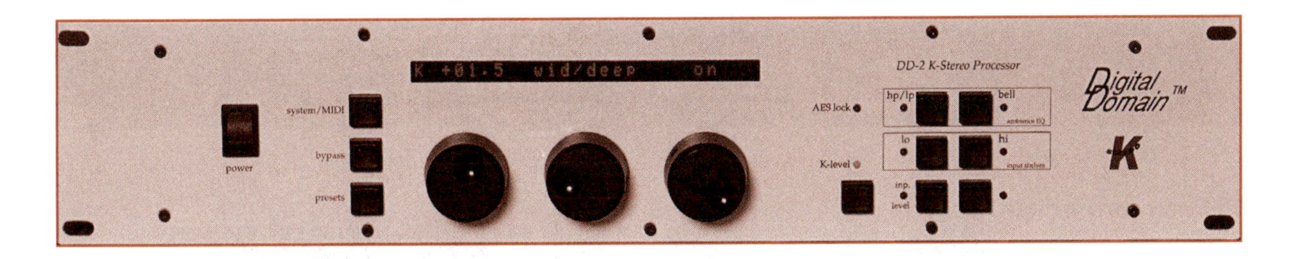


## K-Stereo

**K-Stereo.** The patented **K-Stereo**™ and **K-Surround**™ processes enhance the depth, ambience, space and definition in problem mixes that might otherwise sound small. K-Stereo extracts existing ambience to a stereo result, giving the mastering engineer a handle on reverb returns after the mix has been made. It should be tried before choosing a reverberator, because overall reverberation can muddy an existing mix, whereas K-Stereo selectively enhances elements in a mix which already contain ambience. For example, if a mix has a wet vocal that needs enhancement but also has a dry snare drum, K-Stereo will affect the vocal reverb but not the snare drum. Digital Domain manufactures the **Model DD-2 K-Stereo Processor**. **Weiss Engineering** has licensed K-Stereo for their DNA-1 multi-function unit.

## K-Surround

Z-Systems has licensed the K-Surround process in the model **Z-K6**, which converts 2-channel material to 6-channel. They also manufacture a 5.1 compressor and equalizer as well as the ubiquitous digital routers described in Chapter 2.

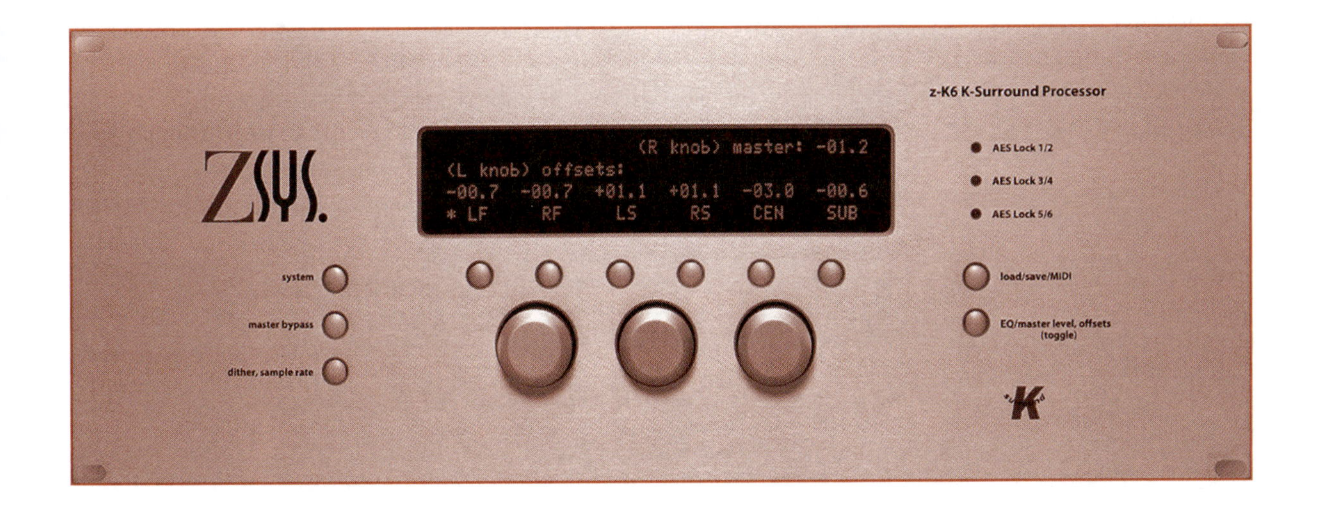

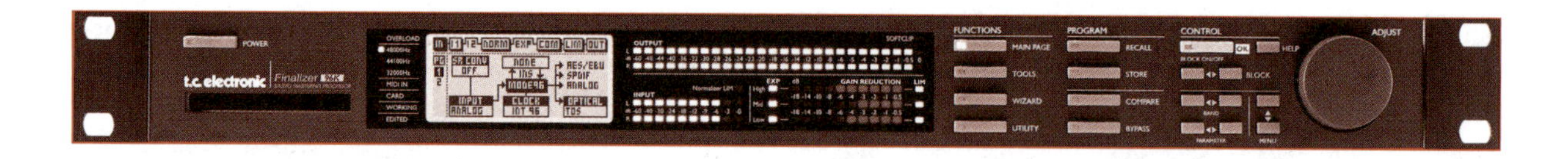


### TC Electronic Finalizer 96K

The Finalizer 96K offers an all-in-one mastering processor.

### TC Electronic System 6000

TC's flagship multichannel product is extremely easy to use, has impeccable sound and is modularly upgradeable. The ICON remote (pictured at right) can control numerous 6000 mainframes at once. Its touchscreen operation is extremely ergonomic. Four 8-channel 96 kHz/48-bit digital engines can perform artificial reverberation with probably the most effective early reflection algorithm (in the VSS4). The MD4 compressor performs excellent normal and parallel compression (see Chapters 10-11). Processing also includes expansion, superb brickwall (oversampled) peak limiting, de-essing, mixing, noise reduction, delay, special effects, and monitor control. The frame also contains a high-quality A/D/A, whose approach to jitter reduction is described in Chapter 21.

### Waves L2

**The L2** was the first hardware product produced by **Waves** and has become an obligatory mastering limiter. It helped spawn the unfortunate philosophy "I can make anything louder than you". Though discontinued, it is the only limiter offered by Waves with a split sidechain option. An exceptional auto-

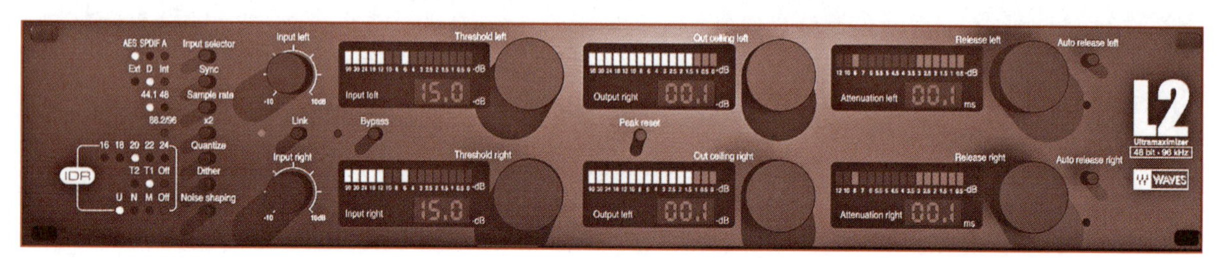


*Waves L2 Ultramaximizer*

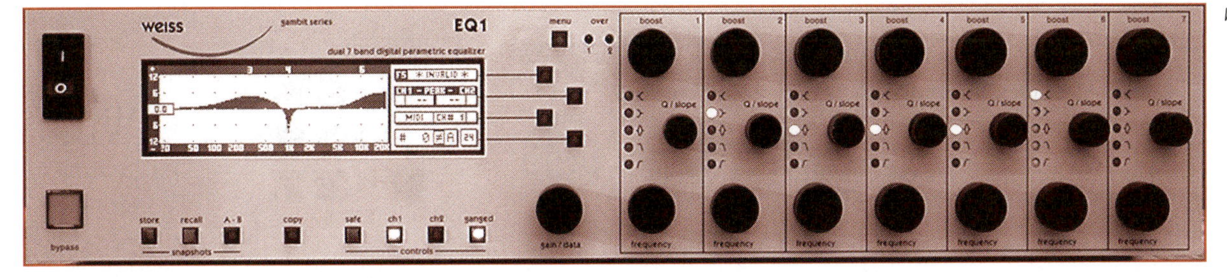


*Weiss EQ1 7-band equalizer*

release and 48-bit processing make the L2 one of the "least damaging" limiters. Yes, this is a left-handed compliment, but the L2 can sound pure and transparent **at low gain-reduction settings**. It contains Waves' IDR dither, among the better-sounding 16-bit dithers, and an excellent 24-bit ADC. The logical successor to this unit is the Waves MaxxBCL, including a Renaissance comp and MaxxBass, but the sidechains are always linked.

### Weiss EQ1-LP

**Weiss Engineering** holds a special place in the hearts of old-time digital mastering engineers (if that's not a contradiction in terms), since they invented the first usable high-resolution digital processing system, still available as the modular 102 series. The Gambit line of rackmount processors is designed for superb ergonomics and sound quality. With a one-knob-per-function philosophy, a Gambit feels just like an analog processor, with the added versatility of memory storage and MIDI remote control. The EQ1 has proved to be an excellent equalizer, with both minimum and linear phase options (see Chapter 8).

### Weiss DS1-MK2

The DS1-MK2 dynamics processor provides downward compression, upward expansion, and parallel compression, which can be wideband or confined to a specific band (see Chapters 10-11) and peak limiting. It is a candidate for "best sibilance controller ever made," ideal for processing sibilant voices within mixed material with minimal artifacts on the instruments. Not all functions are available at once.

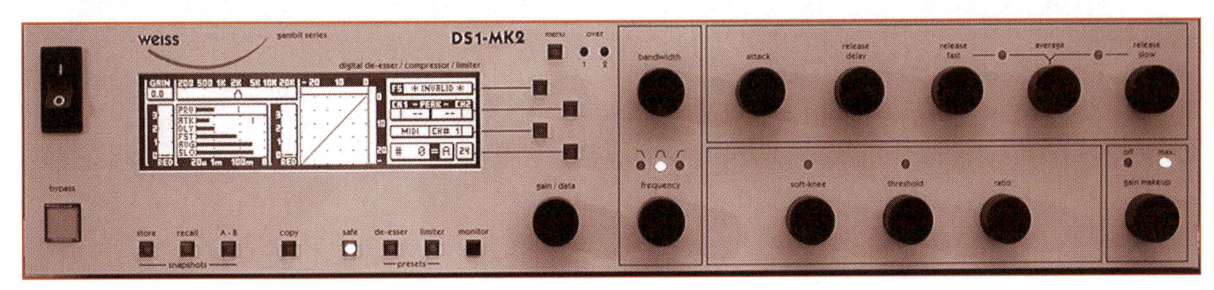


*Weiss DS1-MK2 Compressor/Limiter/Expander/De-Esser*

Z-Systems ZQ-2 6-band stereo digital equalizer

### Z-Systems ZQ-2

**Z Systems ZQ-2** is a 6-band stereo digital equalizer that sounds very clean and relatively *undigital.* I analyze its near textbook-perfect performance in Chapter 17.

## VII. Playback Systems and Recorders

### Slim Devices Transporter

I believe this network music server, and its little brother, the **Squeezebox**, are the future of music playback in many homes. It can play iTunes playlists, and your own music, straight from a computer server connected by wired or wireless ethernet, transparently giving the impression that it is a stand alone music playback device. Internet radio can be played without a separate computer, as long as the home network is connected to the internet. It can play up to 24-bit/96 kHz material while the Squeezebox is a lot more affordable but only goes up to 48 kHz (though it will downsample 96 kHz on the fly). The remote control is extremely ergonomic.

### Tascam DV-RA1000

This recorder is a solution for recording stereo high-resolution audio—up to 192 kHz/24-bit—to inexpensive DVD media. It also interfaces with a computer via USB. Its professional I/O includes balanced XLR analog, AES/EBU and SDIF-3, and it records standard CD-DA, WAVE and DSDIFF files to CD and DVD discs. DSDIFF is the DSD format of

Slim Devices Transporter

*Tascam DV-RA1000*

the SACD. It can also operate as a professional CD-DA recorder, recording standard audio CDs for studio or meeting room installations.

## VIII. Plug-ins and Computer-based Processors

### Algorithmix

Algorithmix has a wide range of products, from high-quality equalizers to noise reduction units. The Algorithmix Red is a 10-band precise linear-phase parametric and shelving equalizer with variable slope shelves. It's the sweetest digital equalizer I've heard and can boost or cut while keeping the instruments in place (see Chapter 8). The Blue is a minimum phase equalizer, whose algorithms emulate various styles of analog equalizers. Renovator is a noise-reduction plugin discussed in Chapter 12.

### AudioEase

AudioEase concentrates on two excellent products: **Barbabatch**, for the Macintosh platform, is a standalone file format and sample rate converter. **Altiverb** is a convolution reverb.

### Audiofile Engineering

This company specializes in Macintosh-based audio processing tools. **Spectre** is an excellent realtime audio analyzer, including K-System metering. **Sample Manager** is an all-inclusive batch file converter that supports metadata.

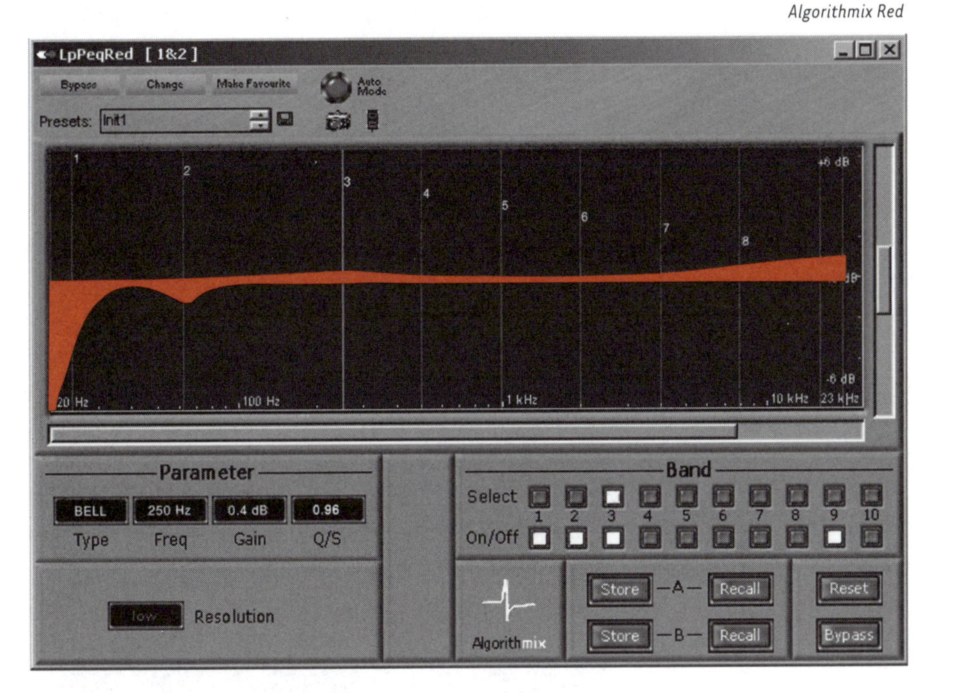

*Algorithmix Red*

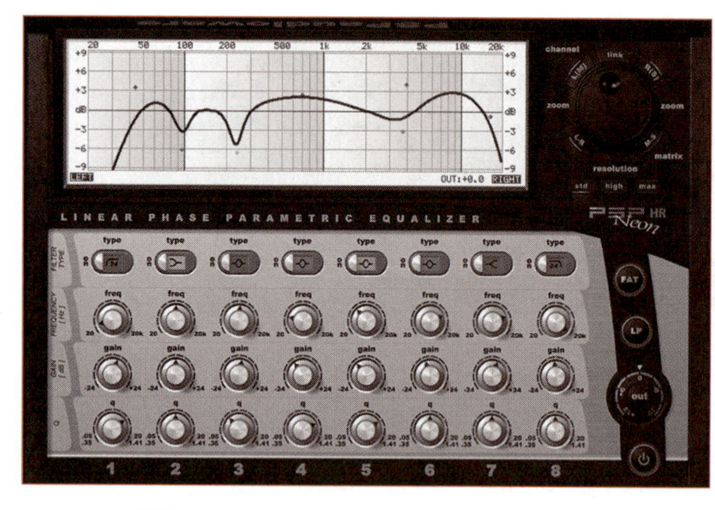


## PSP

The PSP Mastercomp is a quality compressor. The Neon is a linear phase equalizer with less CPU drain than the Algorithmix. The VintageWarmer2 is a saturation processor worth checking out.

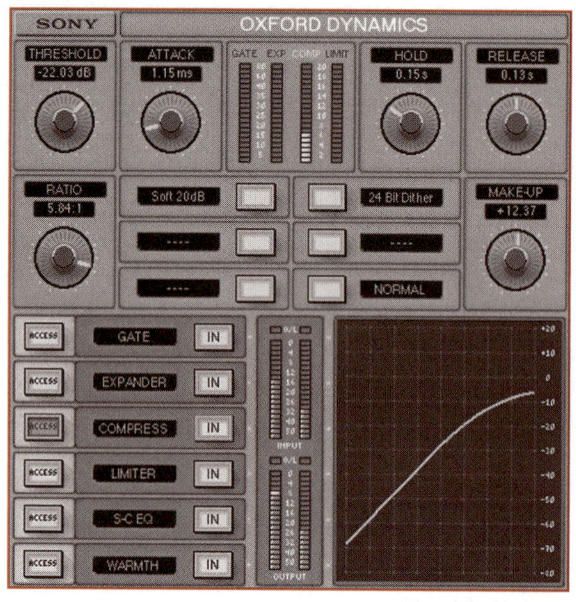


*Sony Oxford Dynamics*

## Sony Oxford

Sony Oxford plug-ins range from EQ's, to dynamics and reverbs. The Oxford limiter oversamples, preventing intersample peaks in succeeding equipment (see Chapter 5).

## TC Electronic Powercore

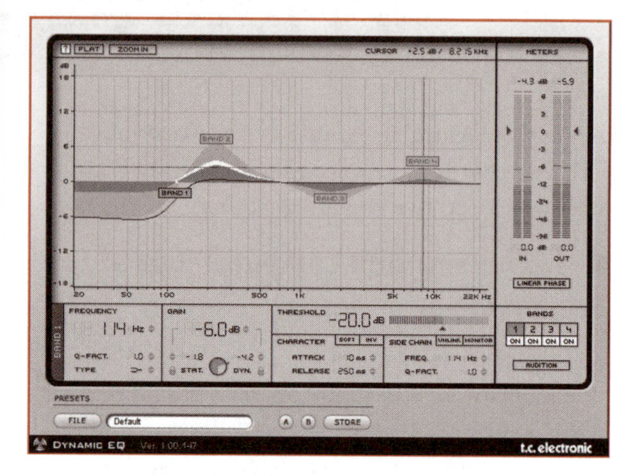


This is a hardware-assisted plugin, using extra DSP on a card to reduce the burden on the native CPU. Multiple units can be used in one system and it has an open platform for 3rd party plug-in developers. The noise-reduction algorithms, particularly the declicker and descratcher, are quite good.

## Universal Audio UAD-1

The UAD-1 provides high-quality hardware-assisted plugins, from equalizers through to reverbs.

### Voxengo

The Voxengo Elephant peak program limiter performs very well. R8Brain Pro is a sample rate converter of extremely high quality.

### Waves

While I prefer to use Waves plug-ins during mixing, many of them are quite suitable for mastering, especially the famous L1, L2 and L3. The Waves IR-1 reverb is an excellent convolution reverb.

The **Q-Clone** is probably based on convolution though not mentioned in the documentation. It is designed to copy the sound of an existing linear

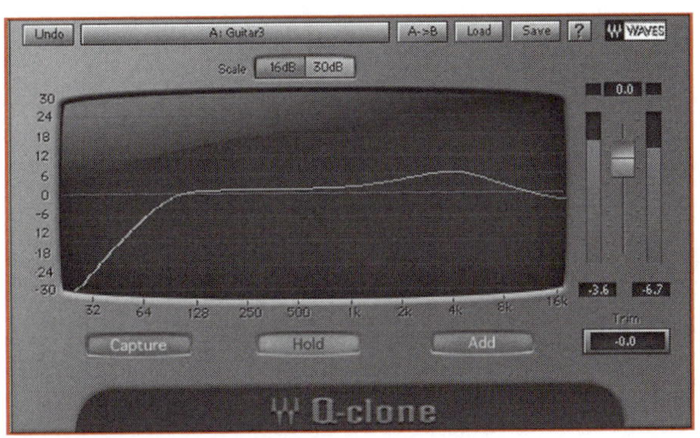


processor such as a hardware equalizer, but is also worth trying for reverberation.

**MaxxBass.** One problem we sometimes encounter in mastering is a bass instrument with inadequate definition or unclear notes. Obviously the best solution is to remix the song with a new bass player, better EQ or compression on the bass, but that's not always possible. **MaxxBass** (pictured at bottom) is designed to help clarify the definition of the bass instrument with minimal effect on the rest of the mix. **It's a form of a dedicated exciter** and a very powerful process that's easy to overuse and requires high resolution monitoring.

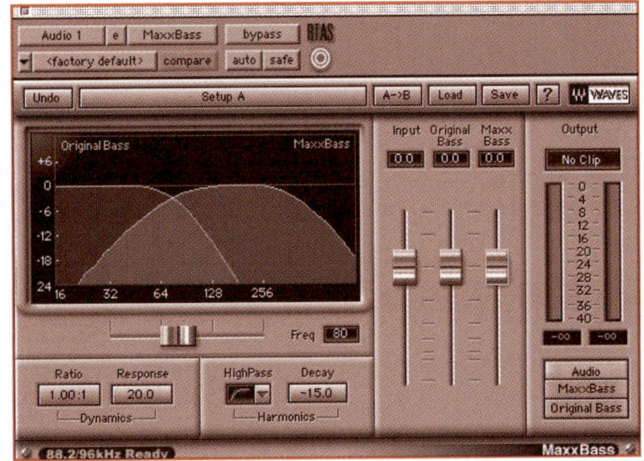

1 The best way to take advantage of multifunction boxes is to load an existing preset, then bypass nearly all the unnecessary and often exaggerated settings that manufacturers habitually toss in, and save the preset as a blank slate. Apparently they can't sell a box to its intended market without presets, but the preset concept is foreign to the way in which mastering engineers work, especially a preset ludicrously named Reggae, Rock and Roll or Smooth Jazz. How can they predict a setting without having heard the recording?

"We'll FIX it in the mastering."

—Anon

# PART III: ADVANCED THEORY & PRACTICE

"Making good sound is like preparing good food. If you overcook, it loses its taste."

—Bob Katz

CHAPTER 14

# How To Make Better Recordings in the 21st Century

## I. The Loudness Race

### History

At the dawn of the 20th century, when making Edison cylinders, everyone had to play loudly to overcome the background noise of the medium. Musicians playing soft instruments had to move in close to the recording horn or they wouldn't be heard and there was very little dynamic range as musicians were taught to never play softly! By 1927, although the electrical recording era had arrived, a vast improvement on the crude mechanicals of the wax cylinder, the low signal-to-noise ratio of the 78 RPM shellac record still kept us from hearing the full impact of the big bands. But by 1950, with the arrival of the Long Playing vinyl record, the noise of the recording medium had been sufficiently reduced so that engineers could achieve an impressive amount of dynamic range and impact. Just ask any record collector to demonstrate some of the best-sounding pop LPs of the 60s through the 80s—you'll be impressed! Ironically, in the 21st century, we have come full circle. While the medium's noise floor is inaudible, we're making popular music recordings that have no more dynamic range than a 1909 Edison Cylinder!

Here is a waveform from a digital audio workstation, showing three different styles of music recording. The time scale is about 10 minutes, vertical scale is linear, +/-1 at full digital level, 0.5 amplitude is 6

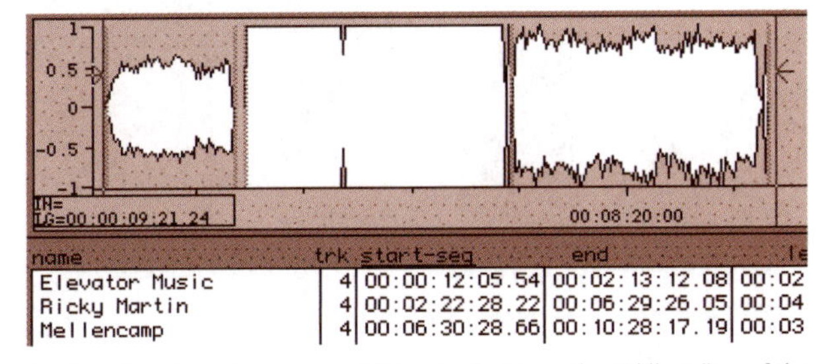


*On the left, moderately compressed "Elevator Music". In the middle, a "top of the pops" selection from the year 1999. At right, a rock and roll record from 1990. Vertical and horizontal scales are the same for all three pieces.*

dB below full scale. The density of the waveform is a rough approximation of the music's dynamic range and crest factor. On the left side is a piece of heavily compressed pseudo "elevator music" which I constructed for a demonstration at the 107th AES Convention. In the middle is a four-minute song from a popular compact disc produced in 1999. On the right is a four-minute popular rock recording made in 1990 that is quite dynamic—for that period. The intrinsic loudness difference between the 1990 and 1999 CDs is greater than 6 dB, though both peak to full scale! Auditioning the 1999 CD, one mastering engineer remarked, "this CD is a light switch! The music starts, all the meter lights come on, and they stay on." Even the elevator music recording has greater dynamics. Are we really in the business of making square waves? Our objective as mastering engineers is to produce recordings that a wide range of consumers can enjoy. Over-loud CDs become fatiguing and eventually unpleasant to listen to; we are mistreating them by engaging in this loudness race. So we need to ask now: why has the average sound quality of popular music gone so far downhill, and more importantly, what can we do to fix the problem?[1]

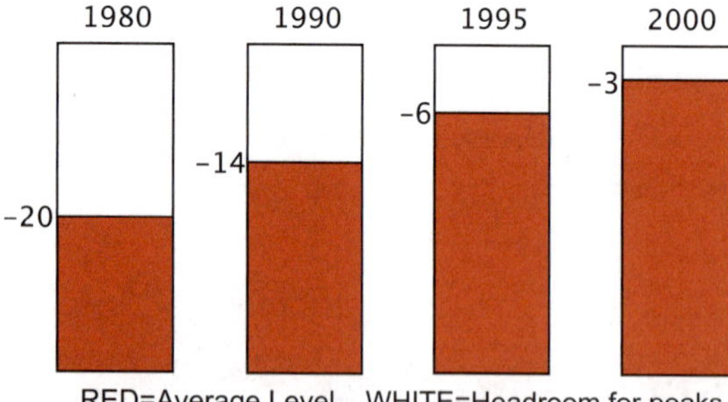


***The Insane Increase in "Hottest" Pop CD Levels***

*The height of the red bar shows the increase in average level of sound recordings during this period. The shrinking white bar demonstrates loss of sound quality, clarity, transient response, impact and dynamics.*

## Peak Normalization Vs. Loudness Normalization

This figure shows the progression of the loudness race in pop music from the introduction of the compact disc in 1980 into the 21st century. The quest for greater loudness is not new; in the days of vinyl, mastering engineers competed to produce the loudest record to attract attention in the jukebox, which had a fixed volume control.[2] The origin of this competition is psychoacoustic: even when two **identical** programs are presented at slightly differing loudness, the louder of the two appears to sound 'better'; **however, this 'improvement' is only a short term phenomenon**. This explains why CD loudness levels have been creeping up until the sound quality is so bad that almost everyone can perceive it. But it does not explain why the intrinsic loudness of CDs has gone up nearly 20 dB in 20 years! Even in the heyday of the LP loudness race, the maximum increase was no more than about 4 dB. The figure (top, next page) explains how the development of digital technology opened up **Pandora's Box**.

In the analog recording system, most engineers used VU meters, ending up with a fairly consistent average level and loudness regardless of the amount of compression used. The way LPs used to be made is similar to loudness normalization, seen at right in the figure.[3] The first bar represents a recording with 20 dB peak to average ratio; its average loudness (in forte passages) is at the 0 dB line. If we compress and loudness-normalize this recording, its loudness remains consistent. However, digital recording allowed us to peak-normalize, seen at left. If we compress and peak-normalize this recording, its average level goes up as we decrease dynamic range. Since mastering engineers can now easily normalize to the peak, compressed material gains an unfair level advantage over uncompressed material.[4] **The Compact Disc became the catalyst for the accelerated digital loudness race. Peak-normalization is the fuel that keeps the motors running.**

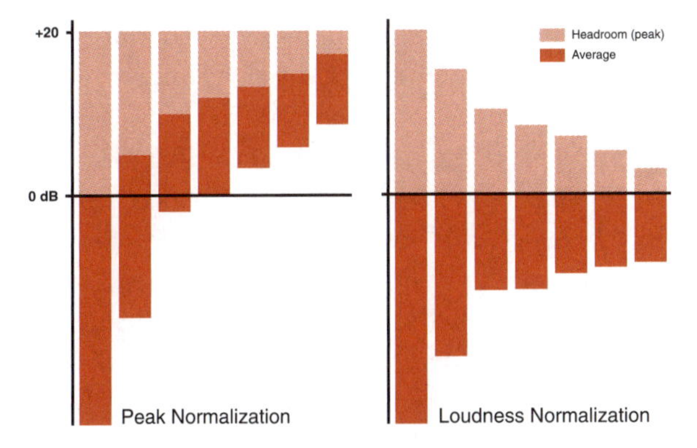


***Pandora's Box: The Fuel For the Loudness Race.*** *Analog recordings were once made as at right. A standardized average level yielded fairly consistent loudness from album to album. However (left side), digital technology lets us normalize to the peak, giving an extreme artificial loudness advantage to highly compressed material.*

## Consistency of Intrinsic Loudness

In the days of the LP, the variation in intrinsic loudness of pop recordings was much more consistent, perhaps within as little as 4 dB. Even at the peak of the vinyl loudness race, I could put on a Simon and Garfunkel LP, or even a Led Zeppelin, and follow that with an audiophile direct-to-disc recording, barely having to adjust the monitor control to satisfy my ears. In the earliest days of the compact disc, before the digital loudness race began, many mastering engineers would dub analog tapes with 0 VU set to -20 dBFS, and leave the headroom to the natural crest factor of the recording. It was not thought necessary to peak to full scale, and so the intrinsic loudness of early pop CDs was much more consistent. However, the inventors of the digital system abandoned the VU meter, which opened Pandora's box. And so the average level began to move up and up. As the digital loudness race accelerated, every mastering engineer began to utilize all the peak headroom, ostensibly to improve the signal-to-noise-ratio, but really to get a little level advantage, but when the intrinsic loudness increased, the headroom and natural dynamics kept on getting squeezed, as pictured.

The figure below shows the insane amounts of dynamics processing required to obtain the hottest levels towards the end of the loudness race. This is the end of the race. There is no headroom left to raise the average level. At the bottom of the figure, more gentle dynamics processing raises the average level without adding peak distortion. At the top of the figure, egregious processing represents mastering engineers' attempts to beat the digital system, to raise the peak level over 0 dBFS, creating severe peak overloads and distortion that translates nowhere (see Chapter 5). Notice how the curve flattens, showing that aggressive processing brings diminishing returns in intrinsic loudness, to say nothing about the sound degradation. Bear in mind that the whole idea of a "loud master" is only a conceit, since the consumer is in charge of their volume control and, more importantly—**"loud" masters sound weak and not powerful when listeners adjust the level for their own comfort.**[5]

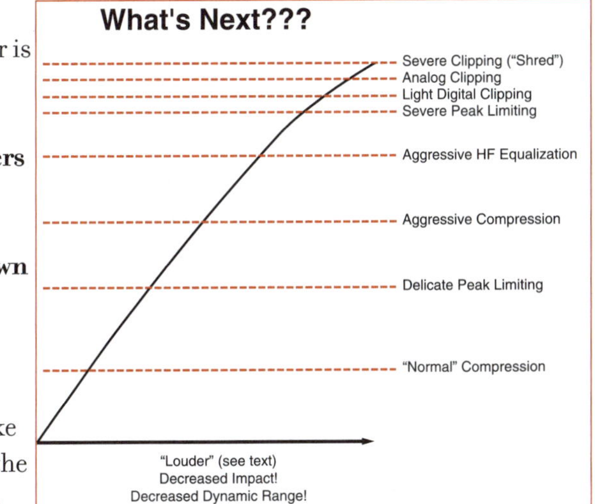


## An Introduction to the Calibrated Monitor

A monitor control is like a water faucet. The greater the water pressure coming into

> *"We have to stop giving the impression that we have any control over the consumer's volume control. In the end we are just making Muzak if we overcompress."* — Clete Baker

the faucet, the less we have to open it to get the same pressure out. Similarly, we can gauge how much pressure must be coming **into** the monitor by how far we have to turn the control and the amount of pressure (sound pressure level) that comes **out**. In a similar way we can draw conclusions about the **intrinsic loudness** of the material by noting the **position of the monitor control** and how loud the music sounds to us (note—not the level of an SPL meter—it is our ears that judge loudness). So by the simple expedient of marking a monitor control in decibels, and using our ears, we can learn a tremendous amount about the incoming signal. A fully open "faucet" would be set to 0 dB, indicating the least pressure coming in. Today's monitor controls should be calibrated, as illustrated in the figure below. In the next chapter we'll learn how to set up a calibrated monitor and later on in this chapter we'll learn how to use it. Putting numbers on the monitor control produces some very sobering observations.

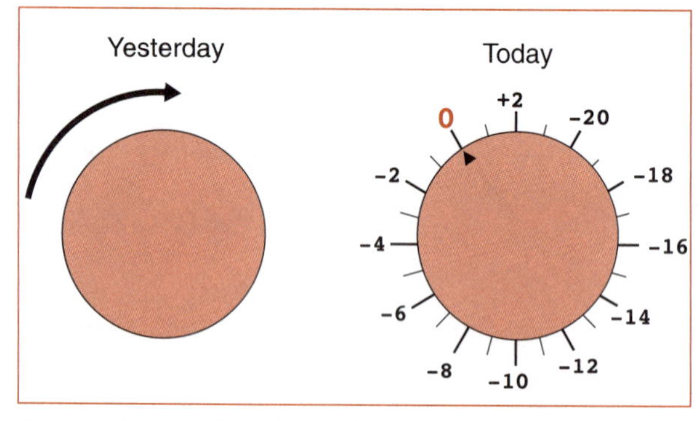


*Today's **monitor control** is marked in 1 dB steps, and the 0 dB position is calibrated to a standard SPL.*

**What happens to relative loudness when we peak normalize?** Let's look at a few recordings (see chart next page). Every recording on this chart has been peak-normalized (to 0 dBFS). The chart indicates the positions of the monitor control that I prefer to audition each type of recording when it has been peak-normalized. The lower the monitor control, the "louder" the recording. Grey bars represent raw mixes, light orange bars mixes which became masters without level processing (other than peak normalization), and red bars processed masters. The chart shows clearly that peak normalization produces unnatural loudness reversals and extreme loudness differences, especially with hypercompressed and distorted masters. Attentive music listeners prefer music played at the natural sound pressure of the original ensemble,[6] and so the consumer is inconvenienced, having to make extreme changes in his volume control. For example, to play a digital recording of a symphony orchestra, I set my monitor to 0 dB. If I then play a harpsichord recording, I have to leap to the control or it will sound 13 dB too loud! Many classical producers know that it's not a good thing to normalize a harpsichord to full scale or, as here, it will sound too loud compared to everything else the consumer puts in his player, but I wish producers from other genres could be so diplomatic! The same is true for a string quartet, which, if peak-normalized, sounds 8 dB too loud compared to a symphony orchestra.[7] I call this difference the **acoustic advantage** and it occurs whenever you peak-normalize a naturally soft instrument or vocal. The other reason the string quartet sounds louder is that its low peak-to-average ratio allows the average level to be raised significantly without causing peak meter overload. The "loudest" master I ever made was of an uncompressed, close-miked solo pennywhistle, which the client insisted on peaking to full scale; I reluctantly obliged. Ironically, this pennywhistle recording is 2 dB louder than the most hypercompressed and distorted rock CDs (based on preferred monitor gain)! Equally surprising is to find acoustic jazz as loud as hip-hop or hip-hop louder than heavy metal. That's what happens in a peak-normalized system with heavily compressed material.

Next, notice how the natural compression of analog tape gives the symphony orchestra a 4 dB loudness "advantage" compared to the digital recording, while removing some transient peaks. This (ironically) brings the preferred monitor gain of the symphony close enough to that of the string quartet to no longer be annoying when the consumer changes CDs. I don't want to be the first engineer to advocate peak-limiting a classical recording—the better consumer solution would be to drop the level of the string quartet. But this will only happen when an ecologically-aware mastering engineer and producer collaborate, though clearly the consumer will benefit if all producers become aware of the limitations of the peak meter.

On the mastering side (red bars in the chart), you can see that around 1990, the majority of popular music (and chamber and jazz) recordings and masters gave rise to a -6 to -8 dB monitor position, **the average monitor position for all types of recordings made between 1900 and around 1990** (if peak-normalized). But 1995 was about the last year the majority of my clients received sonically uncompromised masters. It is no coincidence that in 1990, brickwall digital limiters became widely available. Today, only a small number of clients take notice of the compromises of hypercompression, but more and more people are up in arms about the differences between loud and soft CDs.[8] Still, many masters that I make today are sonically compromised, inevitably at the client's request because they are afraid they won't stand out, and who can blame them for complaining that the master I make for them is 6 dB lower than the "competition"? Is the loudness race client-driven or mastering engineer-driven? I think it is a vicious circle, also driven by that insidious peak meter. With enlightened clients, we try to "back off" our processing, even a little bit, and if we are already at the point where the curve in the figure on page 169 starts turning, 1 dB less can make a big sonic improvement.

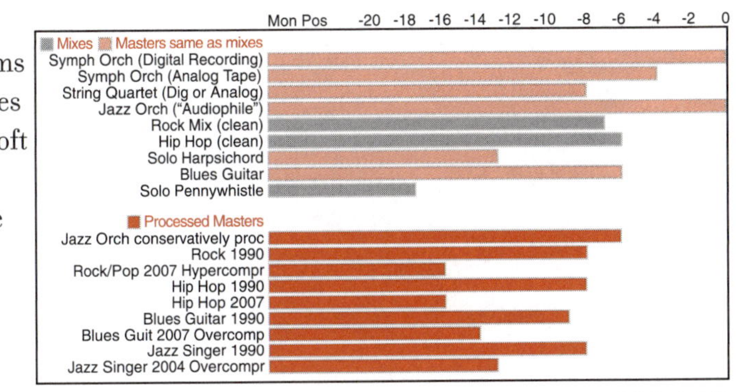

*Full Scale Folly*

**Which type of music suffers first from too much compression?** The first are those with plentiful natural dynamics and live percussion, such as Latin-Jazz and rock with a real drumset (sampled or synthesized sets have a lower crest factor allowing the average level to be raised without using as much peak limiting). Soft acoustic jazz recordings, especially if peak-normalized, do not need any compression for "competitive" loudness because of the acoustic advantage, so it is sad to see that the loudness race has moved into blues and acoustic jazz. The sound of some post-2000 acoustic blues guitar recordings is overcompressed, spongy-sounding,

**Vicious Circle**

1. The acoustic advantage makes peak-normalized acoustic material sound too loud.

2. Then the electric people feel inadequate and start the war all over again.

3. The next salvo—overcompressed acoustic music.

4. Ad Infinitum.

but as the chart shows, they have 6 dB higher intrinsic loudness than an optimally-processed 1990 master. One blues record company expects that level and will simply reject any master that is not compressed enough to achieve it. The 2002 debut recording of an acoustic jazz singer was already strongly compressed, losing that *open* feel, but at least it was listenable. However her 2004 album was hypercompressed, and as you can see on the chart it is in line with the loudness of hypercompressed 2007 rock recordings. Sadly, the 2007 album then becomes the new level benchmark by which our non-enlightened clients measure their own albums.

Fortunately, some clients are still sound-conscious and they know that a more dynamic master sounds better and performs better on the radio. Some other clients are fearful when they compare the level of the car radio to that of their CD, but the radio has its own processing and the two media are not comparable.

*"-6 to -8 dB is the average monitor position for all types of recordings made between 1900 and about 1990"*

**How Much Dynamic Range Do We Need?**

The constant bombardment of heavy compression is fatiguing, and listeners tend to tune it out, treating music like audio wallpaper, which is opposite from what the musicians intended. Dynamic range is the cure for loudness-induced listening stress and we should make every effort to put some life back into our audio. There are specific places where strong or heavy compression is needed: background listening, parties, bar and jukebox playback, car stereos, headphone-wearing joggers, the loudspeakers at the record stores, headphone auditioning at the record store *kiosk*, and so on. Mastering engineer Doug Sax believes that CD playback in cars has driven today's hypercompression, but it has now gone too far even for the car.[9]

**How Low Is Too Low?** The dynamic range tolerance chart (next page, from research performed by TC Electronic) illustrates that our current masters are far less dynamic than the public will tolerate. I'd be happy if we returned to the wider dynamics of a good LP, which corresponds very well with the Living Room bar in the chart. It also shows that if we want to make uncompromised masters for the livingroom, our lower quality reproduction units (like the kitchen CD player) are going to need to apply extra compression internally. We should not master to the lowest common denominator, then good dynamic range can be enjoyed in the environments on the left side of the chart.

**Things Are Looking Up!** There is a direction of movement which might mean that eventually iTunes (and possibly Windows Media Player) take over from the CD changer in many consumer environments. Distributed media systems using iTunes or other music servers, and iPod-based players in homes and cars are better because these systems have both an optional dynamics compressor and a loudness normalizer. This context-based compression is a lot less damaging than universal compression, which compromises recorded music for all listeners. The compressor in iTunes/iPod basically amounts to a low-fidelity replacement for the analog cassette. Every kitchen and car player should have dynamic range compression turned on (though many, like me, will be willing to move the volume control up and down in the car rather than let a compressor try to do it for us).

## II. Using A Calibrated Monitor System for Level and Quality Judgment

### What is a Calibrated Monitor System?

A calibrated monitor system is one that is adjusted to a known standard gain and frequency response. The monitor control is marked in decibels so the monitor gain is easily repeatable and being *calibrated* means that the standard decibel markings on the monitor scale indicate the same thing to any engineer, whether in Calcutta, New York, or rural Suffolk… This will help us to be more consistent in our work, and to produce mixes that will perform together when later assembled at the mastering house. Because the film world already has an absolute reference in the level of spoken dialog, it has this technique down pat. But there is no such exacting reference in the music world, though the chart on page 171 provides some idea of typical monitor positions used for different styles of mixed and mastered music.

By setting a calibrated monitor to its 0 dB position, an experienced engineer can make an excellent mixdown just by listening, without needing a meter or any protective peak limiting! The setup can be done very simply. The monitor level control is calibrated so that the 0 dB position produces 83 dB SPL with a pink noise calibration signal (to be explained in Chapter 15) and the recorded level of this *calibration signal* is set to -20 dBFS RMS. The result is that a comfortably loud average SPL has been set to 20 dB below the peak system level. Since the ear generally judges loudness by average level, and the most extreme crest factor anyone has measured for normal music is 20 dB, this ensures that our peak level will never overload. Typical mixed material has crest factors from 10 to 18 dB, so this mixdown may reach peaks from -10 to -2 dBFS, more than adequate levels for 24-bit recording, as shown in Chapter 5.

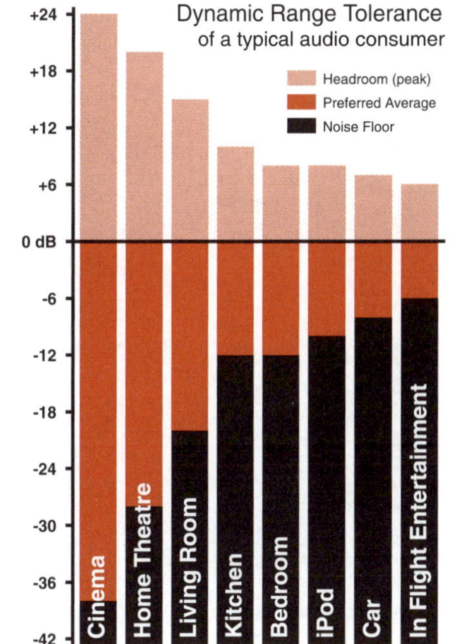


***Dynamic Range Tolerance of a typical audio consumer.*** *Note that at the end of the loudness race, our current compression practices are far more aggressive than what the consumer will tolerate, even in the car. Courtesy of TC Electronic.*

What this means is that a high monitor control position will permit us to produce music with high crest factor. Conversely, the lower we position the monitor control, the more we tend to raise the average recorded level to produce the same loudness to the ear. In the 20th century, we approached this from the opposite way; as we raised the average recorded level, we were forced to turn down the monitor to keep our ears from overloading. **In the 21st century, first set the monitor control to a given position, then start mixing or mastering.**

### Mastering With the Calibrated Monitor

If the monitor control is the key, how do we know what position to set it before we begin mastering? Everyone's room is slightly different, and room

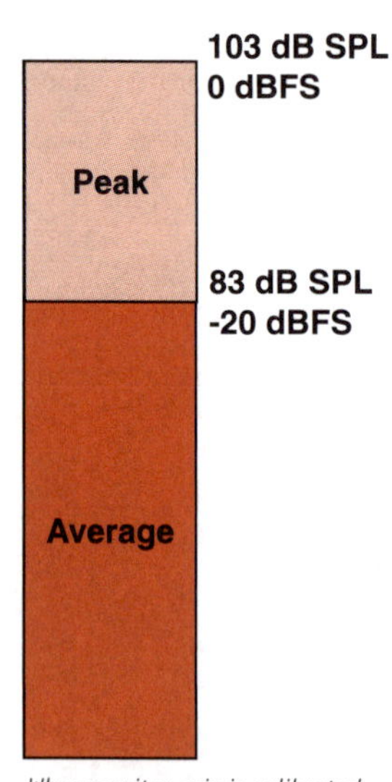


*When monitor gain is calibrated so average SPL is 83 dB at -20 dBFS, and you then mix by the loudness of the monitor, then the music will never overload and you will never have to look at a level meter!*

volume, loudspeaker placement, transient response, and even individual sensitivity at a given time of day affect the apparent loudness of a monitor system, even when identically calibrated. When I first developed the K-System, I determined these recommended monitor control settings in a large stereo mastering room with loudspeakers placed 9 feet from the listener. When I switched to a pair of loudspeakers that have a better transient response, everything seemed about 2 dB louder, illustrating that there are natural variances beyond our control. You should be satisfied to be within 2 dB of the examples shown in this book. In Chapter 15 I'll discuss ways to standardize the monitor controller before mastering.

Let's start with some music which has real drums and potential dynamic *impact*. We start the mastering session by carefully evaluating the raw mix, setting the monitor to produce a comfortably loud level on the forte passages. An experienced mastering engineer should immediately know if this raw mix is up to its full potential: Does it have the sound that the client is looking for? Does it have a good tonal and instrumental balance? Will it benefit from any processing of the many types we've discussed in this book, as invisibly as possible, or should it be left alone? We now turn to the monitor control. If the music was peak-normalized to full scale, and if it has good crest factor and good percussive quality, then it is likely to be sitting in the range from -8 (at an extreme) to 0 dB, and we can probably consider ourselves fortunate to have received a good, *open* sounding mix that can be taken to its full potential without overcooking. Not all music that monitors below -8 falls into the "squashed" or "overcompressed" category; remember that the acoustic advantage may put a peak-normalized soft instrument or ensemble (without percussion) way down on the monitor control.

The next step I call **The Mastering Engineer's Darts, or more aptly The Mastering Engineer's Dilemma. The picture (next page) combines monitor position, quality assessment, and year of release.** While a good mastering engineer already knows the full potential of a given mix, within the loudness race we have to assess the client's level expectations if they do not give us carte blanche to decide the optimal level for sonic integrity. There seems to be a clear division. Some clients are now so used to extreme distortion and lack of dynamics that anything other than a seriously-distorted master doesn't sound right to them! I don't have a problem with particular tastes or preferences for sound, but unless we start loudness normalizing, i.e. dropping the output level of compressed recordings, then hypercompressed and distorted masters end up as much as 6 dB "louder" than a master that is open, clear and has musical impact. A 6 dB difference paints us into an unprofessional corner. It is simply not technically possible to master a good-sounding acoustic jazz album that gives rise to -13 dB monitor gain. But even the most enlightened clients are reluctant to accept a master that's a full 5-6 dB lower than its current competition. So perhaps the best course of action is to ask them if they are looking for a master that is "as loud as possible without compromise." If the answer is "yes", we can have a good time in the studio! To further determine what level they are willing to accept, it helps to listen to some recordings that they like and discuss mutual reactions to their sound compared to their own mix.

Each recording has an optimum level; it is impossible to increase the intrinsic loudness past a certain point without adding distortion and losing depth, transients and dynamics. We can try to turn it into the sound of the current most distorted group (which the client may want to imitate, a sure sign of trouble), but it won't turn their *music* into someone else's. Long experience has taught me that trying to turn a mix into a totally different "sound" from its natural potential usually ruins the music. So we have to make the clients aware of potential compromises, and tell them if their raw mix is not capable of reaching the intrinsic loudness of the latest hit mixed in a multi-million dollar room. We could also remind them—again—that it is not necessary to be that "loud" in order to succeed.

However, if the mix is great, then we can preset our monitor to a given fixed position (hopefully no lower than -8 dB) and go to work. As the average level increases, more compression and peak limiting will be required to keep the medium from overloading and we will find the monitor control moving downward to keep from sounding too loud. The trick is to carefully A/B the mix and the master in the making at matched loudness, and try to make the ideal master, one that has all the virtues of the mix, none of the defects, and which does not degrade the mix quality. Especially for music with transients, the sound quality takes a big dive below about -8 dB monitor position, so if we find ourselves already there, we'll need to try a little less processing; even one dB lower level can make a big sonic improvement.

The loudness race has made it difficult for us to do a good job. Of course, the point where "damage" occurs is subjective and depends a lot on the music and the message. We just have to ensure that the compromise is as small as possible. It helps to point a client to successful recordings in their genre that have made the top of the charts with lower level than other competition. The Honor Roll at **digido.com** can help.

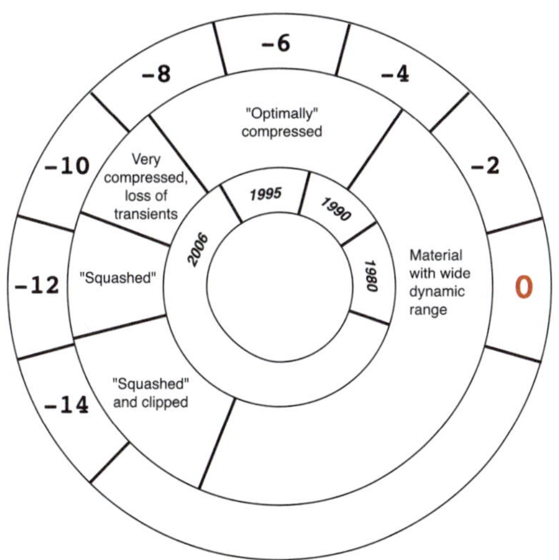


What target are you aiming for today?

## Mixing With The Calibrated Monitor

Mix engineers can also benefit from a calibrated monitor. Here are some tips:

- Using a higher monitor position during mixing encourages making a recording with good clean transients. For example, preset your monitor control from 0 dB to no lower than -6 dB, which will produce a recording that falls in line with the vast majority and has acceptable dynamic range for home and car listening.
- If the mix overloads the peak meter, the monitor is set too low; turn the mix faders down equally, the monitor control up the same amount, and keep on working.
- Never use bus processing during mixing that's

*"I want it as loud as everything else but I don't want to lose the dynamics or add any distortion."*

— CONTRIBUTED BY BRIAN LUCEY

designed specifically to add "loudness"; leave the loudness issues to the mastering session.

- Never use a peak limiter on a mix to "protect" metered level, because there is no reason to protect when you are already in charge of your own level.
- Use subtractive mixing. Instead of turning up the instrument you are concentrating on, look for other instruments that could be brought down or cheated down, especially when driving a bus compressor, which I recommend inserting (if at all) towards the end stage of the mixing process.
- Use only processing that helps achieve the **sound** you are going for. In rare cases this could be a peak limiter, but only if you need it to control **audibly**-disturbing peaks, and even then, peaks can be more invisibly handled with mix automation.

As Bob Olhsson advises:

> I've never heard a compressor or limiter that could beat the sound of manual gain riding in a mix. It's a LOT of work and many people don't have enough time or money but the results sound huge with just a little limiting on some of the peaks in the master.

We cannot restore quality in mastering that has been lost in mixing. An "open" mix produces a better-sounding master, as punch and impact come from microdynamics. You will still be able to be creative with compression and other effects—**a fixed monitor gain is liberating, not limiting**.

## III. The K-System: Integrates Dual-Scale Meters with The Calibrated Monitor

In an ideal world, all mastered programs would conform to a single monitor control position—0 dB and recordings would have a more consistent loudness and better sound quality. This setting (or gain) is currently used in the film world, home theatre, very dynamic symphonic works, and certain rare audiophile music recordings. We would simply lower the level of compressed material by 6 to 8 dB or more to conform with this monitor gain. As we now know, peak-normalization is the culprit for loudness variance, but until the system discourages that practice, we may have to accept a compromise approach that accepts a loudness variance of no more than, say, 8 dB.

### The Origin of the K-System Meters

In the 20th Century we concentrated on the *medium*. In the 21st Century, we should concentrate on the *message*. That's why we should avoid meters which have 0 dB at the top—which makes operators think that they have to peak to full scale. Many mastering engineers use a traditional analog VU meter. But because of the wide range of average levels on current pop CDs, they need a variable meter attenuator to prevent the VU from pinning or reading out of range.

Around 1994, I installed a pair of Dorrough meters which enable me to view the average and peak levels simultaneously on the same scale. These meters use a scale with 0 "average" (a quasi-VU characteristic we'll call **AVG**) placed at 14 dB below full digital scale, and full scale marked as +14 dB. Music mastering engineers often use this meter scale, since a typical stereo 1/2" 30 IPS analog tape has approximately 14 dB headroom above 0 VU.

The calibrated monitor is marked so that a position of 0 dB yields an SPL of 83 dB per speaker with a -20 dBFS RMS signal. But in music mastering, when we attenuate the monitor by 6 dB, it reduces the SPL for -20 dBFS to 77 dB per speaker. On the Dorrough Meter this falls at **-6 AVG**, so the **0 AVG** point falls at exactly 83 dB SPL (pictured at right), which means we're really running average SPLs similar to the theatre standard. Despite that, the perceived loudness is a little less than in the home theatre because we have 14 dB less headroom, and 6 dB of transients have been removed by compression and/or limiting. Our compressed pop presentation is probably about 2 dB less loud than a film even if both are presented at 83 dB SPL. On a flat meter, a brighter or more transient program sounds louder even if they measure the same on the flat **AVG** scale.

Concentrate on the average level. There is no need to normalize the peak to full scale. Regular peaks to full scale are not natural, they only occur with extremely processed recordings. But during the transition period away from peak-normalization, we will need more than one average level standard, and so I developed 3 different meter scales designed to be integrated with the calibrated monitor, called the K-System. Think of the K-System as a **coordinated attenuator for both the averaging meter and the monitor gain** making this **the first integrated metering and monitoring system**.

The K-System meter (see figure next page) shows two levels simultaneously. A bar represents the average level (true RMS, which is more accurate than a simple average) and a moving line or dot above the bar is the current instantaneous (1 sample) peak level. So we always know the crest factor (peak to average ratio). We try to set the average level on forte passages to around **0 dB, which is also a reference loudness.** If we see the average level of a highly percussive piece exceed 0, we might be using too much compression. There are three different K-System scales, with 0 dB at either 20, 14, or 12 dB below full scale. 0 dB on every meter represents 83 dB SPL per channel. These scales are called K-20, K-14 and K-12, and for **one channel** should correlate with monitor attenuations of 0 dB, -6 dB and -8 dB, respectively. But since two or more channels play more loudly, it's not uncommon to run 2-3 dB more monitor attenuation. So let's say that for stereo, a K-14 meter corresponds approximately with a monitor position of -6 to -8 dB. I have not designed a K-10 or less meter, because it would encourage hypercompression and move the loudness center away from the 70-year average. Even the presence of a K-12 meter encourages the peak-normalizing mindset. In this imperfect world, if we must run higher average levels to please a client, we should use the K-12 or K-14 meter and let the average level go into the red zone no more than necessary.[10]

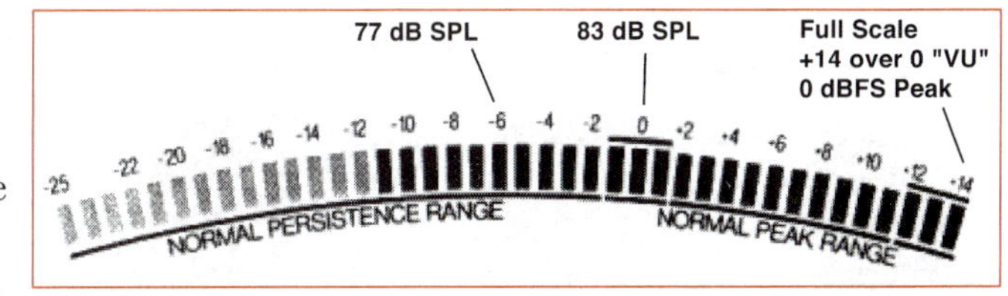


*The Dorrough Meter. With the monitor control's position set to 6 dB below the film reference, 77 dB SPL lands at -20 dBFS, or -6 AVG on the meter. Not by coincidence, this corresponds with 83 dB SPL at the meter's 0 AVG point, revealing the obvious correlation between a mastering engineer's meter ZERO AVG and 83 dB SPL.*

Note that full scale digital peak level is **always** at the top of each K-System meter; it does not change. The average level and the 83 dB SPL point slide relative to the maximum peak level.

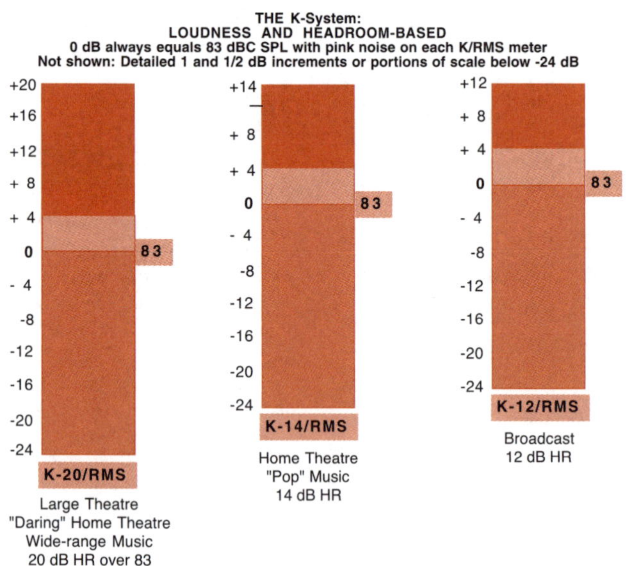


*K-System Meters. For a color image, please see the Color Plates section, Figure C14-1.*

### Peak and Average Calibrated the Same

The peak and average scales are calibrated as per IEC61606:1997, which means that they will read the same with a sine wave signal. Unfortunately, several DAWs do not follow the IEC standard, and their averaging meters read 3 dB too low. If in doubt, test the meter with a sine wave—the peak and average levels must read the same. Analog voltage level is not specified in the K-System so there is no conflict with using -18 dBFS in Europe as the analog reference point.

## IV. Using the K-System

### True Loudness Metering Still In Our Future

Several methods of measuring loudness exist, of varying accuracy (e.g. ISO 532, LEQ, Fletcher-Munson, Zwicker and others, some unpublished). The extendable K-System will accept a "true loudness" meter when a fast enough computer is available. The flat, true RMS meter is closest to the classic VU meter, which is notoriously inaccurate.[11] Operators know that the VU meter lies: bassy material sounds softer than the meter reading, and bright material sounds louder. So the ear must be the final judge, the meters should be a rough guide, the calibrated monitor is really the key. The best option is to adjust the calibrated monitor first to the recommended position, then choose the meter that corresponds.

When working in an unfamiliar room or if uncomfortable with depending on the calibrated monitor control, the meters become our main guide. Wide dynamic range material probably requires K-20, medium range (most pop work) K-14, and work for analog broadcast K-12. It will become a joy to banish hypercompression, and monitor control positions will be consistent within about an 8 dB range instead of the current unprofessional 16 dB.

**Using the Meter's Red (Fortissimo) Zone.** This 88-90 dB+ region is used in films for explosions and special effects. In music recording, naturally-recorded (uncompressed) large symphonic ensembles and big bands reach +3 to +4 dB on the average scale on the loudest *(fortissimo)* passages. Rock and electric pop music take advantage of this *loudest zone*, since climaxes, loud choruses and occasional peak moments sound incorrect if they only reach 0 dB *(forte)* on any K-System meter. If engineers find themselves using the red zone all the time, then the material is very likely going to be overcompressed. If a fortissimo is losing its impact in a compressed structure such as K-14 and especially K-12, pushing the level to +3 or +4 is not going to help, because we are working against mastering processing that softens transients. We should reduce the compression, peak limiting and/or lower the average level, which means moving closer to a K-20.

**Equal Loudness Contours.** Mastering engineers are more inclined to work with a constant monitor gain. But music mixing engineers often vary their monitor gain to check the mix at different SPLs. I recommend that mix engineers use calibrated monitor attenuators so they can easily return the monitor control to a standard position for the majority of the mix. This will result in more consistent mixes and more consistent tonality as well because the equal loudness contours tell us that a program will be bass-shy if the mix engineer monitors very loudly but the final consumer uses a lower level.

**How Low Can You Go?** You'd be surprised how low a level is tolerable, even in pop music, for a short period of time. As I mentioned in Chapter 9, I use a noisy room to evaluate whether a soft passage is not too soft for the consumer. Context is the key—a new track that begins softly after one that ends with a loud climax is far more problematic than a track that ends softly or has a soft passage in the middle. The chart on page 173 shows that the K-14 meter corresponds exactly with a living room environment, and that occasional average levels as low as -20 should be tolerable.

**Tracking/Mixing/Mastering.** For multitracking, the K-System will probably not be needed—a simple peak meter is sufficient. For mixing with the highest sound quality, I suggest using K-20 and save K-14 for the calibrated mastering suite. Using K-20 during mixing encourages a result that will help achieve loudness, all the way to the radio, even if converted to K-14 or K-12 during mastering. K-20 is also a bonus if you are thinking of selling the music to a motion picture; it will sound great in the theatre without need for any program compression.

**When the K-System is not available.** Analog mixing consoles equipped with VUs are far less of a problem than digital models with only peak meters. Using a test tone, calibrate the mixdown A/D gain to -20 dBFS at 0 VU (sine wave), and mix normally watching the console VUs.

**Adapting large theatre material to home use** may require a change of monitor gain and meter scale. Producers may choose to compress the original 6-channel master, or better, remix the entire program from the multitrack stems. With care, most of the virtues and impact of an original K-20 production can be maintained at K-14. We should try to fit this custom reduced-range mix on the DVD as it always sounds better than compressing the theatrical mix.

**Full scale peaks and SNR.** As we've discussed, it is not necessary to peak a 24-bit recording to full scale and there is no obligation in the K-System to peak a recording to full scale. A program's perceived signal-to-noise ratio is determined by its loudness, which sets the position of the listener's monitor level control and the monitor gain determines the loudness of the system noise. That's why the harpsichord recording can be reduced as much as 13 dB without worrying about signal-to-noise ratio. The listener will still set his monitor control to a normal position (shoot for -6 dB); he doesn't turn it up as long as the harpsichord recording sounds loud enough. Remember that even at 0 dB monitor position, system hiss is inaudible with

*{ "The K-system is not just a meter scale, it is an integrated system tied to monitoring gain." }*

*"A good K-20 mix works great in the motion picture theatre, without any additional compression."*

proper gain-staging, good D/A converters, modern preamps and power amplifiers.

As long as two similar music programs reach 0 on the K-system's average meter, even if one peaks to full scale and the other does not, both programs will have similar perceived SNR and loudness. Use the averaging meter and your ears and with K-20, even if the peaks don't hit the top, the mixdown is considered normal and ideal for mastering.

**Multipurpose Control Rooms.** When the calibrated monitor becomes universal, engineers can feel a lot more comfortable when working in a strange control room. Multipurpose production facilities will be able to work with wide-dynamic range productions (music, videos/films) one day, and mix pop music the next. A simultaneous meter scale and monitor gain change accomplishes the job. Operators should be trained to interpret the meaning of the calibrated monitor and log the gain (control position) used on the take sheets.

In ***Color Plate Figure C14-2*** is a picture of the K-20/RMS meter in close detail, with the calibration points. In ***Color Plate Figure C14-3*** is a picture of a K-14/RMS Meter in operation, as implemented by Metric Halo labs in the program **SpectraFoo™** for the Macintosh computer. SpectraFoo includes full K-System support and a calibrated RMS pink noise generator. The SpectraFoo meters also have the option of showing "peak RMS", which indicates the crest factor of the recording at its highest RMS level. Many other K-System meters have been developed, including: **Pinguin**; **RME** (whose Digicheck includes a reconstructing peak meter); **UAD** (in their limiter); **University of York**, **MRC Level Meter** plugin devised by Dave Malham; Spectre, by **Audiofile Engineering**; **Voxengo**; and **Wavelab 6**. The **DK** and **Dorrough** hardware meters nearly meet K-System guidelines but be sure to use an external RMS meter for calibration since they use a different type of averaging. In practice with program material, the difference between RMS and other meter averaging methods is imperceptible.

#### Multichannel

There's good news for audio quality: 5.1 surround sound. Current 5.1 mixes of popular music sound open, clear, beautiful, yet also have impact. Six speakers allow for much more headroom and sound output than two, so when working by the monitor gain, the channel meter levels tend to run a bit lower. What became clear while watching the K-20 meter is that the best masters are using the peak capability of the 5.1 system for headroom, the way it should be.

We hope that these techniques will result in a far more enjoyable and musical experience than we had at the end of the 20th century.

## V. Broadcast and DVD

Pre-production/mastering with the K-System will help by standardizing the levels and sound quality of sources before they reach broadcast or record distribution (CDs, DVDs, Digital Downloads).

### Analog Broadcast

Program material designed for standard analog television broadcast has special requirements, depending on the broadcaster. In order to keep older analog distribution systems from overloading, most analog broadcasters now require that video tapes arrive with average levels of -20 dBFS and maximum peak levels of -10 to -8 dBFS and will reject material that exceeds these levels. To master for analog broadcast, amplify the input to a K-12 meter by 10 dB, and use a peak limiter that guarantees no peaks above -10 or perhaps -8 dBFS.[12] Perhaps this practice will be liberalized as these systems are replaced with digital distribution and broadcast.

### Loudness Normalization in Digital Broadcast

The advent of HDTV and digital radio broadcasting is a big step forward. **There are two possible approaches digital broadcasters will use for loudness normalization after our program reaches the broadcast suite.** The first is to employ loudness normalization at the receiver end, using metadata (to be explained). The alternative is to employ loudness normalization at the transmit end, but instead of compressing, the broadcaster adjusts program levels via an intelligent loudness algorithm.

### Metadata—
### Loudness Normalization at the Receiver End

**Metadata** is *data within data*, that is, control data embedded in a file or in the subcode of the digital audio stream. This data can be used for many purposes, but the most attractive is **loudness normalization**. Readers should visit Dolby Labs website or consult a textbook on DVD authoring for setting metadata parameters such as **dialnorm**, **dynamic range control**, and **downmix coefficients**. Here we will concentrate on how dialnorm affects the esthetics of our listening experience:

**Dialnorm**, one of the metadata settings, was invented by Dolby Laboratories for use in AC3 (Dolby Digital) but has taken on wider use. Also known as *dialog normalization*, *volume normalization*, *loudness* or *replay gain*, it is meant for digital television and radio, DVDs (employing AC3 coding) and also in iTunes as a means of adjusting the loudness of programs to produce a more consistent listening experience. Program level is controlled at the decoder. The amount of attenuation that the decoder should employ is individually calculated for each program and carried as a command on the metadata word. At each program change, the receiver decodes the dialnorm control word and attenuates the level by the calculated amount, resulting in the "table radio in the kitchen" effect. In a somewhat unnatural manner (like analog radio), average levels of sports broadcasts, rock and roll, newscasts, commercials, quiet dramas, soap operas, and classical music all end up at the loudness of dialog. It's a rather strange effect, but it works well for casual listeners, and especially in the car and kitchen.

Dialnorm is a simple gain change, maintaining the crest factor and dynamic range of the studio mix and avoids the need for heavy compression used in analog radio and TV today. In theory, the program compressor will be banished from the broadcast chain, or at least its use will be greatly reduced.

Since DACs overload if gain is applied, dialnorm is an attenuation system. The standard level for spoken dialog is -31 dBFS (average) and the object is to make all programs sound the same loudness as dialog. The very "loudest" programs can be attenuated as much as 30 dB, and a program that is as soft as dialog would not be attenuated at all. Dialnorm values range from -1 to -31. To calculate the attenuation, add 31 to the dialnorm value of the program. For example, normal spoken word would be assigned a dialnorm value of -31. Adding 31 results in an attenuation of 0 dB. Presumably a heavy metal producer would assign his program a dialnorm value of -1, resulting in an attenuation of 30 dB. But many producers try to beat (abuse) the system and assign music for DVD a dialnorm value of -31 dB to "attract attention." However, not only does this make the music sound much too loud, it also affects the proper operation of other metadata such as dynamic range control. This is why broadcasters are working on a foolproof system of assigning dialnorm via an automatic loudness meter.

### The ID3 Tag in Soundfiles

The ID3 tag is incorporated in sound file formats including mp3 and FLAC, and includes replay gain, so any intelligent music server can read the metadata and adjust relative loudness of each song. **Slim Devices** consumer music servers, the Squeezebox and the Transporter, can be set to use or ignore the replay gain.

**iTunes implementation of ID3 tag** (called **volume adjustment** but really means gain adjustment) is a potential step forward but a consumer-based approach that does not meet professional standards. Dialnorm should be program-based, and with music, a program is an album. We don't want to make the ballads as loud as the rockers, unless we're playing music casually, or perhaps at a party, or on the road. Since iTunes is song-based, manually entering "volume adjustment" for an album is highly inconvenient, so it should be set up on a per album basis. Moreover, iTunes' system allows gain as well as attenuation, which is a foolish idea as most peak-normalized CDs will automatically distort the DAC when gain is applied. The fader has no fine increments, no marked steps, and is very unergonomic to adjust. So only the most dedicated consumers will bother to preset the levels for each tune.

Apple's **Soundcheck** is much more useful as it scans all the songs in the computer, assigns them individual "volume" adjustments and to my ears appears to apply a compressor. While not perfect, it's a big step forward and would help many situations that are anathema to mastering engineers, such as background listening, cars, record store kiosks/ceiling speakers, changers, and parties. I'd love to see Soundcheck employed at A&R listening rooms and radio station PD's offices. As I said before, things are looking up; iTunes is popular enough and all it will take is a bit of enlightenment to reduce the pressure for the loudness race.

*If properly applied, dialnorm has potential*. But in addition to issues mentioned above, there are questions whether the consumer equipment can be made smart enough to use broadcast and media metadata properly. Consumers are often presented with confusing menus, especially in the setup parameters of DVD players, and multi-branded digital radios and TVs, not all of which may be implemented properly.

### Loudness Normalization at the Transmit Side

Many broadcasters are concerned that metadata will not achieve a consistent listening experience. They want to guarantee control over program levels when broadcast—in other words, *what you hear is what you get*. This is the approach of ITU specification BS.1770, loudness normalization at the transmit end. In this system, the peak level is **never** normalized to the top of the scale. When program material is ingested into the central broadcast server, an intelligent algorithm calculates the program's "center of gravity" based on an analysis of complex types of programs which can include music, dialog, and effects. Referring to the Dynamic Range Tolerance chart (page 173), the target dynamic range is the third bar, labeled "Living Room", suitable for HDTV. This corresponds to a K-14 lowered so that the average level is at -20 dBFS; this K-14 material must be auditioned at K-20 monitor gains. Therefore, programs we master that already conform with K-14 will be broadcast with no sonic alteration other than a digital level shift.

If the incoming material has too much dynamic range for the target (e.g. cinema material or widerange classical music), broadcasters will perform program compression/limiting on ingest to the server, reducing it to the characteristics of the third bar. Material to the right of the third bar, material which is overcompressed, or material designated for "lower class" platforms, will have its average level lowered to the same as all the others. Some intelligent upward expansion on ingest might help this material if it is to be played on HDTV. The end result is a single monitor control setting of 0 dB for all venues and programs, for serious (not background) radio and TV listeners.

The pitfalls of this system is that it may fall apart if digital meter readers see the extra 6 dB or more of headroom as an opportunity to make their program "louder" than the competition. The meters which they use should not display the peak level, a simple overload light is ergonomically sufficient. Whether broadcasters settle on receive or transmit loudness normalization, it's a bumpy road ahead.

### DVDs

As with the CD, a loudness race has already begun with the DVD and since almost everyone cheats on the metadata, and there is no broadcaster to control the level, the future of good sound quality on DVD is somewhat bleak. All we can do is hope that music and video servers will replace dumb CD and DVD players; that an intelligent form of loudness normalization without compression will develop for home-based servers. Then, and only then, will program producers lose the motivation to hypercompress.

### HD-DVD and Blu-Ray

It's extremely disappointing that the designers of these media did not institute a loudness-normalized system. They are doomed to repeat the devastating loudness race of CDs all over again.

{ *"A fixed monitor gain is liberating, not limiting."* }

## VI. In Summary

The designers of the compact disc never anticipated that an all-digital recording system would accelerate an alarming loudness race (worse than ever took place in the days of the LP) and create seriously distorted music. The K-System provides audio engineers with a common language of loudness, integrating monitoring and metering to help produce more consistent and cleaner masters for CD, DVD, Home Theatre, digital download, and broadcast.

1 I see an interesting analogy of the loudness race and the migration of pitch since the 16th century. Musical pitch seems to be just a little more sharp than the previous generation, so that an A played on an instrument tuned to previous standards is now the G or G# of today, which ultimately produces a problem of transposition. Unfortunately, audio systems cannot accommodate an infinite loudness rise.

2 There, I said it, **volume control!** When referring to a consumer piece of gear, I will occasionally use the colloquial word *volume control*, though we know that there is no formal audio definition of the word *volume*. The jukebox's volume control was hidden on the back. See Chapter 5.

3 This is an approximation. The average level of the LP did not remain steady, it did go up by about 4 dB between 1950 and 1985, because of mechanical and electrical improvements and the efforts by mastering engineers to cut hotter LPs. The left hand bar would not have existed in analog tape, since the system only had about 14 dB headroom at max. In addition, we must make allowances of perhaps 3 dB since the VU meter is not an accurate loudness meter. But the general slopes of these curves remain the same. In the analog days, even those recordings which were peak-normalized did not differ as much because the analog PPM is much slower than the digital peak meter.

4 This figure also illustrates how complete albums must be mastered. The song which is most dynamic and requires the greatest crest factor determines the maximum intrinsic level of the album. For example, if one song is very "open", represented by the left-hand bar, then the rest should not be raised above its level, unless intended. We can cheat, try to raise the "open" song with peak limiting, but only if we can tolerate the sonic degradation.

5 Visit the articles at digido.com for a video demonstration of the loudness race.

6 The late Gabe Wiener's classical recordings noted the SPL of a short passage, encouraging listeners to reproduce the "natural" SPL of the ensemble. I used to second-guess Wiener by adjusting monitor gain by ear, then checking against his listed level. Each time, my monitor gain was within 1 dB of Wiener's recommendation.

7 Composers have a similar dilemma with the acoustic advantage. They mark chamber music scores with the same dynamic markings as symphonies, but a forte mark in the symphony comes out much louder. But it sounds wrong to reproduce the chamber group at the same loudness as the symphony.

8 Consumers jump when playing a hypercompressed soundtrack CD after the theatrical release DVD. Reportedly, the soundtrack CD of *Yellow Submarine* is squashed and "loud" compared to the music on the theatrical DVD.

9 The analog voltage output levels of components were long ago standardized based on some standard modulation. Since the loudness wars, the average level of compact discs has been going up so much that the standard by which car manufacturers align their radio, CD and cassette levels has gone with the wind. This ensures that the current crop of crushed CDs will sound much louder in the car than the radio at the same position of the volume control.

10 I invented these K-(N) terms because there has been no shorthand way to describe crest factor and intrinsic loudness. The non-commercial K-System is available at no charge to developers who wish to produce meters.

11 One of my first lessons in VU meter inaccuracy was in 1972, with William Pierce, voice of the Boston Symphony, speaking clearly and distinctly in Channel 24's noisy control room, yet hardly moving the needle. Operators have to learn how to interpret this measuring instrument.

12 Alan Silverman explains why analog broadcasters are following this strange practice:

> It comes from the original adoption by SONY of -20 dBFS as the "0VU" audio level for digital video combined with the legacy 12dB headroom for the typical analog VTR. You're talking about exchange of material between an analog VTR and a digital VTR and the poor headroom design of the analog VTR. I guess that MTV is still using analog VTRs in some part of the chain. So really, the MTV requirement should be expressed as follows: When working with a digital machine and a VU-meter, set the VU for -20 dBFS with sine wave, and do not let peaks exceed 12 dB above 0 VU = -8 dBFS. In addition, if you lay back audio that peaks much above -8dBFS to a D1, D2 or Digi-Beta deck, the analog audio output of the deck will be driven far into distortion.

# COLOR PLATES

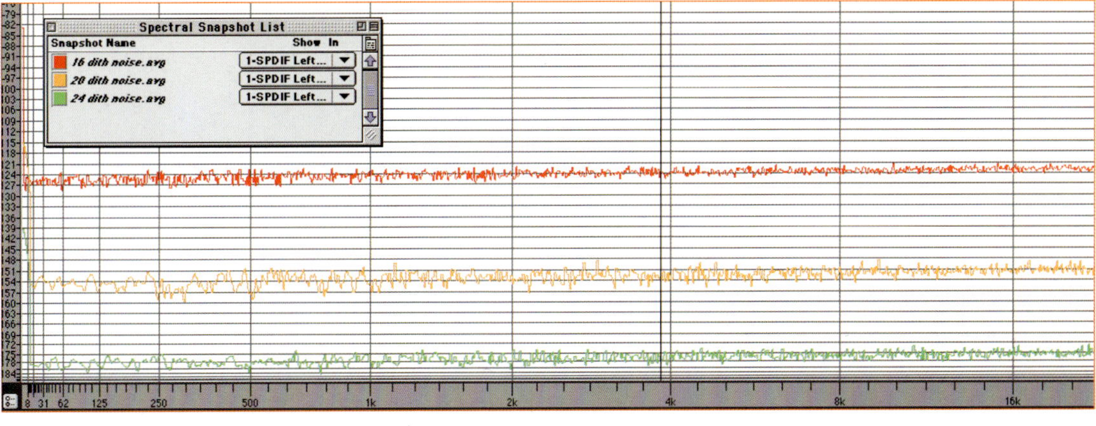


***Figure C4-1:*** *Comparing 16, 20, and 24 bit flat-dithered noise floors (red, orange, green traces, respectively). Though the individual bins of the 16-bit dither measure approx. -127 dBFS, their RMS sum is approx. -91 dBFS.*

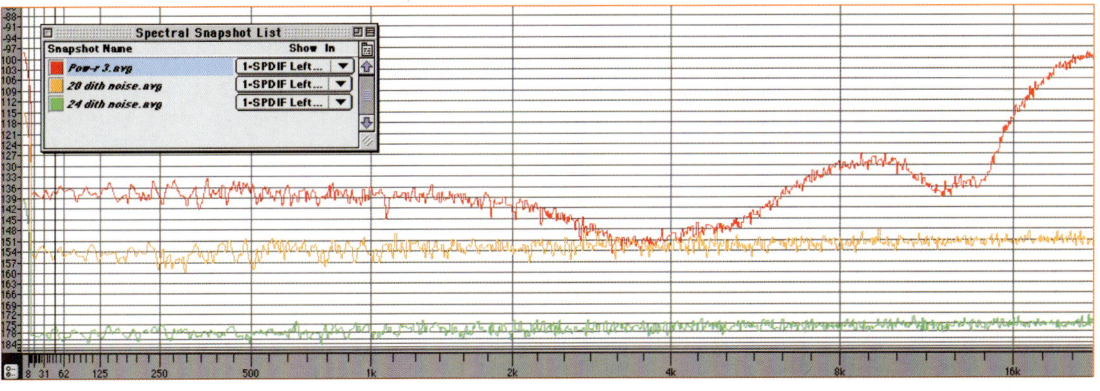


***Figure C4-2:*** *POW-R type 3 at 44.1 kHz/16-bit (red trace), with 20-bit flat dither (orange) and 24-bit flat dither (green) for reference.*

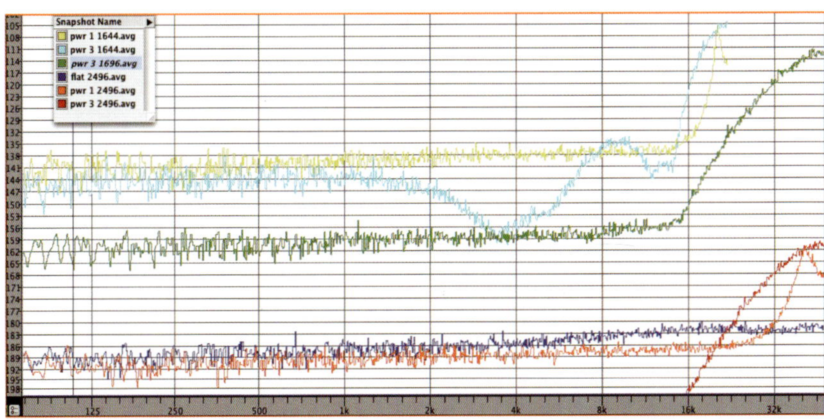


***Figure C4-3:*** *Dithers from highest to lowest level. At 16-bit/44.1 kHz: POW-R 1 (yellow) and POW-R 3 (turquoise). At 16-bit/96 kHz: POW-R 3 (green). At 24-bit/96 kHz: flat dither (blue), POW-R 1 (orange) and POW-R 3 (red).*

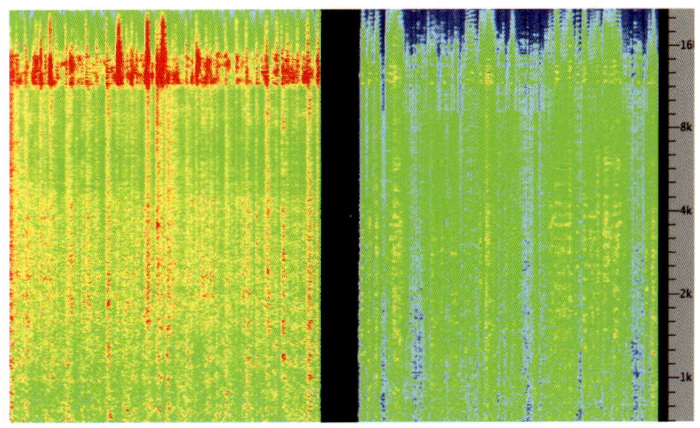


***Figure C5-1:*** *A spectragram of the remnant distortion from an mp3 conversion of two different CDs. At left, a clipped and hypercompressed CD, at right, a hot CD made with brickwall limiting and without clipping.*

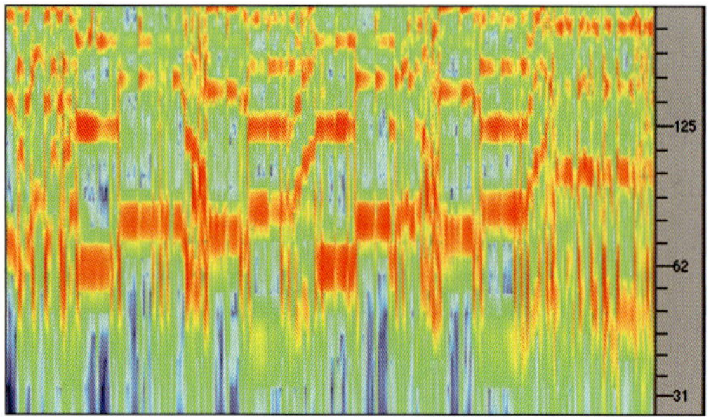

**Figure C8-1**: *SpectraFoo™ spectrogram of the bass frequencies of several measures from a rock piece . Read it like an orchestra score, time runs from left to right. Red represents the highest levels. The bass runs in the 62-125 Hz fundamental range are paralleled by second and third harmonics.*

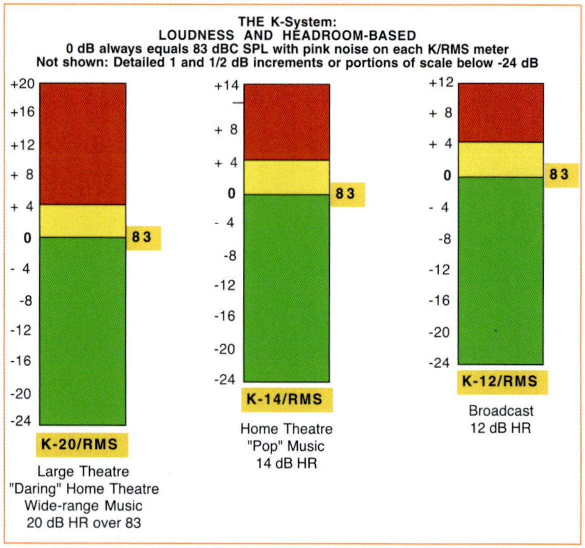

**Figure C14-1**: *The three K-System meter scales are named K-20, K-14, and K-12. The K-20 meter is intended for wide dynamic range material, e.g., large theatre mixes, "daring home theatre" mixes, audiophile music, classical (symphonic) music, "audiophile" pop music mixed in 5.1 surround, and so on. The K-14 meter is for the vast majority of moderately-compressed high-fidelity productions intended for home listening (e.g. some home theatre, pop, folk, and rock music). And the K-12 meter is for productions to be dedicated for broadcast.*

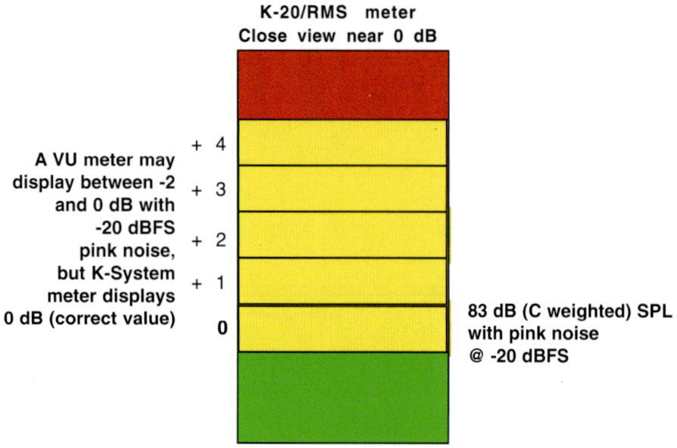

**Figure C14-2**: *A K-20/RMS meter in close detail, with the calibration points.*

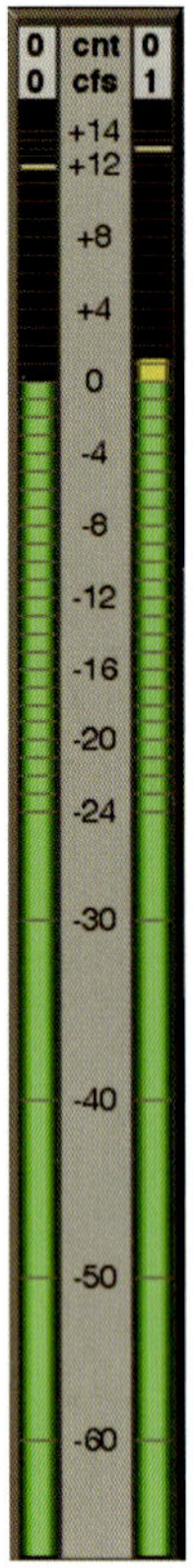

**Figure C14-3**: *A K-14/RMS Meter as implemented in Spectrafoo*

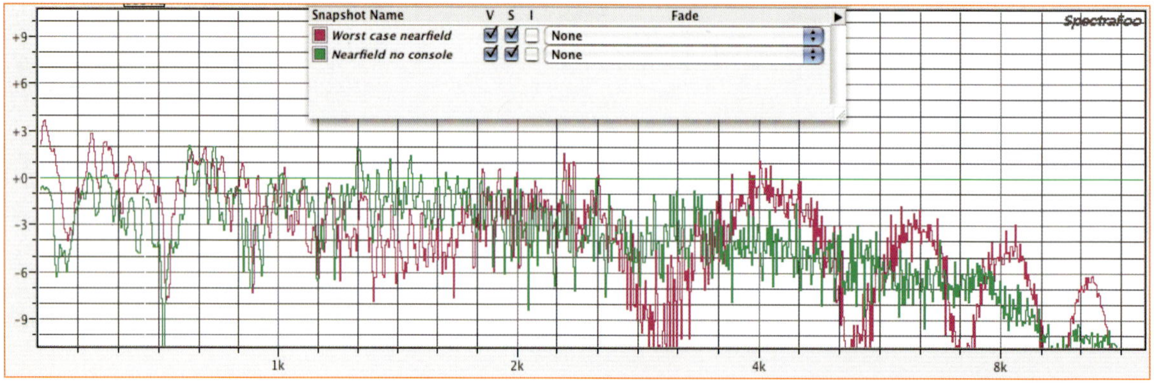

**Figure C15-1:** *Frequency response of a simulated audio desk at nearfield position in purple versus removing the desk, plotted in green.*

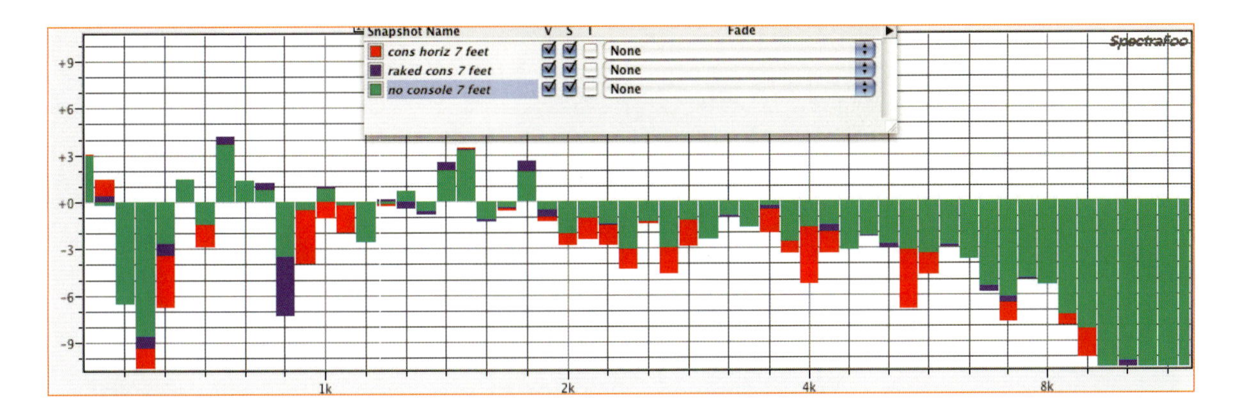


**Figure C15-2:** *Frequency response of a simulated audio desk, where in green is a measurement without a desk, in red is the desk that's parallel to the floor, and in blue the improvement by simply tilting the desk so the back is about 4 inches higher than the front.*

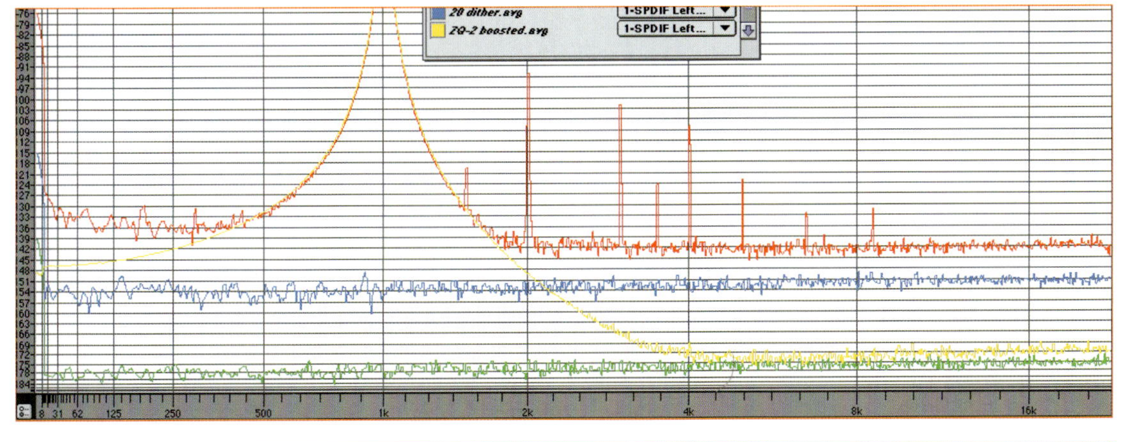


**Figure C17-1:** *Distortion and noise performance of Millennia Media NSEQ-2 analog equalizer in tube mode (red), 20-bit random noise floor for reference (blue), 24-bit noise floor (green), and Z-Systems ZQ-2 digital equalizer (yellow).*

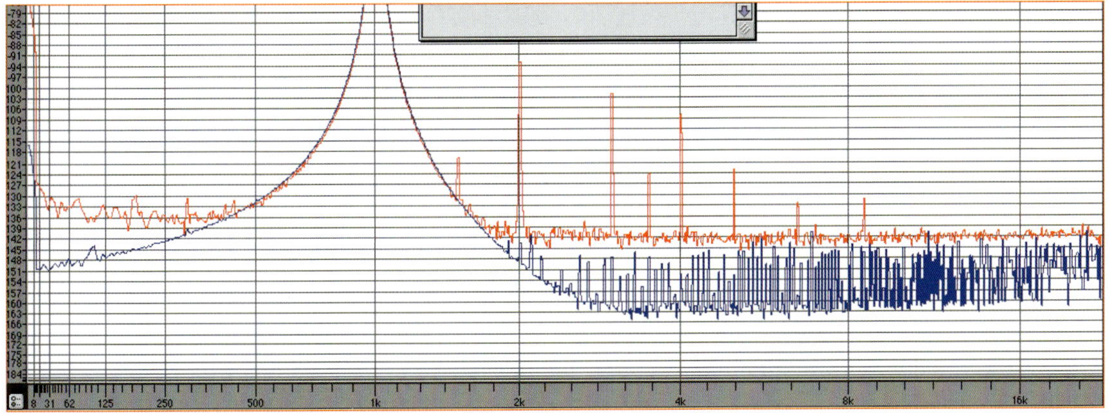


**Figure C17-2:** *Distortion and noise performance of analog Millennia Media NSEQ-2 (red trace), versus Digital Z-Systems set to truncate at 20 bits, no dither (blue trace).*

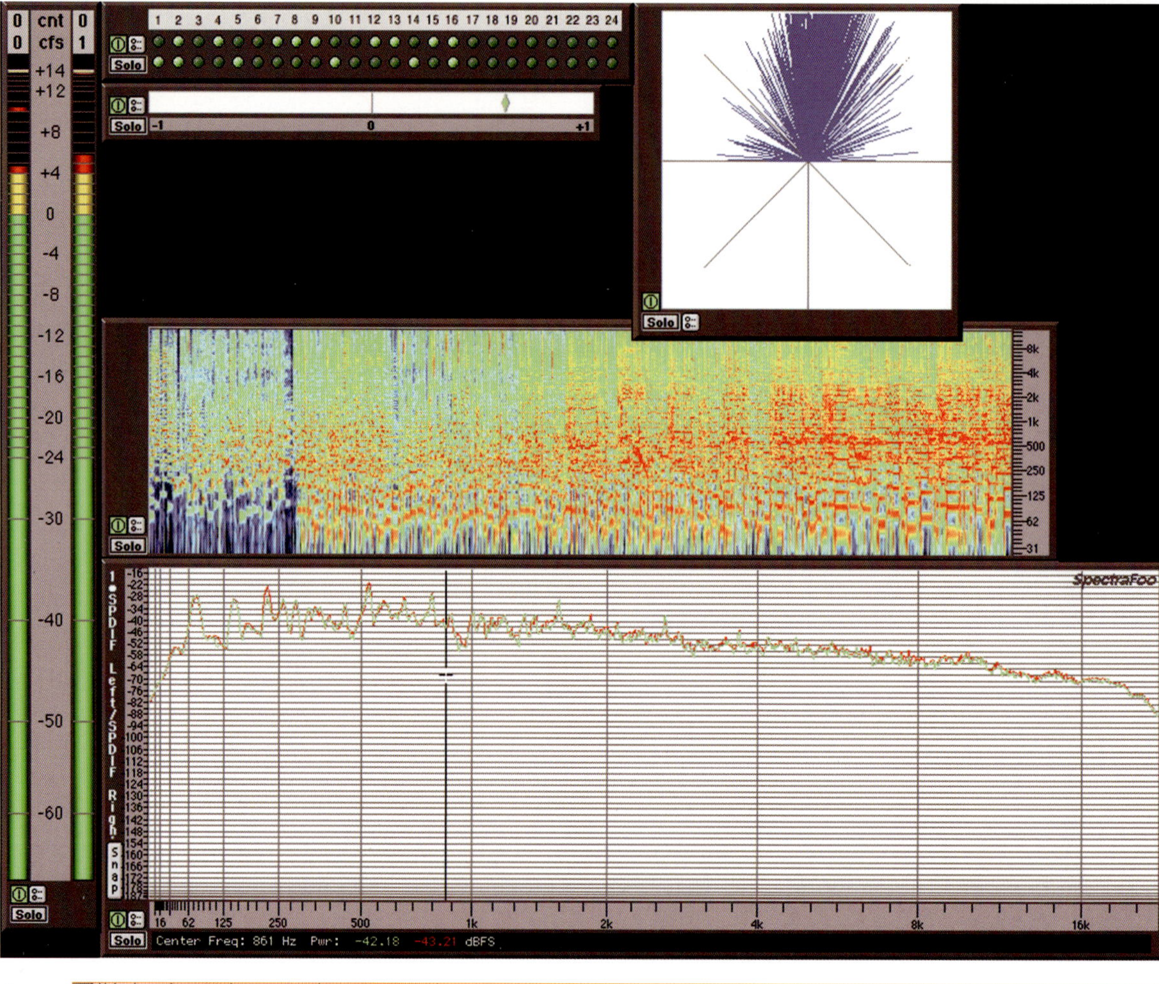


***Figure C17-3:*** *SpectraFoo during a moment of musical action. From left to right at top: K-14 Meter, bitscope, and stereo position indicator. Directly below the bitscope is a phase/correlation meter. In the middle of the screen is a Spectragram, quiet section at left part, then the song begins. At the bottom, the Spectragraph, left channel in green, right channel in red.*

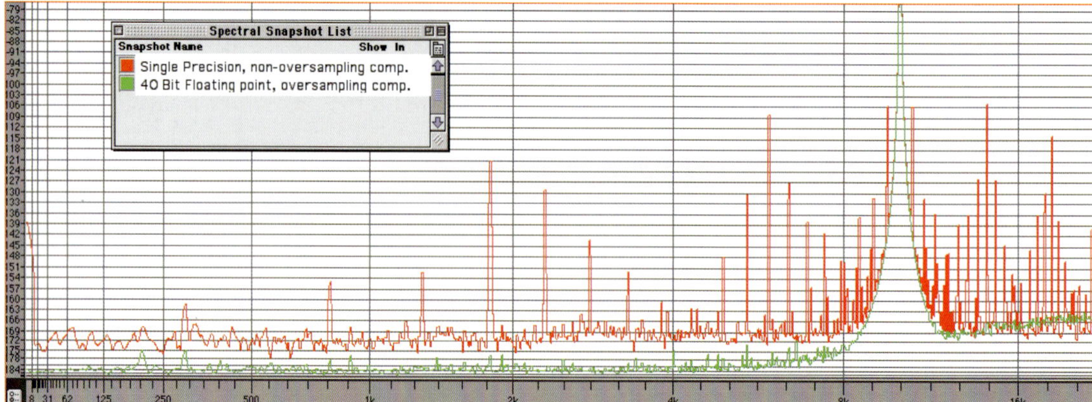


***Figure C17-5:*** *Comparing two digital compressors, both into 5 dB of compression with a 10 kHz signal. Red trace: Single Precision, non-oversampling. Green: 40-bit floating point, double-sampling and dithered to 24-bit fixed level.*

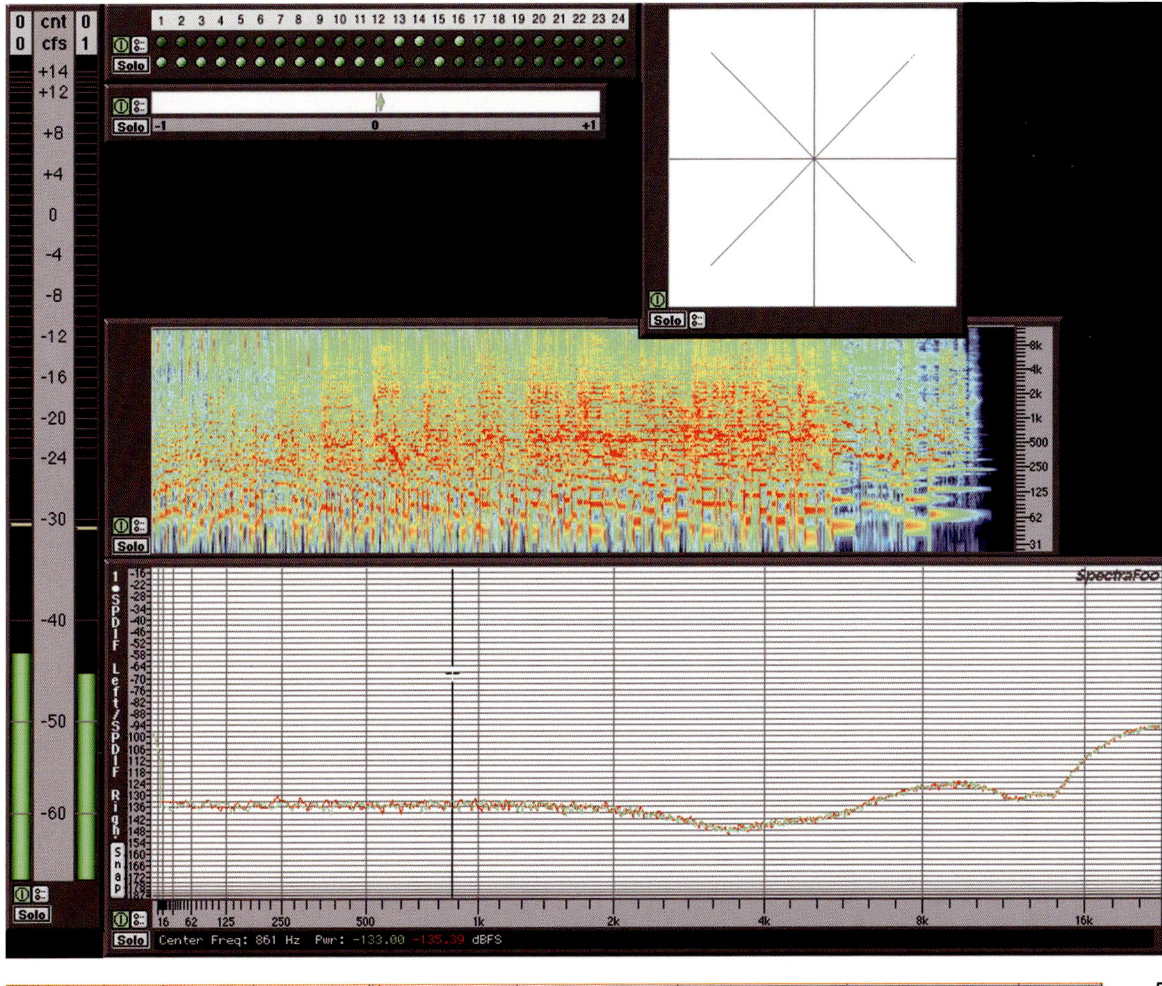


***Figure C17-4***: *SpectraFoo during a pause in the music. Only the bottom four bits are toggling on the bitscope, and the characteristic curve of POW-R dither type 3 is revealed on the Spectragram. The last notes of the music "fading to black" can be seen at the right of the timeline on the Spectragraph.*

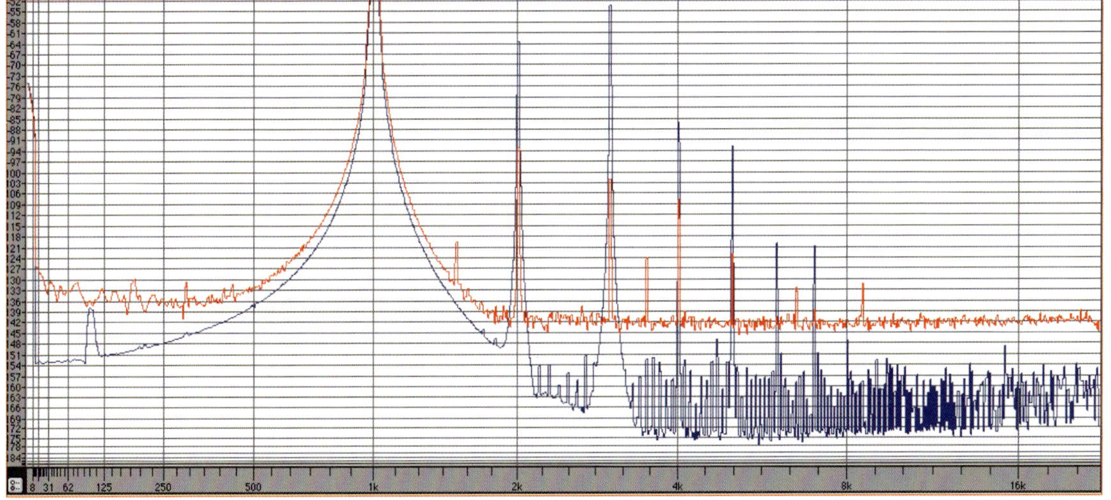


***Figure C17-6:*** *Comparing Cranesong HEDD-192 digital analog simulator (blue trace) to NSEQ (red).*

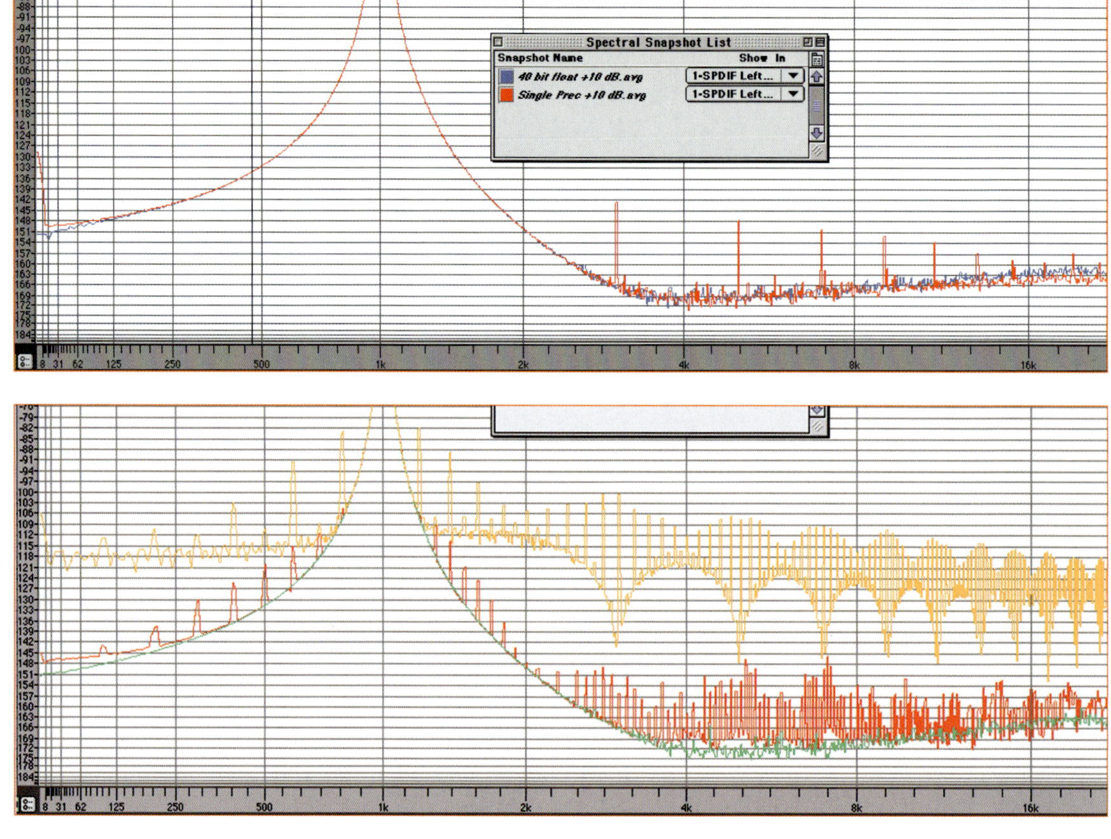


***Figure C17-7:*** *A simple 10 dB boost applied in two different types of processors. In red, a single-precision processor, whose distortion is the result of truncation of all products below the 24th bit. And in blue, the output of a 40-bit floating point processor which dithers its output to 24 bits.*

***Figure C17-8:*** *Compares two excellent-sounding digital dynamics processors, the oversampling Weiss DS1-MK2 (green trace), which uses 40-bit floating point calculations, and the standard-sampling Waves L2 (red), which uses 48-bit fixed point. The switchable safety limiter of the Weiss, which is not oversampled, is shown in light orange.*

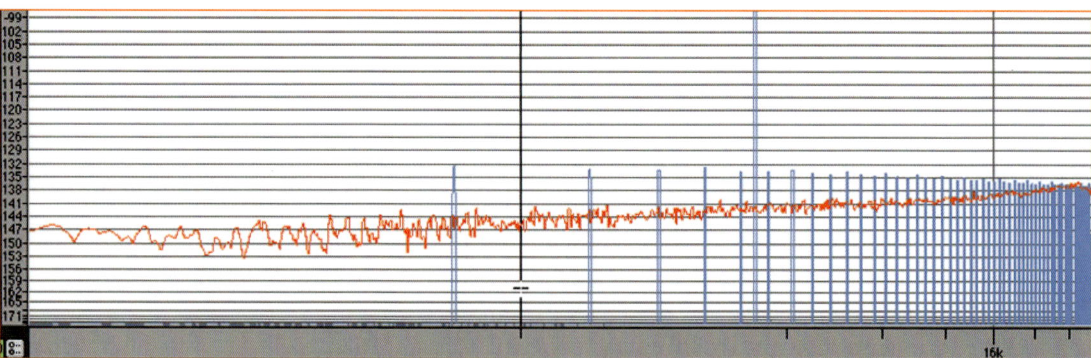


***Figure C21-1:***

*Jitter testing: 16-bit J-Test signal (blue trace) overlayed with the Noise floor of UltraAnalog A/D converter (red trace) which together define the limits of resolution of my jitter test system.*

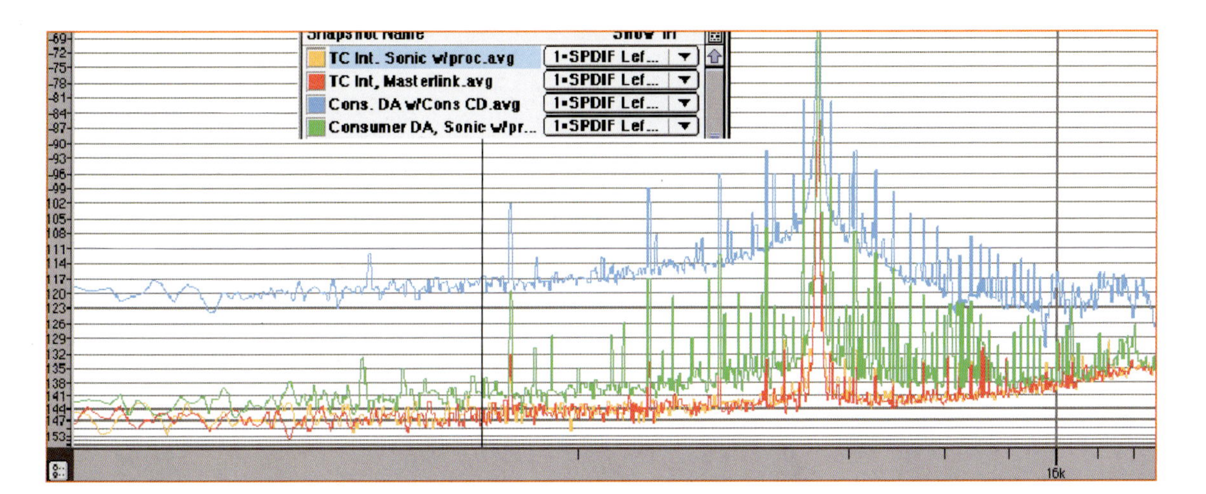


***Figure C21-2:***
*Jitter measurements with J-Test signal:*
*Light Orange Trace: TC DAC jitter on internal sync, fed from Sonic Solutions.*
*Red: TC DAC jitter on internal sync, fed from Masterlink.*
*Blue: Consumer DAC fed from consumer CD Player.*
*Green: Consumer DAC fed from Sonic Solutions.*

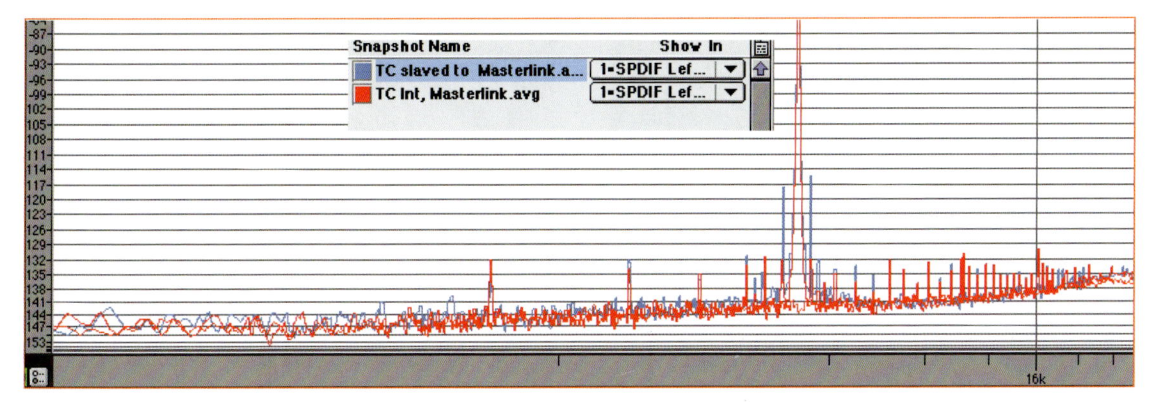


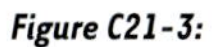

***Figure C21-3:***
*Jitter measurements, demonstrating how different clocking methods may produce different sound with the same source transport.*
*Masterlink transport feeding J-Test Signal to TC D/A.*
*Blue: TC D/A slaved to Masterlink transport via AES/EBU.*
*Red: TC D/A on internal sync.*

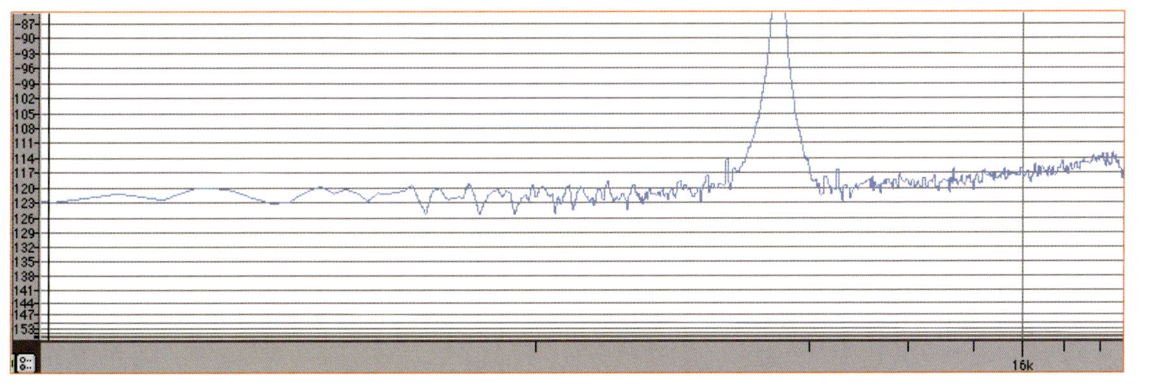


***Figure C21-4:***
*Jitter Measurements:*
*J-Test signal feeding Weiss DAC on AES/EBU sync*

***Digital Domain Studio A**. In front of the listening couch: Rolling rack with two Cranesong Trakkers, Weiss DNA-1 Noise Reduction Processor with K-Stereo, Weiss EQ-1 LP Equalizer, Pendulum OCL-2 Compressor, and two Weiss DS1-MK2 dynamics processors. In back, Lipinski L-707 loudspeakers on sand-filled stands, Pass X-250 Power Amplifier, JL Audio Fathom f112 subwoofers, Mackie Control Universal, Real Traps Mondo Traps. At the couch, Cranesong Avocet Monitor Controller, and the TC Electronic Icon Controller for System 6000.*

CHAPTER 15

# Monitor Setup

## I. Why 83 dB?

### The Magic of 83 with Film Mixes

In 1983, as workshops chairman of the AES Convention, I invited Tomlinson Holman of Lucasfilm to demonstrate the sound techniques used in creating the Star Wars films. Dolby systems engineers labored for two days to calibrate the reproduction system in New York's flagship Ziegfeld Theater. Over 1000 convention attendees filled the theater center section. At the end of the demonstration, Tom asked for a show of hands. "How many of you thought the sound was too loud?" About four hands were raised. "How many thought it was too soft?" No hands. "How many thought it was just right?" At least 996 audio engineers raised their hands. This is testimony to the principle of calibrated monitor gain. Wouldn't it be nice if popular music playback was as consistent as motion picture sound?

The film world's choice of 83 dB SPL has stood the test of time, as it permits wide dynamic range recordings with little or no perceived system noise when recording to magnetic film or high-resolution digital. 83 dB also lands on the most effective point of the Fletcher-Munson equal loudness curve, where the ear's frequency response is most linear. When digital technology reached the large theater, the SMPTE attached the SPL calibration to a point 20 dB below full scale digital instead of 0 VU. In the appendix I'll explain how *85 dB became 83 for a while but then reverted officially to 85*, and why for music production I chose to stick with 83.

## II. Setting Up and Calibrating the System

### Summary of Essential Tools

Now that we know the benefits of having a calibrated monitor, let's enumerate the tools we need to construct a mastering-quality, calibrated monitor system.

- A great room, with proper dimensions, wall construction, layout, and interior treatment. There should be minimal obstructions/reflections between the loudspeakers and the listener, with low noise and good isolation from the outside world.

- For surround sound, five matched full range or "satellite" loudspeakers and amplifiers with flat frequency response (preferably good down to 60 Hz), each capable of producing at least 103 dB SPL with less than 1% distortion. As we said in Chapter 6, high headroom monitors are necessary to make proper sound judgments: if our monitors are compressing, we cannot judge how much compression to use in the recording.

- One, preferably two, subwoofers, capable of extending the low frequency response of all the satellites down to at least 25 Hz, and producing at least 113 dB SPL at low frequencies with (ideally) less than 1% distortion, but some authorities feel that up to 3% THD is acceptable for a subwoofer.

- A low distortion monitor matrix with versatile and flexible bass management, capable of repeatable, calibrated monitor gains, down mixing and comparing sources from 7.2 through mono.

- A monitor selector to feed the matrix, with both digital and analog inputs.

- Measurement/calibration equipment:

**For initial alignment of the room, the most critical ingredient is: knowledge. A trained acoustician should help with the first-time setup.** He will examine the dimensions and construction of the room and recommend loudspeaker placement and trapping, if necessary. Once the room has been set up, he will perform near-anechoic and early-reflection analysis, adjust subwoofer levels and crossovers, and help solve room response errors. If the speakers are free-standing (as opposed to soffit-mounted), he will help find an optimum position that results in the flattest frequency response and best stereo imaging.

**For level calibration:**
**Preferable:** A calibrated 1 octave real time analyzer (RTA) and microphone.
**Alternate (less accurate):** A high quality SPL meter with calibrated microphone, selectable filters and response speed.

**Test Signals**: For level calibration, if using a SPL meter, RMS-calibrated sources of **band-limited pink noise, 500 Hz to 2 kHz** should be used. If a 1 octave analyzer is available, then it is possible to use wide-band RMS-calibrated pink noise.

**Placing the Main Loudspeakers and the Listener**

The ideal reproduction system should have no obstacles in the path between the loudspeakers and our ears. The side walls should be far enough apart to have no effect on the response of the loudspeakers and if necessary, get treatment to reduce reflections to the listener.

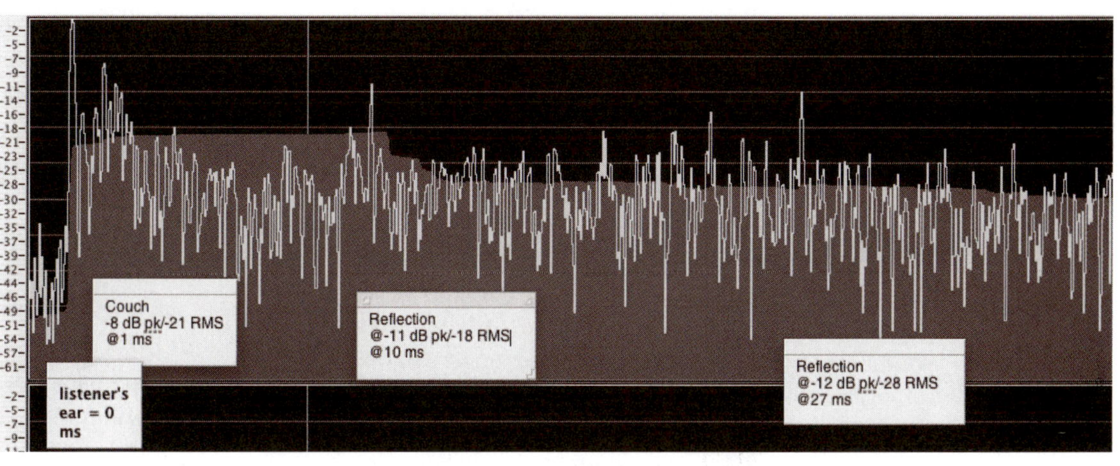


A trained acoustician can test for the effect of obstacles in the studio environment. Reflections within the first 15 to 20 milliseconds after the direct sound hits the ears, should be below about -15 dB. This is called a *reflection-free-zone (RFZ)*. Above is Spectrafoo's level vs. time display for Studio A, showing measurements of reflections in the environment, both RMS (in the red area) and peak. Studio A has very few obstacles between the loudspeaker and listener other than the carpeted floor and a small rack. If the peak level of the specular reflection visible at 10 ms (~5 feet from the mike) becomes a problem, that object should be removed or covered with absorbent foam. The display does not determine the location of a given reflection, so when in doubt, cover a suspect surface with an absorber, and remeasure to see if the reflection disappears. Since the reflection from the couch at 1 ms (6 inches from the mike) is behind the listener's ear and covered by his body, it should not be a perceptual problem.

Time-domain response translates to frequency response. While constructing our Studio B, I built a simulated audio desk out of plywood to measure its effect on frequency response in different positions. Pictured below is the fake desk in nearfield position, which we measured just for fun.

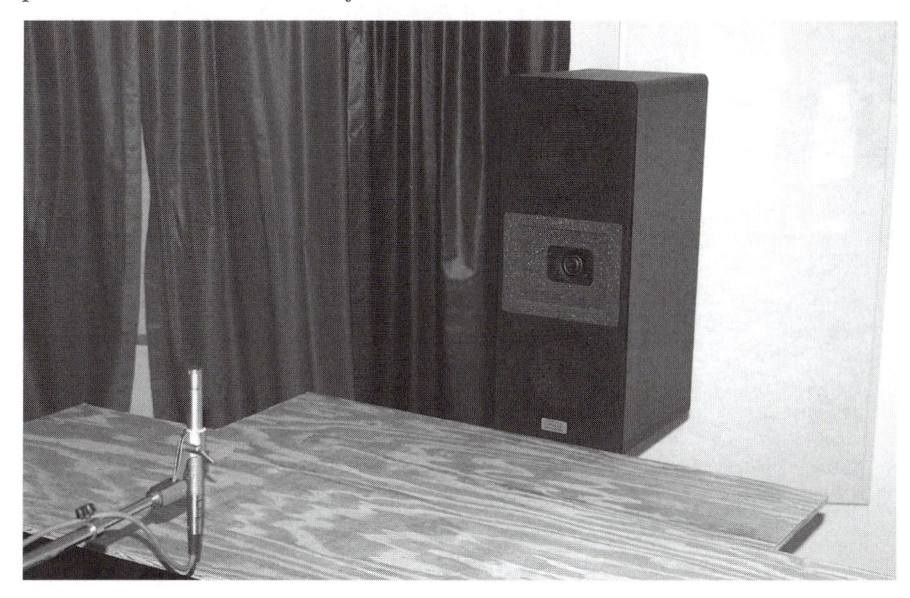

Nearfield position produced the frequency response in purple in ***Figure C15-1*** in the color plates, which should discourage engineers from this practice. Removing the desk produced the response in green.

Lastly, the simulated desk was placed in its proposed position, the loudspeakers about 9 feet away, tweeters at ear height, and its height set as low as possible, just above the listener's seated knees. An angled, low desk surface helps to direct the first reflection into the listener's chest instead of his ears, and reduces comb filtering. We then took three comparative measurements (see ***Figure C15-2*** in the color plates). The response in green is without a desk, in red with a desk that's parallel to the floor, and the improvement (in blue) by simply tilting the desk so the back is about 4" higher than the front. Tilting helps nearly every frequency range except around 800 Hz, but overall, the improvement is significant. Our other choices are much less practical:

- No desk at all
- Stand up to do the mastering
- Use a smaller desk
- Cover some of the desk with absorption or
- Put some large, irregular objects on the desk to enhance diffusion and reduce specular reflections

The next decision involves finding the optimum distance from the walls for the loudspeakers and listener, based on calculations of the standing waves in the room, which again is best left to an experienced acoustician. However, once the optimum distances have been determined, we can adjust the angles and precisely match the loudspeaker distance. With the room interior acoustics reasonably treated, place the loudspeakers and listener at their proposed locations and do a preliminary listen to some stereo material. If we hear a nice image and depth, then this is indeed a good starting point. Now, we drop a plumbline[1] from the tip of our nose to the floor, make a mark, measure back 5 or 6 inches for the location of the ears, and place a real mark on the floor. We reconfirm that this mark is exactly centered between the two side walls, then put a nail in the floor or carpet at this spot. All angles and distances will be measured from this point.

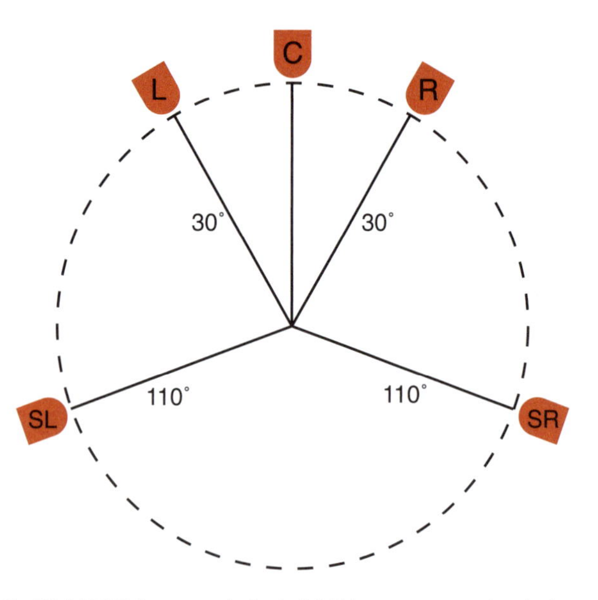


*The ITU-R BS.775-1 recommendation for 5.1. Tolerance on surround angles is between 100-120°.*

### Lasers and Time Delays

Our goal is to adjust the loudspeaker angles and precisely match the distances to conform with the ITU 775 recommendation[2] (previous page), or a variance that we may prefer (see Chapter 19).

The ITU 5.1 setup, (pronounced "five point one", five main channels plus LFE) is a circle, with each loudspeaker at identical distance from the listener. ITU's recommended surround speaker angle is a compromise between good localization or good ambience. Surrounds at 90° at an elevated height would produce an excellent sense of ambience, but 110° at ear height produces better localization. The ITU recommends between 100-120°. Rest the indent of a laser chalk on the nail in the floor. Shine a laser line along the floor to the front wall. Rotate the laser line until it is precisely perpendicular to the front wall, then put a narrow piece of tape on the front wall at that center spot to sight the laser line in the future. Now place the center of a surround speaker alignment chart[3] over the nail and place the laser chalk over it as in this photo (right). With the laser line pointing at the center tape on the front wall, adjust the chart until it's centered and tape it down so it doesn't rotate.

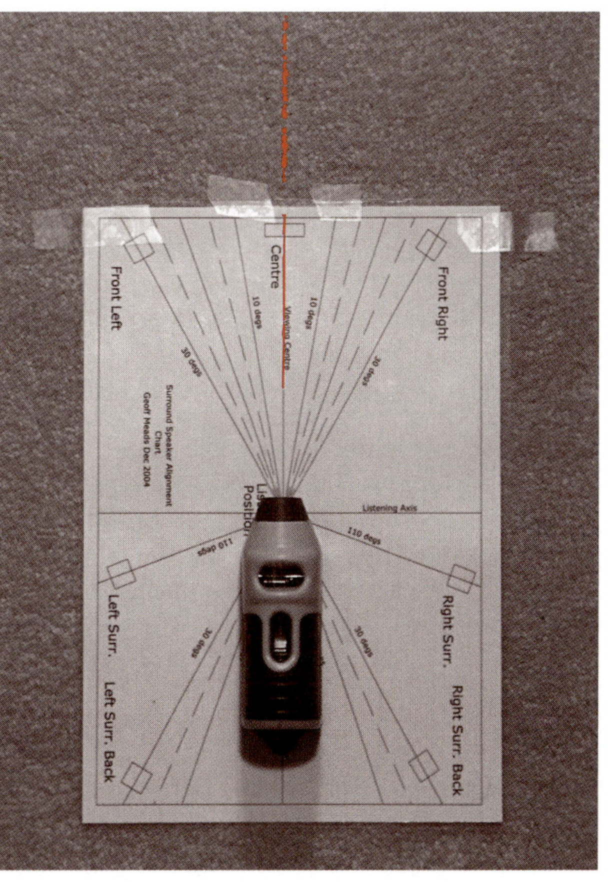


Now point the laser line at each of the speaker positions and place the acoustical center of the loudspeaker at that angle and at the distance proposed by the acoustician for the flattest response in the room; if it helps, drop a plumbline from the speaker so it lands on the laser line. It helps to "toe in" each loudspeaker slightly so the front of the speaker is perpendicular to the laser line. Later, this angle can be tweaked during listening; the toe-in helps off-center listeners to hear the opposite side speaker, but too much toe-in can defocus the phantom center between the left and right loudspeakers for center-located listeners.

The next tool is an electronic distance calculator, the delay measure of a time-domain (FFT) analyzer; if it takes identical time for sound to travel from each speaker to the microphone, the speakers must be at identical distances. SpectraFoo reports the delay in samples, milliseconds, and equivalent distance, which in this image is 9.02 ft or 2.72 m.

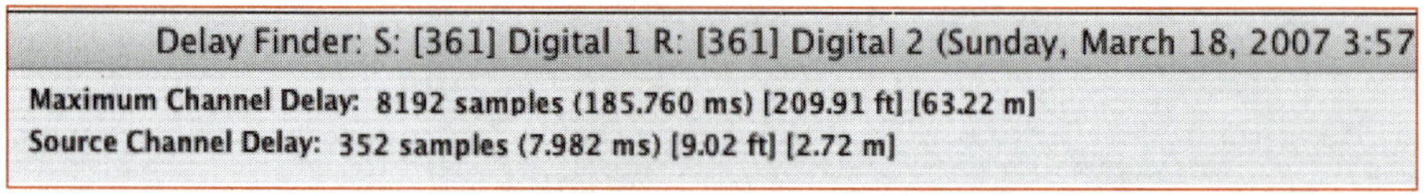

Delay Finder: S: [361] Digital 1 R: [361] Digital 2 (Sunday, March 18, 2007 3:57

Maximum Channel Delay: 8192 samples (185.760 ms) [209.91 ft] [63.22 m]
Source Channel Delay: 352 samples (7.982 ms) [9.02 ft] [2.72 m]

Adjust all the distances until they match. If it's not possible to put the surround speakers at the same distance from the listener as the fronts, then it may be necessary to insert a time delay on the appropriate speaker pair.

For the moment, physically place the subwoofers just in front of, and slightly outside the centerlines of the satellites. Later we will "tweak" the position of the subwoofers for the flattest response at the listening position and best integration with the satellites.

**Connecting and calibrating the system levels**

The 5.1 monitor system has six outputs, which should be connected to the inputs of the corresponding loudspeaker/amplifiers. We're going to construct a system using true stereo subwoofers, which turns the main speakers into full range. Separate subwoofers are an advantage, because we can place them for the flattest bass response while leaving the satellites in the position that gives the best depth and stereo image. Many mastering engineers use full range loudspeakers without additional subwoofers except for the LFE signal. We must not confuse the "point 1" or "LFE" channel with the concept of bass management. Bass management is a crossover between two loudspeakers that each covers part of the frequency range. Strictly speaking, LFE is not part of bass management; though it may be incorporated into a bass manager, LFE is a separate channel, the sixth channel of a five point one system. LFE is meant for special effects, not for the main audio. Many music mixers eschew the LFE entirely (see Chapter 19).

Stereo subwoofers avoid the compromise of a mono subwoofer for several reasons:

- Stereo subwoofers provide a greater sense of envelopment than a single woofer, even when reproducing mono material. There is also evidence that subwoofers can be localized to some degree.[4]
- Stereo subwoofers help create the same effect as full-range loudspeakers.
- Stereo or multiple subwoofers can average out and help to reduce nodal buildup in the room.
- Frequency response is always a compromise with a mono sub, it will never be correct for all sources. This is because low frequency levels are different when combined electrically as opposed to acoustically (in the room). For example, two channels of a center-located in-phase bass instrument combined into a single (mono) subwoofer, results in a 6 dB increase, whereas with two separate subwoofers, the sum is between 3 and 6 dB depending on room acoustics and the distance between the speakers. If only a single (mono) subwoofer is used, we suggest lowering the sub level about 1.5 dB below the individual channel measurement. Then bass instruments panned in the center will only sound a little loud, and uncorrelated bass information or bass instruments panned to one side will sound a little low.

**High Pass on the satellite.** From the point of view of simplicity, I am not a fan of adding a high pass filter to speakers which are already band-limited. I prefer to let the satellites roll off on their own, and then adjust the subwoofer to take over from the satellites. This removes one active crossover component from the main system, which enhances transparency. However, high pass filtering may be required if the satellites distort when receiving out-of-band material.

One way to interconnect a 5.1 system with stereo woofers is to use the two separate inputs within most subwoofers, as illustrated at right. The main speakers receive a full range signal; the sub may perform a low pass on the LFE signal.[5]

We choose the low-pass setting on the subwoofer which produces the most seamless "splice" to the satellites; ideally as low as 40 Hz, but some systems need as high as 80 Hz. Start at the published low frequency response of the satellites.[6] Initially set the woofer polarity to normal and the phase setting to 0 degrees (if the woofer has a continuous phase control). Now we check the integrity of each connection. **Turn the monitor gain control down all the way!** Feed a calibrated, **uncorrelated,**[7] **5-channel** pink noise source at a level of -20 dBFS RMS into all digital inputs of the system monitor control, advance the monitor gain and the trim adjustment on each loudspeaker just a small amount to verify that it's operating. Then solo each output in turn and verify that it's getting to the correct speaker. An RMS-calibrated, uncorrelated stereo WAV file is available at www.digido.com.

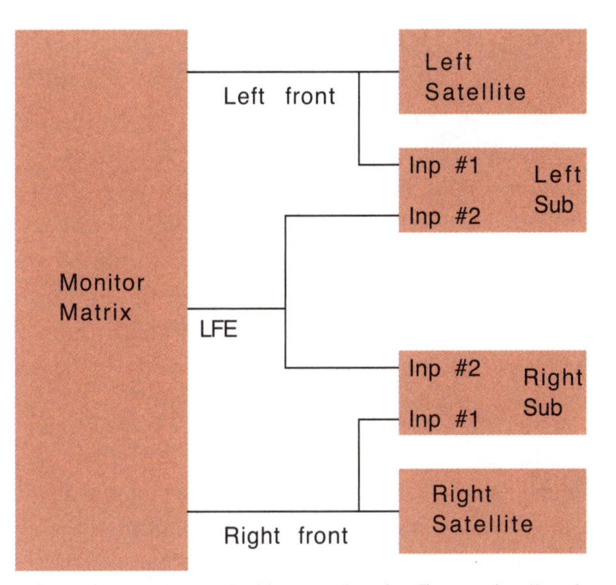


*Connecting a monitor matrix with stereo subwoofers. There are three Y-cords, one for each front channel feeding satellite and sub, and one LFE into both subs. By using the dual inputs of each sub, we can still have a mono LFE signal (the .1 channel) and stereo bass from the front main speakers.*

## Level Calibration: The Best Way

The subwoofer is not needed for the reference level calibration; turn it off or turn its input gain down all the way. We'll be producing some loud test signals, so put on some earplugs. Place a calibrated measurement microphone pointing directly upward, at ear height at the central listening position. We connect this to our 1 octave RTA or FFT analyzer, set to an averaging time between about 3 and 10 seconds, and wait at least that long before taking any reading. **Turn all the loudspeaker trim controls down all the way!** Set the master monitor level to the 0 dB (reference) position. Now, solo **only** the left front loudspeaker. Slowly turn up the left front trim gain until the 1 kHz band reads 73 dB

SPL. If each band were flat at 73 dB SPL, they would sum mathematically to 83 dB SPL.[8] We use the 1 kHz band to avoid measurement errors due to variations in microphone off-axis response, low frequency room resonances, filter tolerances, and so on. Then we inspect the octave band display for a general smooth shape with peaks and dips less than 2 dB. Since octave band display is the most forgiving one, if any band has a significant peak or dip, it's time to bring back that acoustician! The high end should measure a gradual rolloff.

Repeat this procedure for each of the 5 main loudspeakers, sending pink noise one channel at a time. Note that the theatrical standard adjusts the surrounds each to 3 dB lower than the fronts, but for music production, all five loudspeakers should have the same gain.

### Alternate Level Calibration

An accurate alternative is to use a calibrated wideband sound level meter and feed the speakers a band-limited 500 Hz to 2 kHz signal calibrated to -20 dBFS RMS. Set this to read 83 dB SPL. A much less accurate method is to use full range pink noise and a wideband SPL meter set to read 83 dB SPL, C weighting, slow response.

### Total Sound Level

Leave the subwoofers off. Verify that when all five main speakers are operating, the SPL in the midband rises about 7 dB (+/- 1 dB). If not, one or more cables may be wired out of polarity, speaker distances or level calibration could be off, or a component is defective.

### The Fudge

As mentioned in Chapter 14, smaller rooms and closer loudspeakers will produce an apparently louder sound for the identical SPL. If there is a large discrepancy between the examples at the digido.com honor roll and your monitor control, consider using an internal offset for the 0 dB mark. But first ask an acoustician to check for measurement errors.

### Phantom Center Check

Now we check the **phantom center** produced by sending the identical signal in-phase to left and right front speakers, while listening at the central position. This confirms the front main speakers are in polarity and there are no acoustic anomalies. Turn the pink noise off and set the monitor control to about -10. Solo both left and right front loudspeakers. Now remove your earplugs, play a mono pink noise source panned to the middle (or 2 channels of identical signal) and verify the phantom center appears as a narrow virtual image at the physical location of the center loudspeaker. We might need to tweak the angles (toe-in) of the speakers until the phantom image is narrow. If the image is off-center, recheck measurements, speaker distances and verify that the two loudspeakers are well-matched.

Now compare the sound of the phantom center with that of the center speaker itself. The center speaker should sound a little brighter, but the position of the pink noise should not change.

### Bass Management

Bass Management is a fancy word for "crossover"; it is a system for integrating subwoofers and extending the low frequency response of the main speakers. Though we are using stereo subwoofers, when two channels are playing simultaneously, bass response may increase depending on the distance between the subs due to coupling. We strive to make the bottom end response similar no matter how many channels are playing and whether the pink noise is correlated or uncorrelated. If the bass response rises with more speakers playing, try separating the subwoofers slightly.

#### The Right Tool for the Right Job

Below 100 Hz it is impossible to make measurements without taking the room into account, so we must integrate the room response when measuring frequency response. Gating the response to the first 50 ms of sound will produce a fairly accurate measurement down to 20 Hz. A skilled acoustician can precisely tune the level, phase and position of the subwoofers, decide if trapping is needed to deal with room modes and measure the effects of the trapping.

First we try to find the best position for each subwoofer. In principle this is simple: find the position which produces the flattest bass response, but if they end up behind the satellites (further from the listener) we would have to time-delay the satellites to line up the wave fronts. But if we can locate the subwoofers a bit closer, we can use their built-in phase control to compensate for distance. We put our earplugs back on and send pink noise at -20 dBFS RMS to the left subwoofer with the satellites turned off. We start with the display set to 1 octave, moving the subwoofer till the display is flattest, then switch to progressively higher resolution and fine-tune the woofer position. The extreme low frequencies will rise as the sub is moved closer to walls or corners. Now is the time to adjust any other controls the subwoofer may have to flatten its frequency response.

Add the left satellite, which may change the total bass response due to additions or cancellations, and it might be useful to tweak the woofer position further. Adjust the left subwoofer's trim until the FFT shows the low end is in the same ballpark as the rest of the frequencies. There may be some amplitude anomalies near the splice point, indicating some parameters are not yet optimized. Check the polarity of the sub—the polarity that produces the most bass is the correct one; if the result is ambiguous, temporarily set the sub's cutoff frequency as high as possible and recheck the polarity. Now set the FFT to continuous (1-2 Hz) display and inspect the low frequency region to fine tune the amplitude and low-pass frequency. Adjust the phase for flattest combined response at the splice point. Take your time, "focusing" each parameter until the flattest response is obtained at the splice point. If we must compromise, remember, the ear finds peaks more objectionable than dips.

Now we take a spatial average of the response over a few listening positions around the sweet spot, and continue working until we're satisfied the left sub is integrated. This is where art and science interact, for there are no perfect rooms, bass is never absolutely flat, and we may need an elevated or reduced level at the splice point in order to have the most overall flatness throughout the bass region.

Below is the low frequency response of an unoptimized, untrapped small room, presented with a continuous resolution display, with no spatial averaging, accurate to 1.7 Hz.

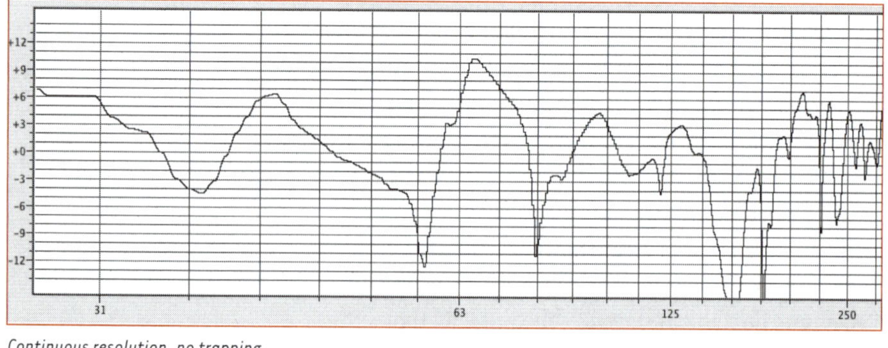


*Continuous resolution, no trapping*

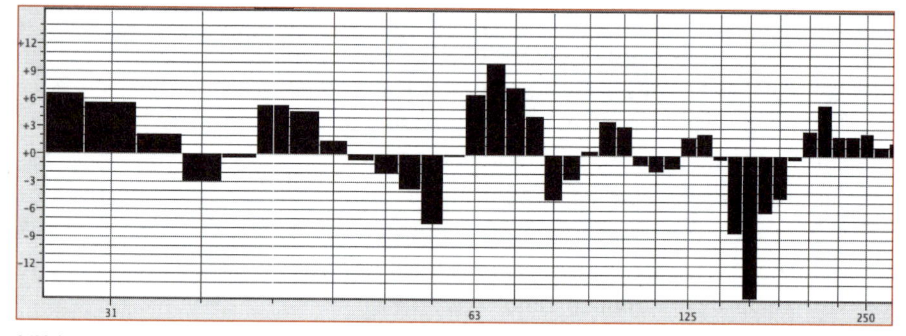


*1/12th octave resolution*

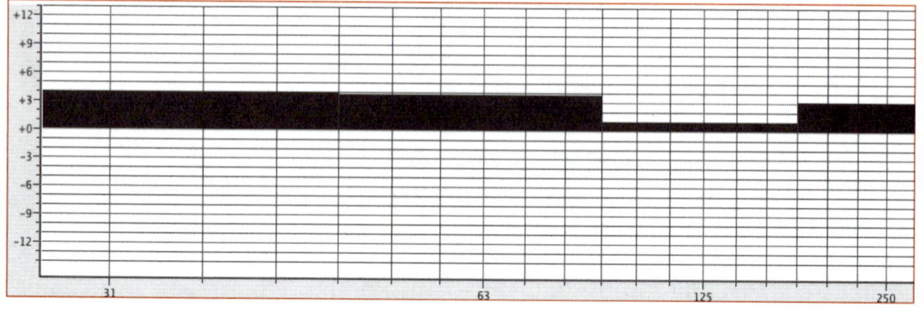


*1 octave resolution*

It's a good thing the ears are more forgiving than the continuous measurement. Continuous resolution is best for precisely analyzing room modes, positioning the woofer, and adjusting woofer to satellite splice points. The various coarse views help us to zero the average of the frequency extremes.

In the middle is the same response presented with 1/12th octave resolution. With 1-octave resolution (bottom left), we can see the general tonal trend, but miss the important peaks and dips that have to be worked on as well.

If the room is symmetrical, it makes sense to place the right subwoofer as a mirror image to the left. Though occasionally, this is not a good idea if the subs both end up at the peak or null of a standing wave between the side walls, which would show up as a significant peak or dip that changes frequency and level with microphone or loudspeaker position. Repeat the above process with the right loudspeaker system. Now send a mono (correlated) pink noise source and play both left and right front channels together (including the subs), turning the master monitor down until the 1 kHz band reads 73 dB. Compare this reading to that of a single channel to see if the subs should be further moved to reduce coupling.

### Subjective Assessment, Stereo First

We have not yet set the bass management for the center speaker or the surrounds, but now that we have two good channels, it's time to listen to some music and confirm our subwoofers are perfectly integrated with the rest of the system. A properly-adjusted subwoofer should not make itself obvious by its presence, only by its absence. Listen to music with the subwoofers on and off. They should not feel "lumpy", they should simply add a sense of weight to the extreme low end. If the crossover frequency is 60 Hz or below, then we may hardly notice a difference except for the additional solidity to the sound. That's the way it should be!

Finding the right recording to evaluate bass is difficult because recordings of bass are all over the map. An excellent way to evaluate a full range system is with a naturally-recorded string bass. My favorite test record is one of my own stereo recordings: Rebecca Pigeon, "Spanish Harlem" on Chesky JD115. This song, in the key of G, uses the classic I, IV, V progression. Here are the frequencies of the fundamental notes of this bass melody:

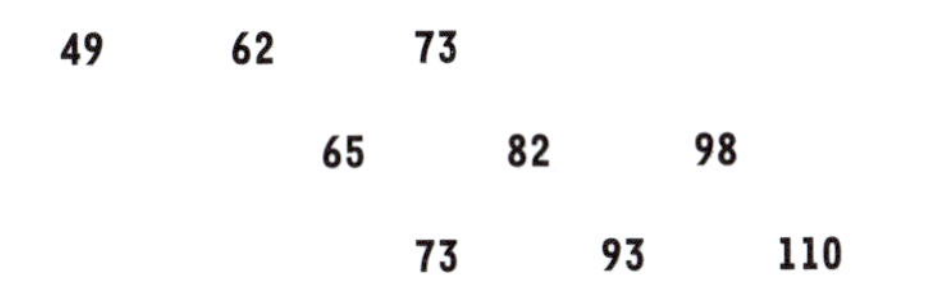


If the system has proper bass response, the bass should sound natural; notes should not stick out too far or be recessed. Start with the subs turned off and verify the lowest note(s) are a little weak. Then turn the subs on and verify they restore the lowest notes without adding any anomalies. The addition of the subs should not move the bass instrument forward or backward in the soundstage or become vague in its placement (an indication the subwoofers are too far apart). Once we are satisfied, we take a break and enjoy Rebecca's performance for its natural acoustic reproduction of voice, string and percussion instruments, and the acoustic depth of a good recording hall. We are off to a good start with an excellent, full-range 2-channel stereo system.

### Bass Management for Center and Surrounds

Our next job is to smoothly extend the low frequency response of the center and surround loudspeakers. To add center and surround information to the subs, each subwoofer needs four inputs, each with its own level control. If not available, we need an external summing box or bass manager. If all main loudspeakers are the same model, we may get away with a single level control in the woofer, otherwise each satellite needs a separate bass level control, and ideally a separate low-pass frequency. We adjust the center loudspeaker for flat response via the bass management level trim of the center (the amount of energy from center redirected to both subwoofers). Then we move to the surrounds and follow similar flattening procedures as with the fronts.

### LFE Gain Setting

The LFE, or .1 channel is an auxiliary channel designed to increase the headroom of the bass channels or for special effects. Its adjustment should be done in the analog domain to prevent overload. To meet the 5.1 standard, the individual RTA bands **for the LFE channel** should read 10 dB higher than the 1 kHz band. **Solo** the LFE output and adjust the level of the LFE channel trim until the 50 or 63 Hz band reads 10 dB higher than the 1 kHz band. If there is insufficient analog trim to raise the LFE level, then we will either need a separate bass manager or to trick out the subwoofers with an outboard set of passive trimpots on each input.[5]

This completes the monitor calibration. Now we sit back and enjoy our calibrated multichannel reproduction system!

## III. Taking it Beyond: Monitor Equalization?

My philosophy is to avoid monitor equalization unless absolutely necessary. I believe that we should do everything possible to fix room-induced problems acoustically, using trapping, absorption and diffusion, and locating subwoofers and/or satellites for the most linear response. Equalization, if performed, should be done by a skilled and experienced acoustician who understands the tradeoffs of electrically equalizing the direct response when a room anomaly is the root cause. There are some modern equalizers that purport to adjust the time-domain response, but this can only be correct for one spot in the room; the best solution is still acoustic treatment. Also consider the tradeoff of additional noise and distortion if an equalizer is added to a system. The X-Curve, which is used for mixing films in large rooms, is not recommended for music to be reproduced in small rooms; using the X-Curve will produce masters which are too bright. After reducing the squiggles with trapping and speaker positioning, we don't hesitate to use our own ears for the last tweak as described in Chapter 6.[9]

1 Plumbline. From the Latin *plumbum* for lead. Attach a small weight to a piece of string to mark the position of a suspended object.

2 International Telecommunication Union, specification ITU-R BS.775-1.

3 Readers can construct their own chart, or obtain a kit with laser chalk and alignment chart from Arcam or one of its dealers. Thanks to Geoff Meads for coming up with the simple but elegant idea of using the "modern technology" of a laser chalk and chart.

4 Braasch, Jonas; Martens, William L.; Woszczyk, Wieslaw (October 2004) Modeling Auditory Localization of Subwoofer Signals in Multi-Channel Loudspeaker Arrays. *Journal of the AES* Preprint Number: 6228 Convention: 117 (October 2004). Griesinger, David. Speaker Placement, externalization and envelopment in home listening rooms.

5 A clever engineer will realize that a typical subwoofer already contains most of the elements to become a bass manager. The inputs to the subwoofer go to an internal summing amplifier, so we can regulate the individual input levels by adding passive level control for each input in an external box. Many subwoofers have both unbalanced and balanced inputs for the "left" and "right" inputs. This means four inputs are internally summed. For even more inputs, a clever engineer will recognize that a balanced input is really a summing amplifier for two unbalanced sources. Pins 2 and 3 of an XLR can be turned into a two input unbalanced summer by inverting one of the sources coming into pin 3. In this fashion up to 6 isolated inputs can be leveled, mixed and managed with a typical subwoofer.

6 If the satellites are good down to 40 Hz, so much the better, because the stereo imaging will probably be more coherent with a lower crossover frequency. The flattest, widest sound will come from the 40 Hz crossover, but to be compatible with consumer bass-management systems and THX recommendations, it is important when mastering, to audition with a mono crossover at 80-100 Hz to test for translation to consumer bass management systems. Many authorities recommend a $4^{th}$ order (24 dB per octave) low pass on the woofer and a $2^{nd}$ order (12 dB per octave) high pass on the satellites.

7 Uncorrelated means there is random, or no continuous relationship between channels. Correlated means there is some relationship. If the same, mono source is fed to all channels, then they are 100% correlated. TMH Laboratories makes test tapes and CDs with calibrated band-limited test signals.

8 The formula for summing any number of uncorrelated sources which are all at equal levels is (level of one source) + 10 log (number of sources).

For example, 5 identical loudspeakers, each playing uncorrelated pink noise at 83 dB SPL will sum to
83 + 10 log 5 = 90 dB. The same formula can be applied to calculate the wideband level based on the level of any single band of pink noise, assuming all bands to be equal. For example, 10 bands of 1 octave pink noise fit between 20 and 20 kHz. 10 bands, each band at 73 dB SPL, sum to 83 dB:
73 + 10 log 10 = 83.

9 While near-anechoic measurement is the best method for adjusting the subwoofer splice and measuring room modes, a wider time window (up to 50 ms) is necessary to interpret how the ear judges timbre. An experienced acoustician must also take into account the loudspeaker's directivity and the treatment in the room. All we know for sure is that the frequency response shows a rolloff at the high end, but how much rolloff sounds right depends on the size and treatment of the room. So I use the subjective method described in Chapter 6.

CHAPTER 16

# Additional Mastering Techniques

This chapter takes us from required basics to advanced mastering techniques including how we "make it louder" if required.

## I. Basic "Objective" Techniques

### Mono Check for Loudspeaker Integrity

Before we begin mastering, during the first listen of the day, it is a good idea to check our loudspeakers by putting the monitor into mono and playing a wide range musical selection. If the center image is tight and unwavering, this confirms that left and right channel loudspeaker drivers have not drifted apart or a tweeter or crossover component has not drifted on one channel. If we suspect a problem, then we take proper measurements as in Chapter 15.

### Stereo Balance of the Program Material

Music feels much better when the stereo balance is "locked in", which can be as small as an 0.2 dB level adjustment in one channel. It is generally unhelpful to use meters to judge channel balance because at any moment in time, one channel will likely measure higher than the other. I've seen songs where one channel's meter (peak or VU) is consistently a dB or so higher than the other, but the balance sounds exactly correct. Since lead vocals are usually centered, this is a good guide, but there are always exceptions. Proper balance should be determined by ear; a stereo position indicator (see ***Figure C17-3*** in the Color Plates) usually just serves as eye candy.

### Fixing Interchannel (Relative) Polarity

**Phase is a concept that operates within the time dimension**. The so-called *phase switches* on consoles are misleadingly named as they **do not shift phase** at all but instead invert signal **polarity.** If two

> *"If it takes me more than three or four hours to master a project, there is something wrong with it."*
>
> — Glenn Meadows

sources are 180° out of phase at all frequencies (or a large band of frequencies), then we say they are *out of polarity with each other*, and so the polarity of one of the channels must be corrected to compensate. This error imparts a hollow quality to the stereo image, can reduce the bass response and even move the image behind the listener. If the correlation meter (see ***Figure C17-3*** in the Color Plates) shows a large phase difference approaching 180°, check for interchannel (relative) polarity errors by switching the monitor to mono and inverting one channel's polarity. The position that gives the most bass or the loudest sound is the correct one. At the mastering stage we can only correct a relative polarity issue if the entire mix has a problem. If, for example, the percussion drops out in mono but the vocal remains, then a remix is required.

### Absolute Polarity Verification

In the real world, some musical instruments create very asymmetrical waveforms. The shape of the waveform should ideally be preserved from microphone through to the loudspeaker. The standard is that microphones produce a positive-going voltage on pin 2 when excited with an acoustic compression, which should produce an upward-going waveform in the DAW, and translate to a forward movement of the loudspeaker when reproduced. This means that the system has correct **absolute polarity**. Is this phenomenon audible? On some systems it is extremely subtle, but it is possible to hear on certain highly coherent loudspeaker systems.[1] I produced an absolute polarity test for Chesky Records, using a solo trumpet recorded in a natural space with a Blumlein microphone pair. When the polarity is incorrect, the trumpet appears (to most listeners) about a meter further back. This is evidence that incorrect absolute polarity can affect how we mix and master.

When mastering, look at the polarity of the waveform of certain instruments. Most instruments produce waveforms with ambiguous polarity, but bass drum attacks generally begin positive, and a solo trumpet on a held note produces a distinct, negative-going waveform. Other than this direct evidence, all we can do is experiment with both polarities to see which sounds better.

### Fixing Phase Shifts and Azimuth Error

Digital consoles can manipulate signal timing (delay). A small timing error between two sources is a genuine *phase error*, which can cause comb filtering especially if the material is combined to one channel. The procedure for correcting small phase shifts between left and right channels requires a keen and experienced ear. Switch the monitor to mono, then, using a timing control calibrated in samples, increase the delay on both channels equally in order to allow advancing one channel relative to the other. Then increase and decrease the relative timing of one channel a sample at a time. Use the timing control like the focus on a camera, searching for the greatest high frequency response and minimum comb filtering at the center of focus. One sample accuracy is pretty coarse, especially at 44.1 kHz, which is why Cedar's digital azimuth corrector makes timing adjustments in sub-sample increments. It is accurate to 1% of a sample. A similar procedure is used by engineers to align spot microphones with the main mikes (see Chapter 17).

#### DC Offset Removal

Sometimes poorly-calibrated ADCs or poorly-implemented DSP processes can add a DC offset, which means that the centerline of the waveform at rest is not exactly 0 volts. When the offset is excessive, overall headroom is reduced in the direction of the offset; so raising gain would cause the audio to clip prematurely in either the positive or negative direction. When using digital limiters, however, slight loss of headroom due to DC offset is not a problem. The best way to determine if a DC offset is a problem is to repeatedly play and stop the material. If we hear a click or a pop when starting or stopping, the DC offset should be repaired. The best solution for DC offset is a very steep linear phase high-pass filter below, say, 20 Hz.[2]

## II. "Subjective" Techniques

#### Achieving Dynamic Impact (Punch)

How do we create a **punchy master**? Here are some opinions:

> If you heard the unmastered mixes, you'd probably find them considerably punchier than the mastered version.
>
> —John Scrip

> Punch is... the right ratio of transients to well-timed compression. Too much of either and the punch is lost. EQ can clean up the punch already in a mix.
>
> —Brian Lucey

> Punch is first captured at the recording stage. If done right it is retained at the mixing stage. If done right again it is retained at the mastering stage... [Only] if it's in there can I enhance it.
>
> —Larry DeVivo

As we described in Chapter 10, when using dynamics processors, punch can be retained and sometimes improved when their attack and release times are optimized to permit both transients and sustained sounds while maintaining or enhancing the overall dynamic shape of the song.

Does this mean that a punchy master can't be produced from a MIDI'ed rhythm section based on 808 kick, synthesized sampled bass and sampled handclaps? Frankly, that's an uphill climb, it requires a team of engineers with ability and knowledge, and it wouldn't hurt to have at least one focused lead or rhythm line played on a standard instrument.

#### Fattening With Tubes

Everyone has his favorite tube processor. Although many of them are one-trick ponies, this is not the case with the more versatile Pendulum OCL-2. By nature it has a very transparent tube circuit, but by adding a drive control and a passive output attenuator, it is possible to overdrive the output tube and obtain a controlled amount of fattening or sweetening with no "digititis". It managed to transform an old digital recording from harsh and edgy to pleasant and "vintage".

### Sample Rate Converters in a Mastering Chain

We discussed upsampling in the mastering session to improve purity of tone, but we need to examine the tradeoffs. Many processors, for example the Weiss, internally double sample only when fed single sample rates. This means that if digitally processing a 44.1 kHz source, if it passes through three processors and then back to the DAW, it has been upsampled and downsampled three times in a row. The improvements from upsampling may be offset by the coloration of the extra DSP and filtering involved. However, if we place a high-quality upsampler in front of the first processor and a downsampler after the last one, there is only one up and one down step. Though the differences may be subtle, in a cumulative chain less DSP usually sounds more transparent or warmer. When using analog processing, the ADC can be the upsampler.

### Pitch and Time Correction

It is impossible to fix the relative pitch of a vocalist once he's mixed with other instruments, so mastering engineers are not often called upon to correct pitch. However, in isolated moments when a singer is *a capella*, or when an entire section of a tune is off-key due to an edit, we can make corrections in mastering. Pitch and time correctors (e.g. Autotune, Melodyne) are now quite sophisticated and we can successfully use one for short periods; however, none are transparent and some degradation can be heard in a high-resolution environment.

**Pitch and speed correction altogether.** Many engineers forget that the easiest and most transparent method is to change **both** the speed and pitch at once, like playing an analog tape recorder faster or slower, avoiding the severe manipulation of a repitching device. It's usually acceptable to the artist as well. We perform a sample rate conversion, then reinsert the material of the "wrong" sample rate into the EDL—this technique can sound excellent if a good SRC is used. SADiE has a high quality resampler for this purpose, it speeds up or slows down the material and makes a new file.

**Pitch correction while retaining the speed.** Not as good sounding but often required. If pitch needs to be raised or lowered the same amount for a long section, I nominate the TC System 6000 VP engine as most transparent. However, it cannot ergonomically raise or lower a single note like Autotune or Melodyne so we may have to sacrifice sound for utility.

**Speed correction while retaining the pitch.** This process, called **time correction,** is probably the worst-sounding. I have not heard a DAW-integrated time-corrector that's as transparent as a two-step process. The first step is to alter the speed and pitch together using a resampler, then to repitch using a good external pitch changer like the TC System 6000.

## III. "Remixing" at the Mastering Session

### Introduction

Everyone has heard the expression, "we'll fix it in the mix"; it's not the right solution for a bad performance, but it happens all the time. Similarly, leaving a problem for the mastering engineer to "fix" is not a good idea, but it happens, too—sometimes because of lack of time, and sometimes because the problem was not perceived during the mix. Given the challenge, we'll find a way to improve, even fix sonic problems. Here are some approaches.

### Vocal Up and Vocal Down Mixes

The lead vocal should be considered "king" in nearly every mix. That doesn't mean it has to sit on top of everything else, but the dynamics of the vocal should help drive the song along with the rhythm; the vocal should not be pulled along by the instruments. A vocalist should neither be so low that she's struggling to be heard, nor so loud as to diminish the impact of the band. A source mix can have a perfect lead vocal level, but occasionally after mastering processing, it might come up or down relative to the instruments. This is why we always recommend that the mix engineer produce vocal up and vocal down mixes, by 1/2 dB. If more than 1/2 dB is needed, then probably not enough attention was paid to the vocal level in the first place.

During mastering we may insert pieces of the vocal up or down mix. For these decisions, we always collaborate with the producer, who may have purposely chosen a vocal level for stylistic reasons or to deal with a pitchy soloist.

### Mastering from Stems

**Stems** are a special kind of submix. For example, a *lead vocal stem* and an *instrumental stem*, which when summed equal the full mix. The best way to produce stems is to make multiple passes of the mix, each time muting different elements. This guarantees that the sum of the stems equals the full mix in level, tonality, and amount of reverb. Generally stems are wet, the vocal stem with its own reverb, and so on. The last thing we want to do in mastering is second-guess what the mix is supposed to be; stems are not supposed to be for remixing or to "fix" the mix, but rather to allow for minor tweaks if the mastering processing changes the mix slightly or to help present the mix in the best light. Stems generally do not work if the mix was made with aggressive bus compression, since the bus compressor behaves differently when fed the full mix than when fed the individual elements. The mix engineer should supply stereo stems, even for mono instruments, to avoid a potential 3 dB ambiguity, for the mastering engineer would not know the panpot law of the console used by the mixing engineer preparing a mono stem. Stems should be sample-accurate and begin at the same timestamp, which disqualifies 2-track analog tape as a stem medium (except for the most adventurous).

> *Learning from your mistakes gives you room to make even bigger ones!*
>
> — Murphy's law of experience

In addition to the full mix, two or three stems may be provided, each mixed in stereo (or surround) incorporating its own reverb:

- Lead Vocal (sometimes labeled *a capella*)
- Instrumental
- TV (instrumental plus chorus)

Sometimes, the instrumental may be split into rhythm and melody stems, or into three stems with the bass instrument separate. Each stem should contain unique elements, otherwise there would be potential for comb filtering or unintended balances. But additional stems should be considered only if the mix engineer has doubts about a particular instrument as it inevitably leads to some remixing at the mastering session. Usually the instrumental stem is not needed in mastering, as the sum of the lead and TV equals the full mix. However, if the client complains that the chorus is too loud, we could add in a pinch of the instrumental and drop the sum of instrumental plus TV. If a full mix and TV are provided, but no lead vocal, it is possible to increase the lead vocal by subtracting the TV from the full mix (add TV with inverted polarity) and then raising the overall level. Given a full mix and a vocal stem we can add or subtract vocal. As soon as we get into these kinds of maneuvers, we must match up levels to get a "base plane" from which to work.

Stems can allow us to perform mastering processing with less compromise. Consider a live recording with a very heavy high hat and ride cymbal due to acoustical issues at the gig. I asked for a vocal and instrumental stem, then processed the instrumental stem with a high frequency compressor which left the vocal untouched.

When clients see what we can do in mastering, they sometimes start to ask for "more keyboards"; but we should discourage them from opening the can of worms of revisiting the mix. Anything may be possible; but our job is to help them maintain perspective and try not to wear the mixing engineer's hat while mastering, for as soon as we start concentrating on "the snare drum is too loud," then we begin to lose our goal of how to best present the sound and feel of the existing mix.

### When Stems are Not Available

Since mixes are rarely perfect, we are often asked to find ways to isolate, raise or lower a particular instrument. Probably the first instrument affected by the attack time of a mastering compressor is the snare drum, which allows us to be somewhat selective on that instrument, especially if we compress in the 1-2 kHz range. We can greatly improve the impact and clarity of the rhythm, particularly the snare, without changing the tonality of the vocal by using upward expansion with a relatively short attack time, or pull the snare back a hair if it's interfering with the vocal, by using a compressor with a relatively short attack time. It's frequently possible to enhance or punch the bottom end of the bass drum without significantly affecting the bass instrument, by using a low-bass-frequency compressor with a relatively long attack time in conjunction with an equalizer. If the bass drum is too loud, as a supplement to EQ, try a narrow-band compressor centered around 60 Hz and stay below the bass instrument. If the bass instrument is too "jumpy" and loud at times, one band of a multiband compressor can do the job, as long as we don't need too much correction; 1 dB gain reduction goes a long way, but more than that can start to suck the life out of the bottom end of the mix.

### MS Mastering

MS stands for *Mid/Side*, or *Mono/Stereo*. In MS microphone technique, a cardioid, front-facing microphone is fed to the **M**, or mono channel, and a figure-8, side-facing microphone is fed to the **S**, or stereo channel. A simple decoder (just an audio mixer) combines these two channels to produce **L**(eft) and **R**(ight) outputs. Here's the decoder formula: M plus S equals L, M minus S equals R.[3] To use just three faders of a mixer as a decoder: feed M to fader 1 up 3 dB and panned to the middle; S to fader 2, panned left; S to fader 3, invert the polarity (*minus S*), and pan it right. Then change the M/S ratio to taste. With more M in the mix, it becomes more monophonic (centered); with more S, the more wide-spread, diffuse, or vague the sound becomes. If we mute the M channel, we will hear a hole in the middle, containing largely the reverberation and the instruments at the extreme sides. Mute the S channel, and the vocalist is dominant; the sound collapses, missing richness and space. There's little separation between M and S channels, but enough to accomplish significant control of a simple 2-track mix. It's great for film work—the apparent distance and position of an actor can be changed by simple manipulation of M/S ratio.

The MS technique can also be used to separate an ordinary stereo recording into its center and side elements for separate processing. There are always tradeoffs, but conservative use of MS tools can turn a good recording into a great one, or save a so-so recording from the dust-heap. For example, a client had mixed in a bass-light room and his bass was very boomy, up to about 180 Hz. At first the vocal level came down a bit by correcting the boomy bass through EQ alone, but via MS processing techniques, I was able to produce a well-balanced master. Given a stereo recording with a weak, center-channel vocalist, first we feed it through our MS encoder, which separates the signal into M and S and we decrease the S level or increase the M level. Listening at the output of the MS decoder, the vocal level comes up, as does the bass (usually) and every other centered instrument. In addition, the stereo width narrows, which often isn't desirable. But at least we raised the vocalist and saved the day! It's also possible to alter the balance between center-located lead vocalist and side-located background singers, even varying the MS ratio between verse and chorus of the song. Some processors have built-in width controls, which internally convert to MS format. The Waves S1 plug-in processor has a gain-compensated width control, which makes life easy as the total level is held constant while the width is being altered. We can accomplish a similar effect, but not as elegantly, by lowering an S control as we raise the M, or vice-versa.

**Automating the MS correction**. MS variation can be accomplished by automating a plugin such as the S1, or directly in an EDL without using any processor. To raise a (centered) vocal, add a duplicate of the material in another stream, with the channels reversed. Add this in at as low a level as tolerable (typically -12 to -16 dB), for if taken to an extreme it will turn the entire material to monophonic. We may add a pinch of a stereo ambience processor to compensate for loss of ambience, width or sense of space, and lower the bass gain to reduce center-channel bass build-up. By contrast, in places where the center vocal sticks out too much, **subtract** a duplicate of the material in another stream, with the channels reversed. In other words, add in a reversed-polarity, reversed-channel duplicate of the source material. This is the simplest type of width enhancer—a true MS encoder/decoder with only four faders. A crossfade into and out of the material in the extra stream is the automation that raises or lowers the level of the center-channel material. Another way to automate this process is to add an MS encode-decode plug-in to the mixer, and automate the panning between the M and S channels on the encode side.

**MS EQ**. We can obtain further isolation and fewer artifacts by using equalization in conjunction with MS processing. Let's take our stereo recording with weak centered vocalist, encode it into MS, and apply separate equalization to the M and S channels. We can raise the vocal slightly by raising, for example, the lower midrange and/or the presence range in just the M channel. This brings up the center vocal with less effect on the other instruments, and less loss of stereo separation.

An equalizer with built-in MS encode-decode, such as the **Weiss EQ-1**, makes equalized MS manipulation a snap. We can spread the cymbals without losing the focus of the snare, tighten the bass image without losing stereo separation of other instruments, and so on. The TC Electronic **Finalizer** 96K's spectral stereo imager is essentially an MS equalizer "on its side"; it's an MS width control divided into frequency bands.

**MS Compression**. Consider a mix that sounds great, but the lead vocal is slightly buried when the instruments get loud. By using MS compression or equalized MS compression (a multiband compressor in MS mode), we can isolate compression to the S channel or to an equalized portion of the S channel; this delicately brings down the sides (effectively bringing up the center) only when the signal gets loud.[4] Or if the vocalist overpowers the band on the peaks, we can compress the M channel and/or expand the S.[5] Equalized MS compression (multiband technique) keeps the bass and treble ranges from being affected by our vocal (midrange) compression. In other instances, we might achieve a better kick drum sound by compressing only the low frequencies of only the M channel. MS parallel compression can be used to enhance and increase the spread of low level ambience. The possibilities are solely limited by our imagination.

**MS with reverb**. Clients have asked us to enhance the vocal reverb without performing a remix. Consider a recording with vocal in the middle and a guitar on each side. Use an MS encoder in front of a reverb send. Or a mono in/stereo out reverb, since the M channel is the sum and it will contain more vocal than guitars.

### Patching Order of Processes

Sometimes it's better to compress before equalizing. For example, if the EQ is being used to enhance the level of some instrument, a compressor after the EQ might undo the effect of the equalizer by pushing the strongest sound downward. However, 90% of the time our equalizer is patched before the compressor, unless there is some emphasized frequency range that causes the compressor to overreact. I almost always put sibilance controllers early in the chain, so they will operate with a constant threshold (sensitivity) regardless of how other devices are adjusted. Parallel compression can work nearly anywhere in a chain before the limiter, but we usually place it early in the chain. The analog processing portion of the chain can go nearly anywhere, except, of course, with respect to the digital peak limiter which must come last. If there is a downsampler placed after the peak limiter, be sure to use a brickwall limiter that takes into account the level rise from filtering.

## IV. Making It Louder with the Least Compromise

### The Red Light District

Probably the most frequently asked question is "how do you make a master sound louder?" My most frequent answer is, "by turning up your own volume control!" But if in a competitive environment the client won't say yes to a master with a lower intrinsic loudness, we can advise on potential negative issues, then apply every technique we can think of to get more apparent loudness at a fixed position of the client's volume control with the least compromise. As we already know, above a certain level the sound quality suffers, with less depth and impact. Many of the following techniques, even used moderately, go completely against the good advice of this book. So: caveat engineer—proceed with caution, and be aware of their limitations even as you may take advantage of some of these techniques.

### Simple Equalization

The Fletcher-Munson effect dictates that high frequency energy produces more loudness than low frequency at the same RMS level. The first thing to try is a high pass filter, the second is a high frequency or presence boost. Note though that both of these can produce the opposite effect when played on the FM radio with its high-frequency preemphasis, so the bright and zippy (and perhaps fatiguing) song that may sound loud in the mastering room will get brought down when it reaches radio.

### Parallel Compression

This is probably the single most potent technique to add loudness and power to a master. If performed well, it can have a singularly positive effect, not just louder sound, but also esthetically better... more *body*, more *punch*. Since parallel compression operates on the lower levels first, we can obtain 2 or 3 dB of additional loudness with little squashing of the higher levels. In a heavy metal or hard rock piece, parallel compression can help supply considerable strength that may have been missing from the mix.

### Cranesong Pentode

This or other harmonic synthesizers can supply edge, depth and a form of compression, unlike a simple equalizer. The down side is that Pentode can easily produce a harsh, thin, edgy sound. See Ch. 17.

### Digital Peak Limiting

Keep in mind that if every mastering engineer is already using peak limiting, every runner in the race has already taken steroids, so there is no performance advantage! 1 or 2 dB of limiting may get the RMS up without taking the sound downhill.

### Clipping

**Clipping can add an edge**, increasing average loudness. Because it is a form of limiting without a defined attack or release time, it can be a lesser evil, with less *clamping effect* than a peak limiter, but like limiting it can take away impact and important microdynamics. At double sample rates or with an oversampled digital processing chain,

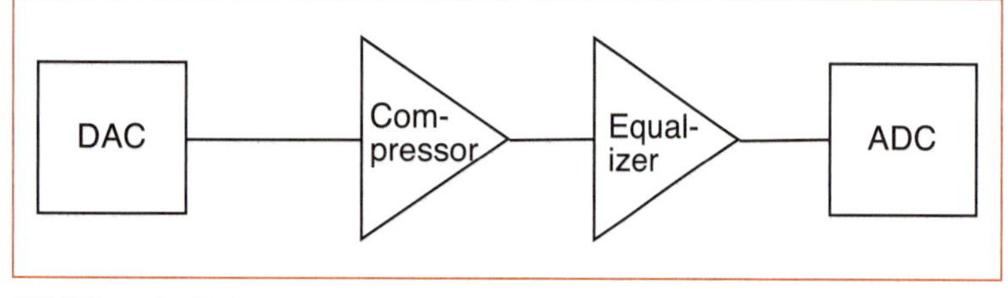


*D/A/D Processing Chain*

**digital clipping** produces less midband distortion and artifacts; the supersonic products will be filtered out when the sample rate is reduced at the tail of the chain. Be aware though, that what sounds "loud" due to clipping in the control room will sound very harsh once it hits a codec (e.g. mp3) or radio station processor; you cannot beat the laws of physics (See appendix 1).

**Analog Domain clipping** is subject to the same cautions, but can sound less harsh than digital clipping. To my experience, an ADC should be run at double sample rate for the fewest clipping artifacts. To ensure the ADC has a controlled amount of clipping, first calibrate the DAC into ADC to unity gain, then add an analog processor in between with a calibrated amount of extra gain. If we receive a clean mix, we **may** get away with as much as three stages of clipping, 0.5 dB of digital clipping, 1 dB of analog clipping, then 1 dB of peak limiting. This is seriously distortion-inducing and so the sonic quality deteriorates extremely fast. Analog domain clipping is most effective if placed just before the digital peak limiter.

We can produce the "loudest" master with the fewest compromises by using a small amount of several processes in a row rather than by engaging just a single process. In all cases, the more *open* and dynamic the mix is to begin with, the easier it is for us to produce a "loud" master. A mix which has already been clipped or limited produces a smaller-sounding, softer master.

1 The jury is out whether the ear truly can detect absolute polarity. We surmise a better loudspeaker, but it may be because the loudspeaker is objectively worse! Objective tests for the ear's ability to detect polarity acuity with asymmetrical test tones have demonstrated that the ear cannot detect a tonal difference between polarities. Perhaps when we do hear a difference it is due to a non-linearity in the loudspeaker driver or magnet structure.

2 DAWs which incorporate DC offset correction may use an inadequate method which centers the average center of the waveform onto the 0 line. The reason this is ineffective is that ADCs tend to drift, so there will still be some remaining offset. This is why I recommend the high-pass method.

3 The formally correct formulas are:

**Encode**: M = 0.5 * (L+R) which is 6 dB less than the mono sum. The encoder sums and attenuates by 6 dB. S = 0.5 * (L-R) which is 6 dB less than the mono difference. The encoder takes the difference and attenuates by 6 dB.

**Decode**: L = M+S. R = M-S. Be aware that an MS encoder and decoder are identical except for the amplitude, and if you use a typical encoder to decode, you will have to raise the level by 6 dB.

4 Remember that a downward compressor brings sound down when it goes over the threshold, so the loudness increase of the compressor is accomplished by raising the gain makeup control, raising average levels but lowering the loudest. In the MS case, just 0.5 dB compression may be all that is necessary to control that "lost" vocalist above the band.

5 If a unit which allows downward compression of M and upward expansion of S is not available, I may compress the M channel in one unit and then upwardly expand both channels in another; when properly adjusted, the net result is the same.

# CHAPTER 17 Analog and Digital Processing

## Introduction

Although mastering equipment is constantly improving, there is still no such thing as a completely transparent audio processor. This means that the mastering engineer must be able to recognize when the interests of the client are best served simply by leaving their recording alone: either because the recording is so good it does not need further work, or where the gains due to processing would not warrant the losses due to the same processing. This chapter is about how we measure performance, and make informed judgments concerning the interaction between objective degradation and subjective improvement.

## I. The Ironies of Perception vs. Measurement

We can use test measurements to evaluate audio processor performance, but we must remember that each single measurement only provides a small part of the overall picture. A processor is like an object inside a house that has a limited number of small windows we can peer through. By seeing the object from each window's unique angle, we can find out snippets of information and add them up, but we can never be entirely sure of what we are seeing, and so we must always leave open the possibility that there may be some aspect that we cannot see, some unknown fact as to why one equalizer sounds "good" and another sounds "bad". For example, here are a couple of "objective" measurements that just don't add up!

### What Makes it Sound Bright?

I've discovered a digital filter that measures "dull" but sounds bright! The TC Electronic System 6000 lets the user choose between different low-pass filters for the ADC and DAC. Some of the filters roll off significantly above 16 kHz (at 44.1 kHz sampling), so you'd think they would sound dull. But instead, the 16 kHz filters called *Natural* and *Linear* sound more *open* and *clear* than the particular 20 kHz filter called *Vintage*. However, there are other converters whose filters extend to 20 kHz and which sound even more *open* than the TC's Linear filter. So clearly, measured bandwidth cannot tell the whole psychoacoustic story. Some of the reasons for this dichotomy can be found in the designs of low-pass filters, which we will investigate in Chapter 20.

### The Fallacy of Single Number Measurements

Equipment in our studio has noise floors which measure from as low as -120 dBFS to as high as -50 dBFS (after A/D conversion). However, much of this equipment is perceptually quiet: if I have to put my ear up to the loudspeaker to hear the hiss, I consider it insignificant. One particular converter whose A-weighted noise floor is -108 dBFS sounds significantly quieter than another converter whose measured A-weighted noise floor is -115 dBFS! This is because the converter which measures better (A-weighted) produces significantly more energy in the region of 3 kHz, but the A-weighted single number was "fooled" by the total noise. An FFT display of the entire spectrum tells more of the story, as does understanding of the critical bands of the human ear and masking effects. Manufacturers' spec sheets need to provide these additional measurements.[1]

{ *"Never turn your back on digital."* — Bob Ludwig }

There are other areas in which traditional measurements do not correlate well with what our ears tell us, particularly in the evaluation of low-bit-rate coding systems. These systems perform adequately according to standard measurement techniques, but once the ear has been trained to hear their errors, we can easily identify unique digital artifacts that analog technology never produced.

## II. Measurement Tools We Can Use While Mastering

### Introduction

FFT stands for *Fast Fourier Transform—a mathematical tool which enables us to move between the time (waveform) and frequency (spectrum) domain*. To really learn how to interpret (and not to misinterpret) an FFT requires a college-level engineering course, and although I cannot claim to be such an expert, I have learned just enough to be dangerous!

### FFT for Music

***Figure C17-3*** in the Color Plate section shows SpectraFoo™ in action during a CD mastering session. High-resolution FFT analyzers such as this are now very reasonably priced, and they provide an essential *early warning system*, a protection from the manifold and varied bugs of digital audio, but are no substitute for the ear. At the middle top is a bitscope, currently showing only 16 active bits, an indication that the dither generator is probably doing its job. Since one of the symptoms of a dysfunctional processor is to toggle unwanted bits, or hold some bits steady when there is no signal, bitscopes can reveal if the DAW or some digital device is malfunctioning. They can also show if there are any truncations caused by defective or misused processors, though they cannot tell us the degree of distortion introduced by a processor, or whether idle bit noise is significant or simply random. At top right is a stereo position indicator, which is frozen at a moment when the information is slightly right-heavy. At left is a meter that conforms to the K-14 standard (see Chapter 14). The meter shows the hottest moment of a rather hot R&B piece (which the client insisted be this hot!). Just below the bitscope is a correlation indicator, which reveals that the material is significantly monophonic. I prefer using a correlation indicator over a lissajoux pattern (another common method of indicating interchannel phase). Meter deflections closer to the center of the scale indicate less correlation between channels which is likely to yield a larger or more spacious stereo image, however with a less defined center. We always use our ears to confirm the image is not too "vague" and perform a mono (folddown) test to make sure the sound is mono-compatible.

At mid-screen is the spectragram, showing spectral intensity over time. This can be useful to identify the frequencies of problem notes, or simply to entertain visitors! At bottom is the spectragraph, whose general rolloff shape gives a very rough idea of the program's overall timbre.[2]

***Figure C17-4*** in the Color Plates shows SpectraFoo™ during a pause in the music, with only the bottom four bits toggling, confirming that the dither is working correctly, since noise-shaped dithers exercise several bits. In this frozen snapshot, bit 15 of the random dither happens to be at zero. The spectragraph shows the curve of the dither noise, which can be identified by its shape as POW-R type 3 or a similar 9th order curve. Using this analyzer, we can often determine the type of dither used by the mastering engineer on recorded CDs.[3] The level meters had not decayed fully when this shot was taken. The correlation meter fluctuates very slightly near the meter's center, showing that the dither is uncorrelated between channels (random phase).

As in geometry, where the shortest distance between two points is a straight line, so too in audio—both digital and analog—the cleanest signal path is the one which is most direct and contains the fewest components. For example, although converters have greatly improved in recent years, we should still avoid extra domain change whenever possible. For mastering work delivered on analog tapes, it's best to do all the analog processing on the way to the first and only A/D conversion. But as mixes are often digital, and as there are a lot of desirable analog processors which the mastering engineer may prefer because they sound more *organic* than their digital equivalents, the subjective benefits of analog processing might well be judged to outweigh the transparency losses of an extra conversion. During the pause, the visual analyzer permits us to see the increase in noise floor caused by the addition of analog processors, which can affect transparency. In general, if the RMS floor of the mastering system remains below -70 dBFS or the FFT components below -100 dBFS, we should not be the least concerned. The quiet mastering room will tell us if the added noise is bothersome.

Wavelab, RME Digicheck, and Sequoia provide similar visualization tools.

## III. Measurement Tools to Analyze your Equipment

As a preventative measure, we can analyze our equipment by sending test tones into the processors and observing an FFT. For example, an FFT can confirm if the bypass switch on a digital processor is truly working. Even though SpectraFoo can only examine 24 bits (the limitation of the AES/EBU interface), it can measure distortion 40 dB below the 24-bit noise floor! This is because it is splitting the energy into smaller segments in the frequency domain. So we can compare the distortion and noise of processors which simply truncate at the 24th bit with others which use 48 bits or so internally and then dither down to 24 bits. When connected digitally, interface jitter (see Chapter 21) is irrelevant to FFT analyzers, which strictly look at data.

### How Many Angels Can Fit On The Head of a Pin?

Some of these measurable differences are audibly important, some are not. The dynamic range of human perception is approximately 20 bits (120 dB), but this varies with frequency. At certain frequencies we can even hear below 0 dB SPL! Given the resolution of the ear, we'd certainly be better off with a 20-bit rather than the CD's 16-bit release format. For intermediate processing however, we need better internal precision than 24 bits to keep digital distortion from becoming audible above the 24-bit level. The major frequency content of digital distortion is **dissonant**, which is less tolerable than **consonant** analog-style distortions, even at louder levels. **Digital processing produces dissonance from harmonic components which beat against the sample rate, producing inharmonic beat or intermodulation products (aliasing distortion).** The term *inharmonic* means that the type of distortion is not part of the integer harmonic series. To keep the dissonant distortions below perceptibility requires high internal processor precision and operation at high sample rates (to be explained). Although the noise increase with digital processing is orders of magnitude less than with analog, calculation errors can accumulate until they reach an audible amplitude and have a deleterious effect.

## IV. Analog versus Digital Recording and Processing

### Accuracy vs. Euphonics

Many people have argued that some digital recordings sound harsh because digital audio is more *accurate* than analog. Their claim is that unlike analog tape, digital recording doesn't compress (and soften) high frequencies, but this *accuracy* reveals the harshness in our sources, which is why we have regressed to tube and vintage microphones. But this is really only a half-truth, since most of these arguments come from individuals who have not been exposed to the sound of good digital recording equipment, that is not only accurate, but can even be *warm and pretty*. Cheap digital equipment is subject to edgy sounding distortion caused by sharp filters, low sample rates, poor conversion technology, low resolution (short wordlength), poor analog stages, jitter, improper dither, clock leakage in analog stages due to bad circuit board design and many others, such as placing sensitive A/D and D/A converters inside the same chassis with motors and spinning heads. It takes a superior power supply and shielding design to make a good-sounding integrated digital recorder.

When it comes to processing, numeric imprecision in digital consoles produces problems analogous to noise in analog consoles, but there is an important difference: noise in analog consoles gradually and gently obscures ambience and low-level material, it is random (uncorrelated with the music) and does not add distortion at low levels. In contrast, numeric imprecision in digital consoles causes errors which increase at low levels, and is correlated with the music, which destroys the body and purity of a mix, creating edgy, colder sound, which audiophiles call **digititis**. Since digital consoles do not make sound warmer, depending on the quality of their digital processing—and the number of passes through that circuitry—it might be better to mix through a high-quality analog console.

Even though good digital equipment is getting cheaper, the need for top-notch converters and processing makes a multitrack digital recording and mix relatively expensive. That is why analog tape and analog mixing remain alive at this point in the 21st century. But I've noticed considerable improvement in all-digital mixes received for mastering, not only because the equipment is getting better, but also because engineers are learning to avoid the pitfalls of digital recording (having too many low resolution plugins in the signal path or pushing them too hard).

### Two Fine Equalizers, One Analog, One Digital

Much inexpensive tube equipment is overly warm, noisy, unclear and undefined, and the common use of "fuzzy" analog equipment to cover up the problems of inexpensive digital equipment is not a cure. Only the best-designed tube equipment has quiet, clear sound, tight (defined) bass, is transparent and dimensional, yet still warm due to subtle harmonic distortion. Modern-day tube designers often make innovative use of low-noise regulated power supplies on filaments and cathodes, which increases transparency and tightens bass, a practice which was impractical in the 50s.

***Figure C17-1*** in the Color Plates section shows the low distortion and noise performance of a well-designed, popular, state-of-the-art analog tube equalizer, the **Millennia NSEQ-2** (red trace). For reference, 20 and 24-bit noise are shown in blue and green, respectively. Notice that the tube noise of the NSEQ is about 10 dB greater than 20-bit, making it a *virtual 18-bit analog equalizer*. However, this performance is dependent on the analog gain structure used. If we drive the equalizer harder, its noise floor will be lower compared to maximum signal, and distortion may or may not be a problem. Since the Millennia's clipping level is around +37 dBu, it is perfectly legitimate to drive it with nominal levels of +10 dBu or even higher, provided the source equipment doesn't overload! Yet even with nominal levels of 0 dBu, as was used for this graph, the NSEQ is extremely quiet. Its noise is inaudible at any reasonable monitor gain and distance from the loudspeakers, demonstrating that

MYTH:
It's a digital processor, so there's no generation loss.

*"Audio processing is the art of balancing subjective enhancement against objective degradation."*

— Bob Olhsson

noise-floor is probably the least of our worries. 1/2" 30 IPS 2-track analog tape has even higher noise, but no one complains about it for popular music.

For this FFT, we set up a DAC, feeding the NSEQ and then an ADC and the FFT. A digitally-generated 1 kHz -6 dBFS 24-bit dithered sine wave feeds the DAC. We adjust converter gain so 0 dBFS is +18 dBu, and boost the equalizer about 6 dB, till just below ADC clipping. The equalizer is coasting at this level, which is 19 dB below its clip point! If we are looking for extreme "tubey" effects, we can drive the equalizer even harder, and also realize a greater SNR.

Notice that the equalizer's distortion is dominated by second, third, and fourth harmonics, which tend to *sweeten* sound. For comparison, in yellow is the performance of the **Z-Systems** digital equalizer, dithered to 24 bits, boosting 1 kHz 5.8 dB with a Q of 0.7. Its harmonic distortion performance is textbook-perfect (no visible harmonics on the FFT). Some engineers use the word "dry" to describe the sound of a component that has little or no distortion. Looking through some other "windows" though, we find that harmonics are far from being the only source of sonic differences. Power supplies and transformers can *loosen* the bass whereas the digital equalizer retains bass tightness. The digital and analog equalizer's curves and phase shift characteristics are also different, though the ZQ-2 does a nice job of simulating the shapes of gentle analog filters. The premium price of both the ZQ-2 and the NSEQ reinforces my point that high-quality analog or digital recording is expensive.

### "Nasty" Digital Processors

Truncation distortion can be fairly "nasty". For example, ***Figure C17-2*** of the Color Plates section shows a comparison between the analog Millennia NSEQ (orange trace) and the digital Z-Systems set to truncate at 20 bits, no dither (black trace). Much of the ambience, space, and warmth of the original source have been lost forever, converted to severe inharmonic distortion and noise.

***Figure C17-5*** in the Color Plates compares two digital compressors, both into 5 dB of compression with a 10 kHz signal. In red is a single-precision (24-bit), non-over-sampling compressor, and in green a double-sampling compressor implemented in 40-bit floating point. Note how the single-precision compressor produces many non-harmonic aliases of the 10 kHz signal, especially in the critical midband. Nasty-sounding compressors are still common in low-cost digital consoles and DAW plugins. When a double-sampling compressor is not available, the sound will be better if the project is recorded and mixed at 88.2 or 96 kHz.

### The Magic of Analog?

Static distortion measurements don't tell us every reason why some compressors sound *excellent* and others hurt our ears. There are analog processors which are *magical* because, although they are not transparent, they add an interesting and exciting sonic character to music. Analog tape recording is a perfect example of this type of process; although objectively measured it is distorted, subjectively those distortion components are sweet-sounding. Noise is another reason why analog tape sounds more *musical* to many people. Low noise must not be our only goal: Maybe -120 dB is better than -144, just enough to cover the ugly parts of the distortion of even some of our best analog and digital gear, so perhaps it's good that our converters aren't any quieter. In addition, noise-free recording media can sound very *sterile* because the nits and cracks and distortions caused by the musicians and their amplifiers are revealed by the quiet media, another case where accuracy is not necessarily desirable and where extra noise can be euphonic by masking less desirable noises. We must remember to consider noise-masking as a tool—distortion or noise may be just what the music needs. In mastering I usually prefer to accomplish this by first passing the signal through the highest resolution electronics, which add little or no distortion, and then add a touch of sauce with a selectively *fuzzy* component. This approach is methodical, controllable, and easily reversible.

As mentioned in Chapter 10, classic analog compressors' signature advantages come from a unique combination of attack and release characteristics (which can be emulated well digitally), zero alias distortion (this type of distortion only occurs in the digital model) and some degree of integer harmonic distortion (which is difficult to emulate well). Certainly the Weiss digital compressor does not sound *digital*, so we know it can be accomplished with programming skill and expensive DSP, but it does not achieve as much punch or warmth as my analog compressors. As my skill at operating the TC Electronic MD4 improves, I often get the sound I want in the digital domain, supplemented—when the recording needs it—by the warmth of a nice analog tube stage used simply as a pass-through.

*"There is absolutely nothing wrong with digital summing other than that it may be very boring."*

### The Summing Amp Controversy

The biggest snake oil currently being sold is the dedicated analog summing amplifier, purported to avoid so-called problems in digital summing by performing summing (mixing) in the analog domain.[4] Many analog engineers who have converted to digital complain that their digital mixes lack separation and depth. But let's get the facts straight: there is absolutely nothing wrong with digital summing, it is essentially perfect, especially since adding numbers is the easiest thing you can ask a DSP to do—equivalent to adding voltages in the analog domain. I am quite sure that summing is not the root of the mix engineer's separation issue because of results obtained from blind tests we have performed at precisely matched gains:

- Analog summing can sound indistinguishable from digital summing provided that transparent analog components and converters are used. In this case, there is no perceived benefit from the analog summing (mix).
- **Any** audio source can gain depth and separation when passed through certain analog components, due to their "friendly" distortion. This is a pleasant "bonus" (or artifact) of the analog chain. Although measured separation may even be less, psycho-acoustically, the distortion appears to increase separation and depth.

**MYTH:** *It's a digital console. It must be better than my old analog model!*

· Only the most superior D/A/D chain exhibits negligible transparency losses sufficient to justify the subjective improvements of the other analog components. To reduce the compromise, expect to spend as much for a superior 2-channel converter as you would have put into 8 medium-quality channels.
· There is no need for the mix engineer to invest in 32 tracks worth of analog summing and 32 D/A/D stages when a single D/A/D stage and 2-track analog module with the preferred kind of distortion can achieve the same result![5] Furthermore, this analog stage can be postponed until the mastering, when it can be integrated and fine-tuned in conjunction with the mastering processing which also affects the sound.
· Not every style of music benefits from analog coloration, either during the mix or mastering. A lot of musical styles (such as classical music and much jazz) are looking for a "clean" approach. Furthermore, during digital mixing it is possible to increase separation and depth (if desired) using techniques described in Chapter 18 or in mastering using specialized transparent digital processors.[6]

### Separation and Timing In Digital Audio

Digital audio presents no obstacle to channel separation or timing resolution. Digital channel separation is literally **infinite**. On the analog side it is limited by noise, so measured channel separation might be 115 dB. Even 20 dB is more than adequate separation; the LP phonograph was lucky to have 15 dB yet sounded quite good in stereo.

*"Digital audio's timing resolution is much finer than the human ear."*

It is a myth that CD's timing resolution is inadequate. In fact, 16 bit/44.1 kHz audio has timing resolution much finer than the human ear, as illustrated in the figure on the next page, it can capture the difference in timing of two waves which are offset by less than a sample period.[7]

### An Analog Simulator—Pick your Color

***Figure C17-6*** in the Color Plates compares the tube NSEQ to the Cranesong HEDD-192, a digital analog simulator of excellent sound quality. The Cranesong (blue trace) has been adjusted to produce a remarkably similar harmonic structure to the NSEQ. For this graph, its levels have been purposely set to produce more distortion than the Millennia was producing. Amazingly, the ear thinks it's hearing an excellent analog processor without any imaging or resolution loss. But the low-level distortion at the bottom of the picture looks mighty suspicious; looking through this "window" you might think the Cranesong was truncating important information. But two important factors ameliorate the result: first, the Cranesong's distortion is about 12 dB lower than that of a truncated device and thus is likely masked by noise and the euphonic harmonics. Secondly, the HEDD has a unique internal architecture that does not alter the original source signal. It clones the original source and sends that to its output, while mixing in the calculated distortion, thereby largely preserving the ambience and space of the original. So the low level distortion in the figure is part of the additive distortion signal and not a result of recalculations to the source. Only the distortion has been distorted! We took this measurement first at 44.1 kHz; then at 88.2 and 96. As you can see in the two figures on the next page, at 96 kHz the unintended distortion is virtually gone, and the Cranesong's distortion is even cleaner, if that's not a contradiction in terms!

### Cooking Better Sound—Naturally

There are certain analog consoles or modules which are highly prized because they add spice, dimension, separation, and some punch to a mix. The necessary ingredients include transformers of a certain type, tubes, or discrete transistor opamps with medium open loop gain. Too much (or too little) distortion can ruin the effect. Some audiophiles feel a well-designed tube circuit can be more linear and resolving[8] than a low-cost solid state circuit, which could be due to the presence of desirable harmonics or the lack of undesirable harmonics, depending on one's perspective. Similarly, some discrete opamp circuits add subtle lower harmonics that we do not see in most integrated circuit opamps. API, for example, has an excellent combination of desirable linearities (like headroom and bandwidth) and nonlinearities (fattening, edge and phase shift, largely from the transformers and some from the opamps). The API modules do a nice thing for rock and roll yet are subtle enough for jazz and classical depending on how hard we drive the stages.

We often master through desirable analog modules. Our role is like that of the master chef who knows what kind of spice and just how much is useful without overseasoning and spoiling the flavor. By the middle of our careers we have collected a sizable spice rack! The Cranesong can digitally mimic three types of naturally-occurring analog distortion, called **Triode, Pentode** and **Tape.** The **triode** control adds a pinch of **salt**, pure second harmonic, which, being the octave, is quite subtle, almost inaudible with some music. It can *clear up* the low end by adding some definition to a bass, but it can also thin out the sound if overused. The **pentode** is extremely versatile; it provides both *salt* and *pepper*. At lower levels it adds third and fifth harmonics, which are extremely seductive, producing at times a warmth, other times a presence boost, and increased depth with little grunge or digititis, especially at 96 kHz as we mentioned. At higher levels, additional odd harmonics add grit and some fatness, like an overdriven pentode tube—a Marshall amplifier in a 1 U rack-mount box! Past the fifth, subtle amounts of seventh and ninth harmonics add a sometimes desirable "edge". It's unpredictable whether the Pentode's warming or presence will predominate until we hear its effect on the source.

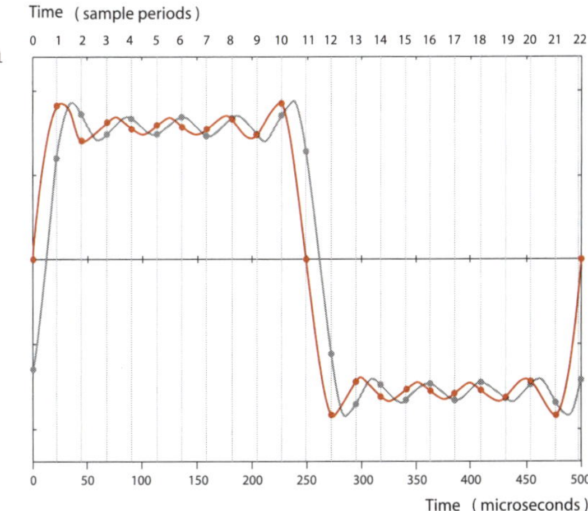


*Sample rate does not limit the timing resolution of digital audio as illustrated here where two waves are offset by less than half a sample period. Image courtesy of Dick Pierce ©2007.*

The Cranesong's **tape** control is the **sugar,** which when mixed in, can sweeten the pentode pepper, yielding colors from red to yellow, green or Jalapeño! The celebrated third harmonic (an octave plus a fifth) sweetens and fattens the sound, much like analog tape. **Tape's** fattening helps to "glue" a mix together. **Tape** can help digitally-mixed sources that may be well-recorded but miss some of that "rock and roll fatness". The control produces largely second and third harmonic distortion, but like the real thing, as it's increased, it produces some additional higher harmonics. Too much sugar gives slow, muddy molasses, rarely desirable, but available if we need it.

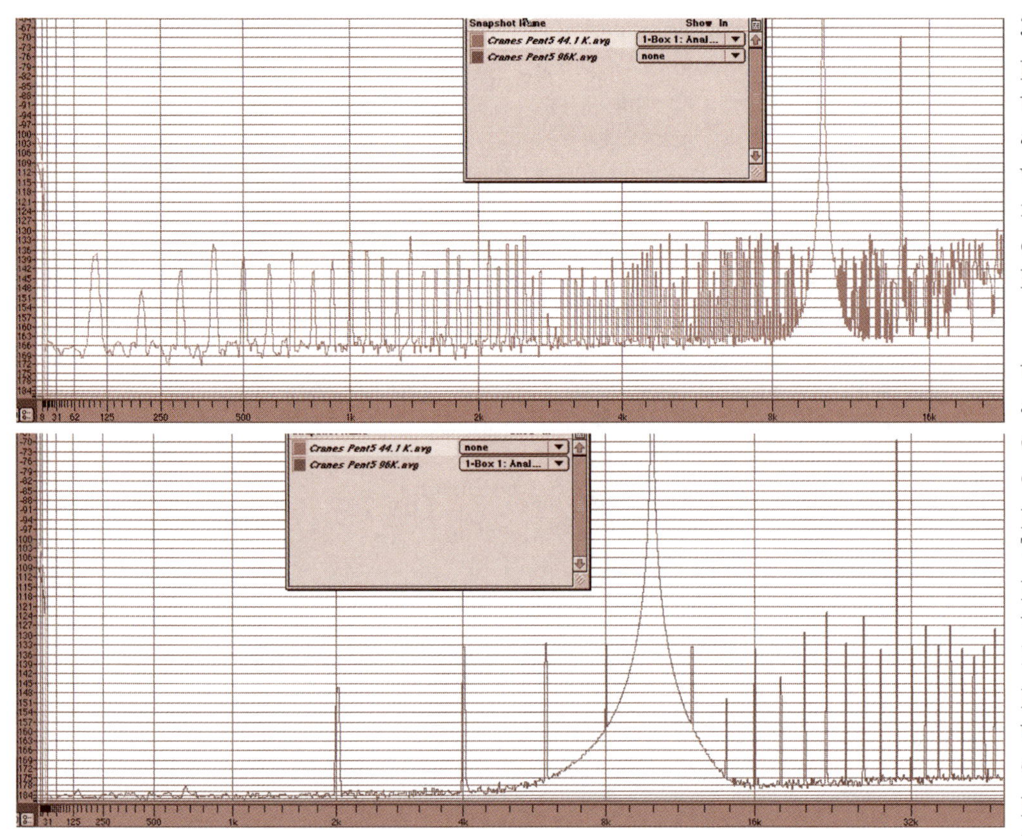


*Comparing Cranesong HEDD 192 in Pentode mode at two different sample rates with a 10 kHz -15 dBFS test tone.*

*At top, 44.1 kHz sample rate. At bottom, 96 kHz. Note the different frequency scales since the higher sample rate displays harmonic frequencies of the audio signal up to 48 kHz.*

## Single Precision, Double Precision, or Floating Point?

First-generation digital processors gave digital processing a bad name. But single precision 24-bit processors are dying out, at least in respectable audio equipment. All things being equal, 32-bit floating point processors are generally regarded as inferior-sounding to 48-bit (double-precision fixed), and 40-bit float.[9] Some newer devices or algorithms are making use of extremely high-precision (80-bit) processing and this may contribute to their sound quality, but just as important is the skill of the designer. For example, Z-Systems has produced a 32-bit floating point digital equalizer using proprietary distortion-reducing techniques that sounds very good and measures as well as some other equalizers using longer wordlengths. The mathematics involved are not trivial, and the designer's choice of filter coefficients can make as much difference as his choice of internal precision.

***Figure C17-7*** in the Color Plates shows that with a single precision processor, even a simple gain boost can ruin your digital day. A dithered 24-bit 1 kHz tone at -11 dBFS is passed through two types of processors, each boosting gain by 10 dB. The distortion of the single precision processor (red trace) is the result of truncation of products below the 24th bit. Nevertheless, the highest distortion product, at -142 dBFS, is extremely low, which would indicate that 24-bit processing or truncation is robust enough for single processes, but not for many processes in series. In blue is the perfectly clean output of a 40-bit floating point processor which dithers its output to 24 bits. I measured similar static performance with a 48-bit (double precision) processor and 32-bit floating point processor, both of which dither internally to 24 bits.

## Double Sampling?

Processing via analog has one distinct advantage. A nonlinear processor (e.g. compressor) manufactures new high frequency energy content, but when this passes back via A/D conversion, the energy above Nyquist is filtered out by the analog anti-aliasing filter. However, when processing digitally, this energy folds back and causes aliasing distortion.[10] This is why the most advanced digital dynamics processors use oversampling technology to raise internal sampling rate to reduce aliasing distortion. High-quality linear phase filters are used in the internal sample rate converters. I'm not certain this has audible meaning for equalizers,[11] but dynamics processors benefit; the higher the sample rate, the less aliasing.[12]

***Figure C17-8*** in the Color Plates compares two excellent-sounding digital dynamics processors, the oversampling Weiss DS1-MK2 (green), which uses 40-bit floating point calculations, and the standard-sampling Waves L2 peak limiter (red), which uses 48-bit fixed point. To compare apples to apples, both processors are limiting by 3 dB (the Weiss is set to 1000:1 ratio) at 44.1 kHz. Note the oversampling processor exhibits considerably lower distortion. However, the switchable safety limiter of the Weiss, which is not oversampled, produces considerable alias distortion even at 1 dB limiting (orange trace). Working at 88.2 kHz and above (not shown), the Weiss safety limiter and the Waves measure better, and double sampling within the processor may not be needed.

Despite these differences, the measurement "window" we've chosen to look through (steady-state sinewave performance) has little to do with the perceived performance of peak limiters, as they operate so fast that the distortion due to aliasing is shorter than the ear's typical ability to detect it. Thus for compressors, which operate slower than limiters, double sampling is a clear choice. Regardless, I notice (and prefer) a slightly purer sound when the L2 is used at double sample rates. Ironically, the reason the L2 produces a louder result than the competition may be its distorted edge (which could be due to fast recovery time, aliasing, or a little of both).

### Better Measurement Methods?

It should be clear by now that we can easily measure simple phenomena that are probably too subtle to hear (such as single tone harmonic distortion near the 24th bit level). But we can hear very complex phenomena that are difficult to describe with measurements (such as the sound quality of one equalizer versus another). What we need to better describe such complex audible phenomena are *psychoacoustically-based* measurement instruments that have not yet been invented. Research based on the ear's masking could lead to noise and distortion analyzers that can discriminate between distortion we can and cannot hear.

### The Bonger—A Listening Test

Since current steady-state (static) sinewave measurements are misleading when measuring nonlinear processors like compressors, a more effective measurement method is by using the **gonger** aka **bonger**, originally developed by the BBC's Chris Travis and available on a test CD from Checkpoint Audio (see Appendix 10). This test is a pure sinewave that modulates through various amplitudes, in the process exercising and revealing any amplitude non-linearities in the signal path. We simply play the bonger through the device under test and listen for noise modulation, buzz or distortion. I run the bonger through my system every time I repatch as it instantly reveals subtle distortion which may be due to clocking errors—distortion that may be masked during normal musical passages. The bonger also mercilessly reveals speaker rattle and sympathetic vibrations in the room.

### Identity Testing—Bit Transparency

A neutral console path is a good indication of data integrity in a DAW. Any workstation that cannot make a perfect clone should be rejected. The simplest test is the identity test, or bit-transparency test. Set the device under test to flat and unity gain, then see if it passes signal identical to its input. This is done through a **null test**: We capture the output of the device to a second file and play the two files synchronized, inverting the polarity of one and mixing the two together. If there is any output, the two files are not identical. Some people scoff at an identity test, since analog equipment could never produce identical output. But this test is important to identify digital distortion-makers. The bitscope is another measure of bit-transparency: a device is likely bit-transparent if we selectively put in 16 bits, then 20, then 24, and get an identical result. We can also watch a 16 or 20-bit source expand to 24 bits when the gain changes, during crossfades, or if any equalizer is changed from the 0 dB position.

### Choose Your Weapon

So, which to use, analog or digital processing? The digital situation keeps getting better each year. If we choose digital, we must be aware of the aforementioned weaknesses and distortion mechanisms. **A mix made through a current-day digital console or DAW may or may not sound better than one made through a high-quality analog console**, depending on several factors: the number of passes or bounces that have been made, the number of tracks which are mixed, the quality of the converters, the outboard equipment, and the internal mixing and equalization algorithms in the digital console. In my opinion, one plugin equalizer (the Algorithmix Red) has the sonic quality of the best digital outboard, and its brother (the Blue) the sonic quality of the best analog outboard, but each uses up so much CPU power that it's currently difficult to use for mixing. It's great for mastering though, where only one or two instances need to be opened. Moore's law will soon overcome even this obstacle.[13]

### Why Is Good DSP So Expensive?

Intellectual property is the most nebulous thing to a consumer. It's easy to see why a two-ton Mercedes Benz costs so much, but the amount of intellectual work that has gone into a one-gram IC is not so obvious. It can take five man-years to produce a good equalizer; there's no doubt that Pro Tools has 500 or more man-years of development, created by individuals each with ten or more years of technical schooling or experience.

### The Source-Quality Rule

An important corollary of this discussion is to illustrate again the importance of the **source-quality rule:** *Source recordings and masters should always have higher resolution than the eventual release medium.* **Start out with the highest resolution source and maintain that resolution for as long as possible into the processing.** When mastering, one consequence of this rule is to reduce the number of generations and copies, and if possible, go back one or more generations when a new process must be added or applied.

It is counterintuitive, but quality of the source counts the most when the end result is an inferior medium. For example, dub to a lossy medium like low-bit-rate mp3 from a high quality source (like a 24-bit mix), and it sounds obviously better than a copy from an inferior source (like an overprocessed 16-bit master). Ironically, you'll never go wrong starting at 96 kHz/24 bit even if the result is destined to be a talking Barbie doll (of course there are diminishing returns).

## V. In Summary

Mastering engineers do not have to think about the meaning of life every time they go to work; many engineers simply plug in their processors, listen, and make music sound better. But I also like to consider why things sound better, because it helps me avoid problems that are not obvious at first listen, and also dream up innovative solutions. I hope this chapter has inspired you to dream up some innovations of your own!

*"The Source Quality Rule: Always start out with the highest resolution source and maintain that resolution for as long as possible into the processing."*

1 The ear's perception of noise is much more than just a frequency response curve, as Jim Johnston explains (in correspondence):

> A single number is ineffective. Noise should be measured separately in each critical band and compared to the ear's threshold for that critical band.

2 As mentioned in Chapter 8, it is not effective to visually analyze a particular type of music and equalize to obtain a "desirable" spectral shape. Every piece of music is unique. For the same reason it is not possible to have an effective compressor preset called "rock and roll"!

3 Most of the SpectraFoo™ screenshots were taken at an FFT resolution of 32K points (32000 "bins") with about 4 seconds average time and Hanning weighting. The actual amplitude of details on an FFT depends on its resolution, so FFTs are only directly comparable if the same methods are used. The levels of the individual bins add up, so for example the bins of 16-bit dither which look like they are sitting at -124 dBFS, add up to a total RMS level around -91 dBFS.

4 It is false advertising to market an analog summing box as "fixing a problem" with digital summing that does not exist, however it is acceptable to market it as "sounding better".

5 We leave open the possibility that the image in an analog mix may sound a bit more "vague" than a single 2-channel module because of the complex crosstalk and leakage between channels in the analog console. Regardless, this is a small contribution to the sound; assuredly, just 2 channels of the right analog gear adds a nice amount of desirable coloration.

6 For the curious, K-Stereo and K-Surround do **not** use harmonic distortion to enhance depth. They use other psychoacoustic principles.

7 Timing resolution of 16-bit/44.1 kHz is the sample period/number of levels. 22.7 µs/65536 = 346 picoseconds, which is more than a million times smaller than the capability of the human ear to detect timing differences.

8 The term *resolving* is meaningless or ambiguous when used so casually.

9 Moorer, J. Andrew (1999), 48-Bit Integer Processing Beats 32-Bit Floating-Point for Professional Audio Applications. *107th AES Convention Preprint Number 5038 (L-3).*

10 Thanks to Dan Lavry for this elucidation, on the Pro Audio maillist.

11 Regardless, the makers of the double-sampling Weiss Equalizer, GML, the Audiocube, and others feel that double sampling is important for equalizers. Some engineers like the sound of accurate high frequency curves that extend beyond 20 kHz, even if that is later cut off when the sample rate is halved at the output of the equalizer. Jim Johnston (in correspondence) states that

> when a digital filter has response extending to half the sampling rate, it can produce some really odd and unexpected frequency responses, indicating that double sampling is important for such type of equalizers.

12 Kraght, Paul (November 2000), Aliasing in Digital Clippers and Compressors. *Journal of the AES Volume 48 Number 11 pp. 1060-1065.*

13 Moore's Law: The empirical observation made in 1965 that the number of transistors on an integrated circuit for minimum component cost doubles every 24 months. Or a correlary, that computing power for the same cost and space doubles every 24 months (or even sooner!).

CHAPTER 18

# How to Achieve Depth and Dimension in Recording, Mixing and Mastering

## I. Introduction

The creation of high quality audio masters requires that some basic acoustic principles be understood by recording, mixing and mastering engineers. But even though we are now in the era of surround recording and reproduction, some mix engineers are repeating their mistakes from two-channel work—panpotting mono instruments to discrete locations, and then adding multiple layers of uncorrelated stereophonic reverb "wash" in a misguided attempt to create space and depth. It's important to learn how to manipulate the surprising depth available from the 2-channel canvas before moving on to multi-channel surround.

### Why True Stereo Recording Yields Better Reproduction

It amazes me how few engineers know how to fully use good ol' fashioned 2-channel stereo, but it is easy to discern the audible difference between simple pan-potted mono versus genuine stereo recordings which create a real sense of depth, by utilizing the natural room acoustics, and reflections from nearby walls. When all the musicians are playing at once, the natural interaction of microphones and acoustics (often called "leakage") helps to make a bigger sounding recording. Without this element, recordings will tend to produce a vague, undefined image; the musical instruments will be obscured and unclear. The ear's computer decodes delay information; a few well-placed echoes solidify and clarify the location of the direct sound, and help to distinguish one instrument from another in a complex mix. This is why a panpotted mono instrument (either close miked, or without stereo-

{ *"Think Beyond The Panpot"* }

phonic early reflections) is hard to locate precisely between two loudspeakers. Its location becomes more ambiguous as listeners move away from the center seat (the "sweet spot") whereas the listening sweet spot is wider and instrument location more precise when it has been recorded in true two-channel stereo.

Many pop engineers know it is possible to simulate depth artificially, using delays or artificial reflections which help localize instruments to sound similar to "the real thing". Engineers need to think beyond the panpot. We need to understand the Haas[1] effect, particularly when implemented binaurally. Most of this knowledge is best applied in recording and mixing, but we can also use it during the mastering session.

## II. Early Reflections and Masking

### Early Reflections versus Reverberation

At first thought, it may seem that depth in a recording can be achieved simply by increasing the proportion of reverberant to direct sound. But the artificial simulation of depth is a much more complex process. *Early reflections* consist of the part of the room sound within approximately the first 50-100 milliseconds of the direct sound, which means the two are highly correlated; think of the early reflections as being *attached* to the direct sound. In a large and diffuse room, after about 100 milliseconds, enough wall bounces have occurred to create random (uncorrelated) *reverberation*, which we can say is *detached* from the direct sound. That is why the early reflections affect our perception of the depth of the sound, giving it shape and dimension.

### Masking Principle

The direct sound from an instrument can mask the direct sound of another; the direct sounds of instruments can also mask the reverberation of the room. **There are three types of masking: amplitude, directional, and temporal. Amplitude masking** is when a louder sound masks a softer one, especially if the two sounds lie in the same frequency range. This is why mixing engineers use equalization and filtering as well as level to separate elements of a mix. If the two sounds happen to be the direct sound from a musical instrument and the reverberation from that same instrument, then the initial reverberation can appear to be covered by the direct sound. When the direct sound ceases, the reverberant hangover is finally perceived.

Mixing engineers can add a small pre delay between the direct sound and the reverberation—a **temporal unmasking technique which** helps the ear to separate one from the other. A good *uncluttered* musical arrangement has built-in temporal unmasking, separating the timing and rhythm of the instruments, making the mixing process easier.

**Directional Masking during Stereo to Mono Reduction**. In concert halls, to some extent in stereophonic reproduction, and to a much greater extent in surround-sound playback, our two ears sense reverberation coming diffusely from all around us, and the direct sound as having a distinct single location; there is little **directional masking**. However, in a monophonic recording, the reverberation is reproduced from the same source speaker as the direct sound, and so as the two sounds overlap directionally, we may perceive the room as being less reverberant. A very live recording hall is bad for mono recording, because reverberation will directionally mask direct sound. This is one explanation for the incompatibility of many stereophonic recordings with monophonic reproduction.

In 2-channel and multichannel recordings we can overcome directional masking problems by spreading artificial reverberation spatially away from the direct source, achieving a recording which is both clear (intelligible) and warm at the same time. One of the first tricks that mix engineers learn is to put reverberation in the opposite channel from the source. But though this can help to unmask the direct sound, it can produce an unnatural effect.[2] More sophisticated techniques use multiple delays or stereophonic early reflections to yield a more cohesive, natural effect than is possible by simply using opposite-channel panning. Cheap reverbs containing no early reflections and a basic reverb "wash" muddy up the sound—perhaps it's better to have no reverb than a cheap reverb. True stereo reverbs produce better depth than mono-in-stereo-out models.

In natural environments, the early correlated room reflections are captured with their correct placement; they support the original sound, help us determine the distance of the sound source and do not interfere with left-right orientation. The later uncorrelated reverberation naturally contributes to the perception of distance, but because it is uncorrelated with the original source, it does not help us locate the original source in space. The better the original room and the miking techniques, the more convincing the sense of space and the less artificial reverberation will be needed in post-production.

## III. Recording Techniques for Depth

### Recording in Natural Rooms

**Balancing the Orchestra with only a few microphones (minimalist).** The loudness of an instrument affects its balance in the mix; softer instruments also sound a bit farther away. But the primary influence on perception of depth and distance is the amount of early reflections and reverberation. A musical group is shown in a hall cross section (pictured next page). Various microphone positions are indicated by letters **A-F**.

Microphone position **A** is located very close to the front of the orchestra. As a result, the ratio of **A**'s distance from the back compared to the front is very large. Consequently, the front of the orchestra will be much louder in comparison to the rear, and the amount of early reflections reaching the microphone from the rear will be far greater than from the front. Front-to-back balance will be exaggerated. However, there is much to be said in favor of mike position **A**, since the conductor usually stands there, he purposely places the softer instruments (strings) in the front, and the louder (brass and percussion) in the back. Also, the radiation characteristics of the horns of trumpets and trombones help them to overcome distance, the focus of the horn increases direct-to-reflected ratio. We take these factors into account when arranging an ensemble for recording.

The farther back we move in the hall, the smaller the ratio of back-to-front distance, and the

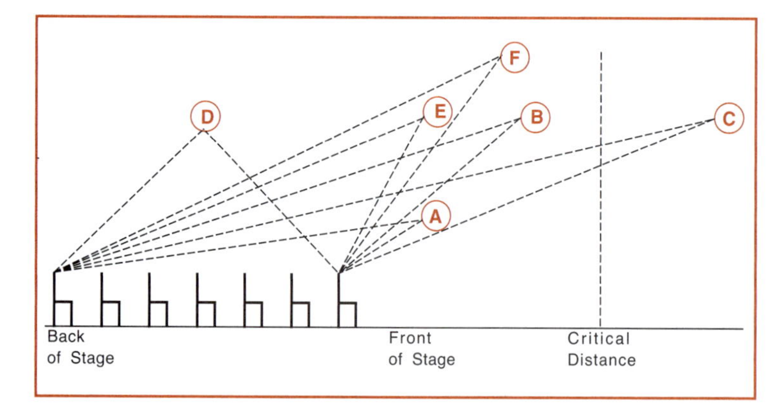


front instruments have less advantage over the rear. At position **B**, the brass and percussion are only two times the distance from the mikes as the strings. This (according to theory) makes the back of the orchestra 6 dB down compared to the front, but in reality there is much less difference, because level changes less with distance in a reverberant hall.

For example, in position **C**, the microphones are beyond the critical distance—the point where direct and reverberant sound are equal. If the front of the orchestra seems too loud at **B**, position **C** will not solve the problem; it will have similar front-to-back balance but be more buried in reverberation.

## Using Microphone Height To Control Depth And Reverberation

Changing the microphone's height allows us to alter the front-to-back perspective independently of reverberation. Position **D** has no front-to-back depth, since the mikes are directly over the center of the orchestra. Position **E** is the same distance from the orchestra as **A**, but being much higher, the relative back-to-front ratio is much less. At **E** we may find the ideal depth perspective and a good level balance between the front and rear instruments. If even less front-to-back depth is desired, then **F** may be the solution, although with more overall reverberation and a greater overall distance.

## Directivity Of Musical Instruments

Frequently, the higher the mike, the more high frequencies it will capture, especially from the strings. This is because the high frequencies of many instruments (particularly violins and violas) radiate upward as well as forward. The ear perceives a brighter sound as closer, overcoming the distance. When the mike moves past the critical distance, we may not hear significant changes in high frequency response when height is changed.

The recording engineer listens and makes changes in mike placement based on these factors. The difference between a B+ recording and an A+ recording can be a matter of mere inches.

## Mike Spacing, Pattern and the Depth Picture

**Coincident Microphones.** The various simple miking techniques reveal depth to greater or lesser degree. Microphone patterns which have out of phase lobes (e.g. hypercardioid and figure-8) can produce a holographic depth quality when used in properly angled pairs. Even coincident figure-8s provide as much or more of a depth picture than spaced omnis. But coincident miking reduces time ambiguity between left and right channels, and sometimes we seek that very ambiguity. With any given mike pattern, the farther apart the microphones of a pair, the wider the stereo image of the ensemble and the greater the hole in the middle. Instruments near the sides tend to pull more left or right, center instruments tend to get wider, more diffuse, harder to locate or focus.

The technical reasons for this are tied in to the Haas effect (to be explained) for delays of under approximately 5 ms. vs. significantly longer delays. Very short delays between two spatially located sources produces ambiguous image location.

**Spaced microphones**. I have found that increased intermike spacing increases the center depth; for example, the front line of a chorus no longer seems straight, instead, it appears on an arc bowing away from the listener in the middle. If soloists are placed at the left and right sides of this chorus, a rather pleasant and workable artificial depth effect will occur. Adding a third omnidirectional mike in the center of two other omnis can stabilize the center image, and reduces center depth.

### Beyond Minimalist Recording

Even after obtaining perfect balance, the engineer/producer often desires additional warmth, ambience, or distance. There may be a more distant mike position with equally good balance, but time is of the essence in orchestral recording and we hesitate to fix what isn't broken. Another call for increased ambience is when the hall is a bit dry. The engineer may try changing the microphone pattern(s) to less directional (e.g. omni or figure-8) but this then also requires a different spacing and angle.

Perhaps the easiest solution is to add ambience mikes, being careful to avoid acoustic phase cancellation, which does not occur when the extra mikes are placed far enough to be in the uncorrelated reverberant field, or by applying the **3 to 1 rule.**[3] When these mikes are mixed into the program, direct frequency response should not deteriorate, we should simply hear an added warmth and increased reverberation.

**Multiple Miking**. While multiple close mikes destroy the depth picture, soloists do need to be heard, and for many reasons they are not always positioned in front of the group. When the soloist cannot be moved, plays too softly, or when hall acoustics make him sound too far back, then one or more *spot mikes* must be added. The depth image may seem more natural when the spot is a stereo pair, than a mono solo mike.

Apply the 3 to 1 rule, listen closely for frequency response problems when the close mike is mixed in. This will (not surprisingly) appear to bring the solo instrument closer to the listener. If this is not overdone, the effect is not a problem as long as musical balance is maintained, and the close mike levels are not changed during the performance.

When mixing a recording made in a live acoustic space with multiple microphones, try to match the panning to the way the musicians placed themselves on stage (unless this produces an awkward balance). Don't get hung up on putting the bass in the middle—if he was playing originally on the left and in the second row, put him there. Since each spot mike also picks up ambience, the sum and spread of these accurate panning positions will enhance the stereophonic depth picture. In addition, leakage will end up in its proper virtual position. For example, if there is drum leakage in the piano mikes, the drum imaging and depth will be more accurate when the piano mikes are properly panned, producing less "smearing" of the drum image.

**Delay Mixing**. Adding a delay to the close mike to synchronize it with the main pair pulls the soloist back and helps to maintain natural depth, but we still need to hear some early reflections around the soloist, which hopefully arrive at the main pair with some strength. Otherwise we should try a bit of artificial early reflections. To adjust the delay of the spot mike(s), start with a delay calculated by the relative distance between the solo mike and the main mike, then focus the delay up and down in 1 ms increments until the sound is most coherent and focused, clarifying the sound of the soloist.

### Dead Studios

Minimalist miking techniques do not work well in a dead studio. In a dead room, simple miking has no advantage over multiple miking with panpots. So when there are no significant early reflections, choose multiple miking, with its attendant post-production balance advantages.

## IV. Adding Depth in Mixing and Mastering

### The Haas Effect

The Haas effect can help increase definition, depth and fullness without causing masking problems. Haas says that very short echos (less than about 5 ms) produce an ambiguous (confused) image. However, echos from about 10 through approximately 40 milliseconds after the direct sound become **fused** with the direct sound—only a loudness enhancement occurs. This is what happens in a real room with the earliest wall and floor reflections. Since the velocity of sound is approximately one foot per millisecond, 40 milliseconds corresponds to a wall that's 20 feet distant (assuming a flat wall perpendicular to the angle of the direct sound).

### Haas Delays in Mixing to Enhance Spatial Qualities

In pop or classical mixing, we can use delays to take advantage of a very important corollary to the Haas effect, which says that fusion (and loudness enhancement) will occur even if the closely-timed echo comes from a **different direction** than the original source. The brain will continue to recognize (binaurally) **the location of the original sound as the proper direction of the source**. The Haas effect allows added delays to enhance and reinforce an original sound without confusing its directionality—just as long as the delay is not too long and the level of the delayed signal is not too loud. When the delay is too long or the delayed signal too loud, it starts to be perceived as a discrete echo; which we call the **Haas Breakdown** point. Long delays maximize the definition of the source, as long as we have not reached breakdown. The Haas breakdown point is shorter for percussive sounds; for example, sometimes only 15 ms is tolerable for a drum hit, while up to 30-50 ms is permissible for strings.

To take advantage of the ear's own decoding power during mixing, generally use panned and leveled delays in the 12 to 40 millisecond range. Haas delays are more effective than equalization at repairing the sound of a drumset which was recorded in a dead room. To create layers in the mix, put single delays on some instruments, multiple (or no) delays on others; try doubler and quadrupler delay plugins with built-in panning, supplemented with the panpots in the console. Mix engineers also use computerized early-reflection simulations found in devices such as the TC Electronic, EMT, and certain models of Sony reverbs. Using a variety of techniques and processors can increase depth and space.

When mixing in surround, it is best to avoid power panning (standard panpots) between front and surround as it produces a very ambiguous image. As with stereo, think beyond the panpot. Virtual panpot positions do not image well—it is much better to use a surround early reflection generator, which produces a more stable image and allows a wider sweet spot.[4]

**Haas and mono-compatibility**. When utilizing simple Haas delays, be sure to check the recording in mono for comb filtering. The more complex and numerous the delays, the less likely that comb filtering will occur.

### Haas In Mastering

We receive recordings for mastering which lack depth, spatiality and clarity for two reasons: 1) the mix engineer did not mix early reflections or reverberation well enough or loudly enough. 2) the recording was made in a dead room and the mix engineer either used no reverberation or cheap reverberation which does not contain adequate early reflections. In the first case, adding further artificial can muddy the sound. But ambience extraction, used subtly, can increase depth and definition. In the second case, we may suggest a remix with better reverberators. Or if the reverb used in the mix has a decent tonality but no sense of depth, we may try adding early reflections from a superb algorithm such as the TC Electronic VSS4, which can add depth to a dead room. However, the depth effect will be more convincing when the original room has some useful reflections which we can combine with the artificial ones to enhance the reality. The VSS4 has a feature called **decrease** which prevents generating artificial early reflections which are already present in the source.

### Using Frequency Response to Simulate Depth

In a natural acoustic environment, the apparent high frequency response is reduced as the distance from a sound source increases. This provides another tool with which the recording engineer can simulate distance. An interesting experiment is to alter a treble control while playing back a good orchestral recording. Notice how the apparent front-to-back depth of the orchestra changes. We can use mikes with differing treble response, or during mixing, change the high frequency characteristics to move instruments forward or backward.

### The Magic Surround

We can take advantage of the Haas effect to naturally and effectively convert an existing 2-channel recording to a surround medium. When remixing, simply place a discrete delay in the surround speakers to enhance and extract the original ambience from a previously recorded source. No artificial reverberator is needed if there is sufficient reverberation in the original source. Here's how it works:

Haas fusion only works with **correlated** material. The ear fuses correlated sources with their delayed replicas (e.g. a snare drum hit) and so continues to perceive the direct sound as coming from the front speakers. But this does not apply to **uncorrelated** ambience—because the ear does not recognize the delay as a repeat, thus spreading, enhancing, and diffusing the ambience between the location of the original sound and the location of the delay. Dolby laboratories calls this effect *the magic surround*, for they discovered that natural reverberation was extracted to the rear speakers when a delay was applied to them. Dolby also uses an L-minus-R matrix and logic elements to further enhance the separation. The wider the bandwidth of the surround system and the more diffuse its character, the more effective the psychoacoustic extraction of ambience to the surround speakers. My patented K-Stereo and K-Surround processes start with and extend these principles.

## V. In Conclusion

### Influence Of The Control Room Environment On Perceived Depth

At this point, many engineers may say, "I've never noticed depth in my control room!" As described in Chapter 15, the widespread practice of placing near-field monitors on the meter bridges of consoles kills almost all sense of depth.

### Listening Examples

Here are some examples of stereo audiophile recordings I've made that purposely take advantage of depth and space, both foreground and background, on Chesky Records. Sara K. *Hobo*, Chesky JD155. Check out the percussion on track 3, "Brick House". Johnny Frigo, *Debut of a Legend*, Chesky JD119, especially the drums and the sax on track 9, "I Love Paris". Ana Caram, *The Other Side of Jobim*, Chesky JD73, particularly the percussion, cello and sax on "Correnteza". Carlos Heredia, *Gypsy Flamenco*, Chesky WO126. Listen to track 1 for the sound of the background singers and handclaps. Phil Woods, *Astor and Elis*, Chesky JD146, for the natural-sounding combination of intimacy and depth of the jazz ensemble.

### Technological Impediments to Capturing Recorded Depth

Depth is the first thing to suffer when technology is incorrectly applied. Here is a summary of some of the technical practices that when misused, or accumulated, can contribute to a boringly flat, depthless recorded picture:

- Multitrack and multimike techniques
- Small/dead recording studios or large rooms with poor acoustics/missing early reflections
- low resolution recording media
- excessive dynamic range compression (which tends to amplify the mono information and bring everything forward)
- improper use of dithering, cumulative digital processing, and low-resolution digital processing

**In Summary: To resurrect the missing depth in recording, mixing and mastering**, use the highest resolution technology, best miking techniques, and room acoustics. Process dead tracks with Haas delays and early reflections, and specialized ambience recovery tools.

1 Haas, Helmut (1951), *Acustica*. The original article is in German. Various English-speaking authors have written their interpretations of Haas, which you can find in any decent textbook on audio recording techniques.

2 Even if unnatural, it can be interesting, nevertheless. Listen to 1960's-70's era rock recordings from the Beatles, Beach Boys, Lovin' Spoonful, The Supremes, Tommy James and the Shondells, where mono instruments or vocals are panned to one side, and often their reverb return completely to the other side.

3 Burroughs, Lou (1974), *Microphones: Design and Application*, Sagamore Publishing Company. (Out of Print). Burroughs quantified the effects of acoustic phase cancellation (comb filtering, interference) with real microphones and real rooms, and devised this rule: The distance between microphones should be three times the distance between each microphone and the source of the sound to which it is being applied. This is particularly important to avoid comb filtering when both microphones are feeding a single channel; when the microphones are feeding different channels (e.g. stereo), the degradation will be much less noticeable in stereo but still be a problem in mono.

4 www.digido.com links to further articles on this topic.

# CHAPTER 19 Surround Mastering

## Introduction

In this chapter we meet four of the most talented and experienced surround sound engineers in the business. Each brings his own motif, but the themes are universal. Three are mastering engineers and one is a mix engineer:
**Dave Glasser,** of Airshow, Boulder, Colorado; **Bob Ludwig** of Gateway Mastering & DVD, Portland Maine; **Rich Tozzoli**, independent producer/mixer; and **Jonathan Wyner** of M-Works, Cambridge, Massachusetts.

## I. The Approach

### Are You Dedicated Primarily to Mixing or Mastering?

**Dave Glasser** Mostly mastering, but in a few projects we mixed and mastered in the same session.

**Jonathan Wyner** I have been doing some surround mixing as well as mastering. The line is blurred between the two. Partly because while mastering, I find myself needing to redistribute the "sound field", e.g. relying on phantom center or creating center channel information. In stereo there is a greater separation between the disciplines.

**Rich Tozzoli** I will never claim to be a mastering engineer. Bob Ludwig did my last DVD project, and he rocked that. Dave Glasser has also mastered a few of my 5.1 SACD discs and he makes a big difference in the final product. We won a Surround Music Award together.

**Bob Ludwig** Occasionally we have to make changes. I was doing a Coldplay project and they needed a clean version. Unfortunately the instru-mental versions still had the F-word reverb bouncing

Dave Glasser

off the rear of the hall. So we had to get all of the tracks to fix it.

If you have to manipulate something, you can stack up groups of six tracks pretty quickly. We worked on a live Foo Fighters project and most of the tracks are from night three, but some were from night two and night one. The DVD sequence was in a different order from what they actually played. By the time you have restored all the applause, you can get 48 tracks or more happening without too much trouble.

I hate when it happens, but there are times we've had to make faux 5.1 recordings. The Police had all their hits remixed for 5.1. There were a couple of stereo demos that ended up being used and the producers wanted to maintain the feeling of 5.1, not have it revert to two channels. I did an Unwrap (TC Electronic) which creates a believable 6 channel presentation and did some further panning. A Brian Ferry 5.1 DVD also used a demo.

**Do you ever get stems?**

**Dave Glasser** The stems that I have gotten were for a live performance DVD, a Grateful Dead movie. The music has one stem and the rest, dialog, behind the scenes elements, which was nice. But usually we do not get stems.

**Jonathan Wyner** I'm more interested in stems when working in surround than in 2 channel where I usually discourage it unless it is critically important, e.g. when a client is really out to sea with a mix. With surround, I'm happy to get stems so that I can more easily redistribute the soundfield. If I want to spread the image a little wider it's much easier to control when I have access to individual elements. I have a ZK-6 box to help redistributing (Z-Systems).

**Dave Glasser** Some people would mix the vocal, for instance as a phantom center but also as a hard center, and so if you do need to do something specific to the vocal, you could always work with the center channel. So depending on how it's mixed, you can work around those limitations. I probably wouldn't want stems unless it was from a producer who really had their act together.

**Rich Tozzoli** I do six-channel mixes, no individual stems. It ends up as a continuous six-channel mix with all the audience between songs. Then I assemble and cut a sweetened six-channel master of one non-stop performance. However, I also bring some multichannel audience to mastering, where you can fly it in as necessary. Or we often create two six-channel tracks and have one six-channel song feed into another, then "glue" them together in a final layback to Pro Tools or a video deck.

**Dave Glasser** Actually I might prefer that if somebody was mixing totally in the box, which more people do—why not bring the whole session in? It could be a can of worms, but if it's a producer or an engineer who knows what they are doing and who has a sense of perspective, it might be easier than doing stems, as long as the producer understands this is not another mixing day. But if you envision "a little more vocal", or "I need to pan these guitars back a little bit further", then you have that option.

**Bob Ludwig** We seldom get stems in 5.1 except on large live shows where we get sent the entire session as it was mixed "in the box".

**Are most mastering sources Pro Tools sessions?**

**Rich Tozzoli** They're always Pro Tools except sometimes we print right to HD decks for broadcast, as they can take eight channels of audio. I'll bounce to disc or I print right to the video deck. I also bring the Pro Tools sessions as a safety, on multiple platforms including a DVD disc. If some little problem should arise, you can clean it up real fast.

**What do you discover when you hear your mix at mastering?**

**Rich Tozzoli** You actually hear positive information and the engineer helps your mix as that is what you go there for. I find a good mastering engineer makes things 20% better. I am very careful with my mixes, I QC them in headphones before they get to him. That is another tortuous process because an hour and half concert takes several hours time.

**Does a surround mix take longer to do than stereo?**

**Rich Tozzoli** Yes, because there is less masking—which makes it much harder. Surround brings out imperfections that you didn't know were there. So, if there is a noisy channel, you can't mask it. When I did David Bowie's Ziggy Stardust live with Tony Visconte, the 12-string was clipping on the original 16-track analog master from 1973, which was mostly masked in the stereo mix. In surround, it was very clear on the center channel, so we decrackled it with Waves—carefully judging the quality loss.

**Jonathan Wyner** Noises… that is a fascinating question. They are potentially more distracting in surround, depending on where they come from. When the noise comes from behind as it might for instance in the audience during a live concert, it has the potential to be a great distraction to the listener. What we try to do is keep people in the illusion that we're creating. The whiplash effect or whatever people call it can be extremely distracting [Tom Holman calls it the exit sign effect].

**Rich Tozzoli** In my approach I mix the surround first, followed by the stereo, because that reveals the most imperfections. While working in surround, you set your reverb, EQs, compressions and overall levels. Then it becomes much easier to do the two mix. A separate, independent two mix is not a folddown. You can then get a stereo record done in a day and a half to two days because everything is already set. It becomes a tweak session. However, the stereo is very much of a letdown compared to the surround mix.

Bob Ludwig

**Bob Ludwig** I would like to point out that Rich is the exception. Most engineers with whom I work do stereo first and then spread it out for 5.1.

**Would you call "getting the perspective from song to song" mastering?**

**Rich Tozzoli** Yes, especially with a surround concert broadcast (Dolby E-delivery) or a DVD concert video, which is a huge market. Audience cuts become so revealing that a whole sweetening session often happens after the mix. So, that is the combination of mastering and sweetening at the same time, which is an absolute art form in itself. When editing, if there is a noise in the surrounds and you have to make a cut, the unmasking reveals a lot of sounds. Often, you have to fly in a stereo audience pair to cover it up. So, what we are trying to do is cheat in and cheat out. Even if you do the finest live recording, you still sweeten the audience levels to make it more exciting. In surround it is that much more complex.

Rich Tozzoli

Luckily, I am able to ask a lot of engineers before they go into the live recording, "please put up boundary mikes, audience mikes, balcony mikes." Give me six channels of audience because sometimes they are going to overload, sometimes there are going to be people screaming at one mike. We may have to take out individual claps, as sometimes you can hear one annoying person clearly in surround. And that is something you might not want to spend time on at a costly mastering house. It is definitely what we call sweetening. Unfortunately few people have the money to take it beyond that process for the final mastering step because the budget is in danger. You sometimes have to beg to get mastering money in the world of surround. That is why projects like Blue Oyster Cult were not mastered, which absolutely should have been. They cut the money off and you beg, "Aah, just that one more little bit for the mastering." "Nope! Print it, it's going to tape." Unfortunately, that is what happens .

**Jonathan Wyner** When going from song to song, the most obvious thing that comes to mind is dealing with perspective vis-a-vis the center channel and sub-content. More often than not, I find that the amount of program that's located in the center channel varies wildly from piece to piece. This is especially true in pop music recordings where days, weeks, or months elapse between mixes. The center channel needs to be consistent enough to present a fairly consistent listening experience across an entire record. A consistent balance between phantom center and center channel needs to be struck. Typically I lean towards relying on a phantom center, and use the center channel to shore it up, to anchor the image.... consider the off-chance that a listener doesn't have their center channel up! It is something that mix engineers need to think about when they are preparing their mixes for surround. If you rely wholly on the center for a vocal, you might open up a can of worms.

I often redistribute the energy a little bit. Sometimes it is simply a matter of adjusting center level to side channels and sometimes it requires a little judicious midrange EQ. It's usually a subtle adjustment in order to get a sense of consistent width across an entire record. There have been instances where the issue was that there was no phantom center whatsoever and so I matrix the center channel.

**Dave Glasser** With stereo, every now and then you end up re-balancing the channels because maybe it's a little left heavy. With surround, we're re-balancing the channels quite often. Occasionally you cheat things by taking the front and bringing them into the room a little bit more.

**Bob Ludwig** To me, just as any stereo music needs mastering, the 5.1 needs mastering. On rare occasions we'll get something that comes in that's so good that it doesn't need anything. I've mastered some of Tom Jung's—five microphones direct to DSD, where it called for just re-panning some of those five microphones ever so slightly. It was such minutia.

## II. Surround Monitor Quality

### What level of monitor quality is required?

**Bob Ludwig** A mid or near-field speaker might be fine for a mixing engineer, but you would never want to master with those. We put a lot of effort into the studio and gear.

**Jonathan Wyner** You want something that's truly full range for mastering. I've heard mixes using the JBL 6328 Series self-powered that sound great... they're reasonable sounding speakers for a midlevel application.

I have clients who have mix rooms with fantastic monitoring that still come to me for mastering. I think that simply because somebody has a high skill level and excellent monitoring doesn't obsolete the skill set or the need for mastering.

**Bob Ludwig** It's also a function of how well they know the speakers. Look at all the great mixes I got from Bob Clearmountain. They were done on NS-10s. You could also look at the ITU recommendation for stereo monitoring [links at digido.com].

Jonathan Wyner

**Dave Glasser** Since surround is new territory, I try and get clients to come in with some mixes or send some surround mixes ahead of time. This is even more helpful for surround because there are so many more variables. The material I get mostly has the same problems as with stereo, the mix is too boxy or boomy.... One problem that I come across often is where the front and rear speakers are not integrated very well. I think it is mainly because of how they have their system set up and dialed in.

**Rich Tozzoli** I mix on NHT Pros at home. The reason I prefer them is they are the perfect blend of what the consumers can hear, which is ultimately what we really need to listen to, and what I need to hear as a music professional. I use full range Genelecs and a sub when working in New York City.

### Are Mastering Engineers becoming dinosaurs with high resolution monitors?

**Jonathan Wyner** I don't think so. There are often elements in a recording that are not necessarily audible in most control rooms or to consumers that have implications for what they hear regardless of the resolution of the delivery format. Low frequency 'P'-pops, for example can have a

fundamental frequency well below what most speakers might reproduce, but, if there is an AGC circuit somewhere that gets a hold of that and pulls it down it can create a jarring effect for the listener...and if it's not dealt with, one may find out about it later the hard way, at the most embarrassing moment. My ethic is, if there's a problem you've got to know it's there. You don't always have to fix it—distortion may be acceptable to a producer or artist, but you've got to know it's there. That's part of our job as mastering engineers.

**Bob Ludwig** In post production there is always a place for high resolution. No question about that.

**Do you have a surround-dedicated room?**

**Dave Glasser** My room was designed by Sam Berkow; we designed it so that it would work for surround. Most of the time, it is used just for stereo but the surround monitoring system is permanently set up. When people come in to do stereo, they usually ask, "Is the center speaker on?" and they say, "I swear I hear sound coming from that speaker." As for left-right angle, I find that 60° works well in stereo with my Dunlavys, which people have always said work better spread a little wider anyway.

## III. Level Calibration

**Are you mastering with the music surround calibration or 3 dB down (as in film)?**

**Dave Glasser** Music. Equal level.

**Jonathan Wyner** The same.

**Bob Ludwig** Equal level unless it is for theatrical release where the rears are a row of speakers instead of a single point source. It often seems there are film people for movie theatres, home DVD people, and broadcast people who don't realize there are different needs for each.

**Rich Tozzoli** We are using film surround calibration most of the time for concerts with video. It depends on the place. For music only productions, I would use flat. In broadcast, that is the fine line again where you will have to mix for the consumer. Bob turned my surrounds up 3 dB and I usually print them hot. It was definitely the right call.

**Are you doing full level or broadcast (-10 dB)?**

**Bob Ludwig** If something is going to be done for broadcast, then it is lowered about 10 dB, very often by us. It depends on where that tape is going. So sometimes we send that out and it is marked that this has been lowered for broadcast and other times, we say it is full level for DVD.

**Rich Tozzoli** We need to work towards simplicity, right now it is much too complicated. For the broadcast standard I use the L360, which is Waves multichannel limiter; we pretend -10 is 0.

**Jonathan Wyner** We are typically sending out files at full scale, though we'll drop the levels if we're laying back to video. If a network/uplink receives a program peaked to full scale, they'd have to make their own dub and knock it down, although there are a couple of instances I have heard about where that did not happen with disastrous results.

**Dialnorm and Consumer Dynamic Range Compression**

**Bob Ludwig** When we first started doing DVD video before DVD-A or SACD was invented, DTS was a big player. DTS really got surround off the ground with their 20-bit surround CDs. When the

DVD video format was invented, they pushed to have DTS as part of the specification. In 1997, when we were doing the early surround discs , you had to have something in the LFE or the demo person felt, "I just sold them a $60,000 system, and they can't hear anything out of the LFE. They're going to get worried."

Back in those days when consumers compared the Dolby Digital with the DTS in order for the Dolby to sound as loud as the DTS you'd have to increase the dialnorm to unity gain, -31.

DTS does not have any down mixing nor the "Dolby compression" option that is the bane of my existence. Lots of old Bose systems default to "compression on" unless you purposely turn it off. We did a live DVD and everybody approved the references. The producer of the video went out to the DVD plant to approve the run because he was behind schedule. He called saying "the sound's not right out here." I said, "it's got to be the Dolby Digital compression." He said, "Oh no, it is not that. They told me that it's not on. Plus every player I play in the facility has the same pumping sound." To make a long story short, apparently, up until that day, every Dolby system in their place had the compression turned ON in their decoders! He turned it off and said, "Ah, that's what I remember." So, they actually stickered those particular DVDs, "For best fidelity, in your DVD player's set up menu, set Dolby compression to *off*."

When we first started authoring, all discs defaulted to 5.1 because everybody was trying to push 5.1. So, we used to do dialnorm -31 as a rule. But if we author a disk that's got dialnorm at -31, now we have to make it default to stereo, because if you default to 5.1, you might hear a folddown instead of true stereo unless you take the time to go to a menu and change the audio settings, which a lot of consumers don't bother to do. If somebody listens to a folddown, the Dolby downmix compressors will just go nuts. You never want the listener to hear that. So, we defaulted to stereo so they would hear the dedicated stereo mix, which is usually a PCM stream or a pretty good Dolby 2.0.

This is an authoring, not a mastering problem. You need to have a dialogue between the authoring place and the producer. When James Guthrie did Dark Side of the Moon, he set dialnorm to -24, 7 dB lower in level, because that way it is more universally playable if somebody leaves the Dolby compression on. It is a workaround, which is very sad. So even a very loud group like Pink Floyd or Metallica is willing to forego 7 dB of level in order to have their disc not hit the Dolby compressor in case it's on. When they play a Led Zeppelin disc at -31, it would be 7 dB louder than Metallica now. So, in my opinion, instead of fixing the quality problem, the Dolby compressor situation is often making it worse.

**Jonathan Wyner** The variety of sound that comes out of the DACs in DVD players is astonishing. Sometimes the discrepancy of levels on a single disc is equally alarming. Once we're done with the main program, we might be handing it off to authoring and hopefully they will match levels of menu audio and extra bits to what we provide. It can get to the point where you have to throw up your hands... you just try to make it sound as good as you can.

## IV. Monitoring: Full Range vs. Bass-Managed

**Bob Ludwig** In the control room, I have the full range EgglestonWorks speakers that go down to 13 Hertz all by themselves plus I have a pair of M&K subwoofers, just for the "point one", there is no bass management. It sounds just glorious in there. In my client lounge right outside the door of my studio, is a highly bass-managed Bose Home Theater "Lifestyle" System. We have the movie EQ turned off and the dynamic compression turned off.

Happy to say that when the Bose system is hooked up the way I like, it translates beautifully from room to room. I usually check everything between the two systems. Occasionally I will hear something in the bass-managed system where it treats the bass a little bit differently than I had imagined it. There is a certain range of acceptability for EQ, I'm thinking "should it be a dB hotter or a dB lower at 60 Hz?". Sometimes what I hear on the bass-managed systems will influence that EQ decision. You definitely need to check all 5.1 on a bass-managed system, especially if there are phase problems with the bass.

Cars can be a great place to hear 5.1. In 2006 millions of cars came out with 5.1 for the first time. I hope it continues to grow.

**Dave Glasser** Our Dunlavys are not equal sized. I have model SC-Vs for left and right, four ways with 12 inch woofers and model SC-IVs for the centers and the surrounds. But the 4s for all intents and purposes are full range. I have a Martinsound ManagerMax bass manager that I can insert into the monitor path to check how it works on a bass-managed system. It redirects the low frequency information from all five to the subs. And it does degrade the sound of the speakers a little bit because they are not designed to be rolled off on the bottom. But you do get a good idea of how a bass-managed system is going to behave. We do this because we do not have the luxury of another room.

**Rich Tozzoli** I use the Waves M360 bass management setup. I do mix with bass management. I pop it in and out. You will see channel 6 in Pro Tools, which is the feed to the sub, disappear when you pop bass management out because I barely print LFE. I used to send a lot more than I do now, for example on the kick, I used to kick the LFE way up so it sounded great. But it was way too much, it was muddy.

But in New York I also run through Dolby AC-3 hardware encoders into a consumer home theater in another room.

## V. Are You Using ITU Monitor Layout?

**Dave Glasser** I do not use 110°. I am closer to 130°. That is closer to the NARAS recommendations. ITU is 110° but many people think that it does not work that well for music. I think 110° is definitely not far back enough. What we recommend is something greater like 130° or 135°.

**Bob Ludwig** In my mastering room, where I do mostly rock and pop surround, I have the rears at 135°. At home, where my listening is more classical oriented, I listen at 110° as I feel there is less of a "disconnect" between the front and the back, plus I know that all the European engineers are using ITU recommendations (and note, it is a recommendation, not a standard). Also at work, in one of my production rooms where I have a ProAc 5.1 system set-up, I also use ITU. For our loudspeaker QC pass, having the rears at 110° delivers more acuity in the rears than the more severe angles.

**Dave Glasser** One of the big problems is that it sounds like there are two different things going on. For instance, take a live concert. With the band in front and a couple of audience noises coming from behind, it is not very well integrated. So that is where tweaks end up often.

**Bob Katz** With all due respect, I think 135° is contributing to that dichotomy. That's why they didn't place surrounds behind us in the motion picture theater, because it is most disconcerting.

**Dave Glasser** I just know when I had these speakers at 110°, it felt like there was a big hole in the back. But once you work at it, you can get it to work really well. So I think it is due to people not thinking in terms of creating a sound field. They just see five speakers and decide—okay I'll put something there and I'll put something there, but they're not perceiving those five speakers as one integrated sound field. And that holds true whether you are doing an ambient rear production or whether you are going all out with discrete sources.

## VI. Has the Volume War Invaded Surround?

**Bob Ludwig** Let me state for the record that there is no need to have over-compressed recordings. Simply turn the playback volume clockwise if a recording sounds too soft.

The flat 5.1 mixes we got from Nine Inch Nails were mixed hotter than I would have dreamt of mastering it. The meters just pegged and never came off the peg. They are trying to get it as hot as they can get it, and I guess it works for their music. The mastered version came out at the same level as the mix. With DVD home theater, unlike movie theatres, there is no adherence to the 85 dB calibration standard. Fortunately, most of the people that are doing DVD video work do not have this louder-is-better mentality that some record A&R people have. I have to say that no one is going to tell Nine Inch Nails to bring their record down.

**Dave Glasser** I haven't had problems with overcompressed 5.1 material yet. I'm not doing a lot of rock & roll 5.1 though. I'm doing jazz, classical, and acoustic music. So the volume wars haven't caught up to that.

**Jonathan Wyner** Fear motivates people to compete in a volume war. You've got to stick to your sense of ethics and do what is right and what is best to create work that sounds as good as it can, and if that involves making something that's 6dB lower than some "Crushers" idea of what's supposed to happen then so be it. The real question is how can we keep our livelihoods? Certainly not by producing unlistenable music that's distorted and causes ear fatigue. We need to stick to our guns and produce work that is worth listening to!

One small silver lining in the demise of record labels is the disappearance of a centralized filter, focusing people's attention on any particular body of musical work. There's not so much direct comparison these days and audiences are getting more different kinds of music from different places and having more different kinds of listening experiences. In some cases that allows us to relax levels. I do not think artists are generally dealing with program directors or A&R offices set up with surround systems who are judging one thing better than another because it is louder. Surround just hasn't penetrated that part of the market place.

I have some clients for whom it was important that the surround version of their project have a little bit more impact and be slightly louder than the stereo version, and this is just in terms of SPL in the room. Fortunately, having the multiple channels in the room provides enough additional SPL. It doesn't take a lot for a 5.0 version to compete with its stereo companion simply because there is more energy from the 5 speakers. You'd be surprised.

My contention is that something that is well recorded and well mixed, sounds loud naturally. I personally try not to adopt a practice simply to compete, but focus on the values, ethics and goals manifest in an artist's work.

## VII. Sample Rates

### Do you see any 96 kHz surround material?

**Bob Ludwig** Oh! Totally, we get that and 88.2 kHz and even 192 kHz. The Nine Inch Nails was at 192 kHz. In the past we seldom got 48 kHz masters. Now, where things are pretty much DVD video oriented, I have noticed more projects at 48 kHz. When we were doing SACDs and DVD-As, it was all 96 kHz. We still get a lot at 96 kHz even now.

If a project comes in at 96K, we'll master at 96K and then we will give them six broadcast wave files at 48K for Dolby encoding. We've still got the high resolution 96K files for a future DVD-A or Blu-Ray. If they make files for digital downloads, they can encode down from that.

**Rich Tozzoli** I do. Mostly for audio-only projects. 96 kHz usually ends up on a SACD. That is a very difficult choice right now because there is no other way for non-video music to go in surround.

**Jonathan Wyner** Program comes in at all sample rates. I don't see a future for 96 kHz surround material for consumers right now. We are delivering 44/24 for music, 48K for soundtracks and DVD video.

## VIII. Destination Media

### Is SACD a dead medium?

**Jonathan Wyner** I believe that's right and the players are starting to disappear.

**Bob Ludwig** I don't think SACD is dead, I think it won the format war. In Europe, there are still a lot of indie labels that do hybrid SACDs. I listen to the BBC-3 "CD review" program. Often the new releases they highlight are from labels like Telarc, Delos, Hyperion, Pentatone and Harmonia Mundi and they're hybrid SACD's. That makes so much sense to me. As a classical buyer, I will buy an SACD before I buy a stereo CD. In the US 1.3 million SACD players have been sold in the past 9 months; they are in the PlayStation 3. Today, Arkivmusic.com lists 1,461 available SACD titles. I'm telling you, it's not dead.

### What about Blu-Ray and HD?

**Jonathan Wyner** I think these could be viable formats but the success will not be driven by fidelity…

Let's think back for a minute to the history of consumer formats. At what point in the last 40 years has any format had success in the wider marketplace where it didn't represent an obvious advantage in terms of convenience or economy for the consumer? In terms of fidelity gains, if the internet infrastructure improves so that the majority of consumers have much faster connections than they have now, I am hopeful that at least 44/16 resolution, if not higher, will become the minimum de facto standard.

**What about just plain DVD for a surround music release?**

**Jonathan Wyner** If you mean DVD video, it's a good format from the standpoint of compatibility but audio-only is bound to be something of a vanity project right now as there is not a substantial commercial market for it.

**Rich Tozzoli** It's not that I have abandoned that format by any means, but I have told artists to remain format agnostic right now. Just hold on to your six-channel masters and let the wave ride out. We, the producers, engineers, artists and ultimately the public are suffering because competing formats create a conflicted consumer. The struggle is about the bottom line of a dollar over what is right, and that in itself is wrong. I do hope it gets worked out, even if one format has to disappear like the Beta deck, for us all to have a proper multichannel platform to work with. Consumers love surround sound, and if they could have one logical format to buy, they would buy it!

I always default to the fact that most of my mixing is for broadcast. I feel very strongly about the world of HD television for the delivery of surround. Anybody who thinks they are going to do surround mixing for audio only is sadly mistaken, as it will not happen.

**Jonathan Wyner** Beck's recent release was a surround that doesn't have any video associated with it. It's visual, there is a slide show along with it, but it is not a music video per se, the slide show is pretty psychedelic. It's abstract and seems designed to complement the abstract nature of the aural experience. My hat's off to him. He lives in his ears and his sense of orchestration and arrangement is wonderful.

I think the primary context for working in surround for the foreseeable future will be audio married to video although there will continue to be niche markets representing all kinds of interests: high res, audio only, etc.

I can't get too worked up over exactly what the marketplace will support or is going to allow. It's something I have no control over, so why worry about it. I never saw surround as having a profound impact on the wider marketplace or being something that would drive consumers to a particular format. Example - *Diana Krall: Live in Paris*. Fantastic performance, beautiful video direction, and sound! Al Schmidt did an incredible job. People who are inclined to buy that disc are simply going to appreciate the fact that it looks and sounds great, not that it is surround.

You can produce an economical video, though not on that same level (with 9 cameras and the orchestra she used in a couple of numbers). Test versions of my recent DVD production cost us about $5000. But video and surround are not within reach of a lot of

artists and may not be justified. I think it is important for people to understand that the post-production steps in surround, especially when video and authoring are concerned, are profoundly more complicated and time consuming.

I recently completed two audio-only productions, no video, the exception rather than the rule... One is a piano recording; six pianos recorded at the Brooklyn Academy of Music scored specifically for surround by a composer named Gordon Green. And the other, a project by singer-songwriter Laura Austin, produced by Mark Doyle. It's a beautiful project. They're bound and determined to release it as an audio-only DVD-A in spite of the fact that the market is so small. Otherwise a more typical project is a song by the Click Five that we're doing a surround mix from stems for inclusion in a Disney picture. So, it is eventually going to be incorporated into a visual product.

## IX. Authoring

**Bob Ludwig** We were the first mastering studio to offer authoring in December of 1997. We have D5s and digital NTSC and PAL Betacams. We do just as much work in PAL as we do in NTSC. It is relatively easy for us to supply surround clients with any sort of reference. If you don't have authoring available, you should be able to afford Dolby or DTS software for references. But most of our clients come in with video and want to see references with the video.

**Dave Glasser** For SACD, we can provide the finished cutting master. But not for DVD—maybe some day, but that requires a dedicated "nerd" who knows how to do the authoring and we don't have that.

**Jonathan Wyner** We provide simple authoring services in-house and can build a DVD video or DVD audio ref. And we have a world class authoring facility on the other side of the wall that can make discs jump through hoops. It is fantastic not only given the proximity and their skill level, but everybody over there comes from an audio world with the exception of one guy who is their video and film geek.

## X. Reverb in Surround

**Dave Glasser** We end up using reverb a lot more often than ever comes up in stereo and I have the TC 6000, which is fantastic, and a Sony S777.

**Rich Tozzoli** I have also been using impulse responses since they were first introduced with the Sony S777. Once I heard the unit, I recognized the value. As a surround engineer, part of what you can recreate in the world of surround is the acoustic space. Now, I know how to do my own impulse responses and I will use those in the mix because they work. I was very lucky that I got Waves to capture some of my favorite recording spaces, Trinity Church, New York, and I got AudioEase Altiverb to capture the Club House. So, I have my two favorite acoustic spaces as presets in people's impulse response folders. Software impulse responses make a surround experience much better because surround is about envelopment and realism to me.

**Dave Glasser** What you can do and what I have done a couple of times is, using the TC—in the front channels, add some early reflections and in the rears bring in the reverb.

CHAPTER 20

# High Sample Rates: Is This Where It's At?

## I. Introduction

After curing the wordlength blues—it's time to tackle the sample rate issue. Regardless of the real benefits for the professional and the consumer, the current relentless drive for higher sample rates is lucrative for the hardware manufacturers. Engineers who must replace their expensive high-resolution processors to keep up with the Joneses will spend big bucks.

I've been working with higher sample rates for several years.[1] A great number of engineers think that higher sample rate recordings sound better, pointing to their ability to reproduce extreme high frequencies. They cite the *open, warm, spacious, extended* sound of these recordings as evidence for this contention. But let's offer an alternative explanation. First, objective evidence shows that higher bandwidth cannot be the reason for superior reproduction—since **the additional frequencies that are recordable by higher sample rates are inaudible.** How can our ears detect differences between 44.1 kHz, 96 kHz and even 192 kHz sample rates since most of us can't hear above 15 kHz?

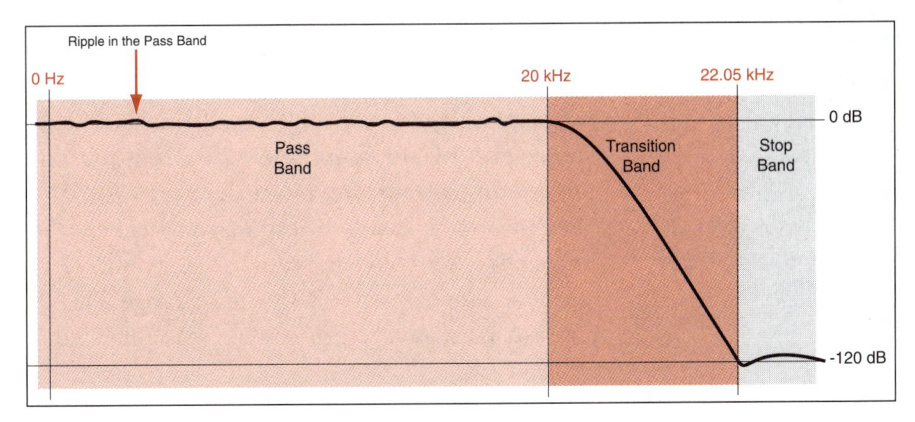


*Low-pass filter terminologies. The **passband** is the part of the frequency response which is not filtered or attenuated, in this example, from 0 Hz to about 20 kHz. This figure shows some **passband ripple** (non-flat passband frequency response). The **transition band** begins at the nominal cutoff frequency, until the **stop band**, where the response reaches the maximum loss of the filter. To avoid aliasing, the stop band must be at or below the Nyquist frequency. The steepness of this transition band is the slope of the filter. In this example, the transition band is only about 2 kHz wide.*

**Nyquist, Sampling and Aliasing**

**Why filter on ADC**
Sampling without filtering will include ALL signals, from the baseband that you want to keep, along with the out-of-band stuff you DON'T want, all the way out to infinity. This folds down (aliases) to the baseband, producing alias distortion, which sounds a lot like ring modulation, especially obvious on instruments like trumpets that have lots of high frequency harmonics. That's why an antialias filter is required every time audio is sampled.

**Why filter on DAC**
The sampled audio stream which is played back contains the baseband and EVERY image of that baseband, all the way out to infinite frequency. That's why an anti-image filter is required when going from sampled to continuous.

I believe the answer to the dilemma lies in the design of digital **low-pass filters**, used in **oversampling A/D and D/A converters and sample rate converters** (see figure on the previous page). Filters of lower quality or which are unoptimized, exhibit tradeoffs such as low calculation resolution, higher distortion, ripple, ringing, and potential for aliasing. The artifacts of ripple are time-smearing of the audio, and possible short (millisecond) echoes. Aliasing is a form of distortion which occurs if the filter does not have enough attenuation in the stop band (see sidebar). To avoid aliasing, we must use either a very steep filter, or a gentle filter with a higher cutoff frequency (which requires a higher sample rate). It is harder to engineer a steep filter with low ripple, but it is perfectly doable; this can be achieved with a large number of filter taps. For the same number of taps, a more gentle filter will have less ripple. Ripple in the passband should be less than 0.1 dB.[2]

*"The filters in a typical compact disc player or in the converter chips used in most of today's gear are mathematically compromised."*

**Oversampling**

One of the biggest improvements in digital audio technology came in the late 80's, with Bob Adams' **oversampling ADC**; this form of ADC has a front end which operates at 64 or 128 times the base sample rate. In other words, for 44.1 kHz operation, a 128X converter operates internally at 5.6448 MHz! The converter's noise is spread around a wider frequency spectrum and shaped, moving much of the noise above the audible frequency range. This high rate must then be digitally downsampled to the destination rate, at which time the supersonic noise is filtered out, to yield as much as 120 dB signal-to-noise ratio within a 20 kHz bandwidth.

Downsampling is accomplished with a digital circuit called a **decimator**, a form of divider or sample rate converter, which must contain an anti-aliasing filter. An oversampled DAC has an **anti-image filter** with an analogous role; though it operates at a higher sample rate, it too must have low distortion and ripple to sound good. While this was once costly to implement, the price of silicon is now infinitesimal compared to the benefits of good filtering. Still **the filters in most converters made today are compromised.** There is no longer a reason for this practice—chipset manufacturers should begin doing it right.

To overcome the limitations of current converter chips, high-end converter manufacturers can either roll their own discrete converters (very expensive) or create fixes using off-the-shelf components. Some manufacturers add filters of their own design to supplement the chipset filters. For DACs, they upsample in front of the chip's own filter, so that filter does not have to work as hard. These hot-rodded DACs operate at 88.2 or 96 kHz regardless of the incoming rate. For ADCs, these manufacturers run the converter at double rate, also easier on the built-in filter, followed by their own high-quality SRC. Supplementary filters would be unnecessary if the manufacturers of the chips used higher quality filters in the first place.

#### An Upsampling Experience

Audiophiles, and some professionals, have been experimenting with digital upsampling boxes which are placed in front of DACs, supplying specious reasons to justify them.[3] In some cases they report greatly improved sound. Although the improvement may be real, in my opinion it can be attributed to the various digital filter combinations, not to bandwidth or frequency response or (especially) the sample rate itself. Remember that 44.1 kHz sample rate recordings, already being filtered, cannot contain information above 22.05 kHz. An upsampler cannot "manufacture" frequency information that wasn't there in the first place.

I've compared the sound of upsamplers against DACs working alone. Sometimes I hear an improvement, sometimes a degradation, sometimes the sound quality is the same either way. Sometimes the sound gets brighter despite a ruler-flat frequency response, which can probably be attributed to some distortion. **Sonic differences have come down to mathematics in this digital audio world**.

> *"The issues of the audibility of bandwidth and the audibility of artifacts caused by limiting bandwidth must be treated separately. Blurring these issues can only lead to endless arguments."* — Bob Olhsson

## II. The Ultimate Listening Test: Is It The Filtering or the Bandwidth?

In December 1996, I sought to systematically find reasons for sonic differences between sample rates, performing a listening test, with the collaboration of members of the Pro Audio maillist. The question we wanted to answer was: *Does high sample rate audio sound better (or different) because of increased bandwidth, or because of less-intrusive filtering?* We developed a test that would eliminate all variables except bandwidth. Other major factors were held constant, sample rate, filter design, DAC, and jitter.

The test we devised was to take a 96 kHz recording, and compare the effect on it of two different low-pass filters. The volunteer design team consisted of Ernst Parth (filter code), Matthew Xavier Mora (shell), Rusty Scott (filter design), and Bob Katz (coordinator and beta tester). We created a digital audio filtering program with two impeccably-designed filters which are mathematically identical, except that one cuts off at 20 kHz and the other at 40 kHz. The filters were designed for overkill, with exemplary characteristics: double-precision dithered, FIR linear phase, 255-tap, >110 db stopband attenuation, and <.01 dB passband ripple.

For the first listening test, I took a 96 kHz orchestral recording, filtered it and laid both versions into a Sonic Solutions DAW for

Continuous is another word for "analog".

The higher the sample rate, the higher the permitted filter cutoff frequency, 1/2 of the sample rate, known as the Nyquist Frequency.

The same basic rules apply to resampled digital streams. In other words, any sample rate converter needs to properly apply anti-aliasing and anti-imaging filtering as it involves re-sampling, very similar to the processes used in ADC and DAC.

Usually, the filtering is built into the resampler, transparent to the user.

**Contributed by Dick Pierce**

**MYTH:** *Upsampling makes audio sound better by creating more points between the samples, so the waveform will be less jagged.*

comparison. I expected to hear radical differences between the 20 kHz and 40 kHz filtered material. But I could not hear any difference! Next, I compared the 20 kHz filtered against "no filter" (of course, the material has already passed through two 48 kHz filters in the converters). Again, I could not hear a difference! The intention was to listen double-blind; but even sighted, 10 additional listeners who took part in the tests (one at a time) heard no difference between the 20 kHz digital filter and no filter. And if no one can hear a difference sighted, why proceed to a blind test?

I then tried different types of musical material, including a close-miked recording of castanets (which have considerable ultrasonic information), but there was still no audible difference. I then created a test which put 20 kHz filtered material into one channel of my Stax electrostatic headphones, and the time-aligned wide-bandwidth material into the other channel. I was not able to detect any image shift—there was always a perfect mono center at all frequencies in the headphones! This must be a pretty darn good filter!

As a last resort, I went back to the list and asked maillist participant Robert Bristow Johnston to design a special "dirty" filter with 0.5 dB ripple in the passband. Finally, with this filter, I was able to hear a difference... it added a boxy, veiled, "gritty" quality that resembles the sound of some of the cheaper CD players we all know.

After I conducted my test, several others have tried this filtering program, and most have reached the same conclusion: **the filter is inaudible**. One maillist participant, Eelco Grimm, a Netherlands-based writer and engineer, performed the test and reported no audible differences using a Sonic Solutions system, yet he and a colleague passed a blind test between filtered and non-filtered using an Augan workstation. He did not compare the sound of the 20 kHz versus 40 kHz filters, so we are not sure if he was hearing the filter or the bandwidth (I suspect the filter). We are not certain, but perhaps the reason Eelco uniquely reported a sonic difference is that the Sonic system produced sufficient jitter to mask the other differences, which must be very subtle indeed! Be aware that two other 48 kHz filters in the chain may have obscured the audible effect of the test filter, so it is very difficult to design a perfect test.

This 1996 test seems to show that a "perfect 20 kHz filter" can be designed. Regardless of whether Eelco's group did reliably hear bandwidth differences, it should be clear by now that differences people hear between sample rates are more likely due to filter design than to supersonic bandwidth. Ironically, it was necessary to make a high sample rate recording in order to prove that high sample rates may not be necessary.

## III. The Ultimate Sample Rate

**Let's be logical: since the human ear cannot hear above (nominally) 20 kHz, then any artifacts we are hearing must be in the audible band.** Audio researcher Jim Johnston,[4] who knows as much about the time-domain response of the ear as anyone, has shown that steep low-pass filters at or near the high frequency limit of the ear interact with the cochlear filter, creating pre-echoes which the ear interprets as a loss of transient response, obscuring the sharpness or clarity of the sound. Jim has experimentally calculated that the minimum sample rate which would support a Nyquist filter gentle enough to elude the ear is 50 kHz, so if he is right, then the 48 kHz professional rate is nearly sufficient.[5]

We also have to consider cumulative effects, for even if an inaudible filter can be designed, will 2, 3 or 4 in series also be inaudible? Perhaps this is irrelevant, as researcher Peter Craven has discovered: **Ringing or pre-echo problems in a filtered system can be completely eliminated by adding a properly-specified gentle slope filter anywhere in the record or reproduction chain.**[6] It seems counterintuitive that such a filter placed at the tail end of a chain can repair previous ringing issues, but Dr. Craven has the mathematical proficiency to prove this, so his paper ought to have a profound effect on how converters and digital systems are designed. His discovery may explain why some digital audio systems sound better than others; it may explain the discrepancy between my listening test and Eelco's. For example, if Eelco was listening through a DAC with a steep filter, and I was listening with a gentle—that could override the effect of the sharp filter under test. Craven's discovery alone is justification for using 96 kHz sample rate, or upsampling to that rate for reproduction. Converter and systems manufacturers must review this research and provide tools for better-sounding digital audio; all it takes is the impetus to do it right.

**How good is 44.1 kHz sample rate?** The answer: A lot better than I used to think: that sample rate conversion to a lower rate drastically reduced sound quality. But with an improved upsampling DAC the audible difference between a 96 kHz original and a 44.1 k result is subtler.

*"A well-designed DAC should exhibit very little audible difference between sample rates"*

Once again we point to the filters as the culprits, not the sample rate or the downsampler (the exemplary Weiss SFC, see figure below). A well-designed DAC should exhibit very little audible difference between sample rates. Can 44.1 kHz ever sound equal to 96 kHz? It may be impossible to find out without building a custom, discrete DAC. A more effective question would be: Is 192 kHz necessary? We'll discuss that in a moment.

We need to perform objective experiments to determine the lowest practical sample rate to use without audible compromise, to help get us out of this expensive, marketing-driven sample rate war. If 192 kHz sounds even marginally better than 96 kHz I can assure you it is not because we have suddenly

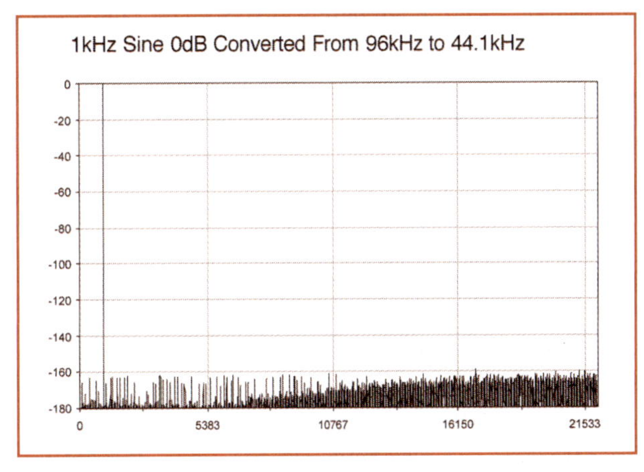


*The textbook-perfect distortion and noise performance of a Weiss sample frequency converter. With this SFC it is possible to convert between non-integer rates with identical measured performance. In other words, no difference between downsampling from 88.2 kHz or 96 kHz to 44.1 kHz.*

*"The filters in a typical compact disc player or in the converter chips used in most of today's gear are compromised."*

developed supersonic hearing. And when comparing 192 to 96, be sure you know what you are comparing; since all modern converters are upsamplers, the only difference between a manufacturer's 192 kHz and 96 kHz converter is probably the decimator. Is 192 kHz necessary? Dan Lavry has studied the objective characteristics of conversion at 192 kHz and found that at those extreme rates, distortion increases and conversion accuracy decreases.[7]

### The Advantages of Remastering 16/44.1 Recordings at Higher Rates

Researchers such as J. Andrew Moorer of Sonic Solutions, and Mike Story of dCS have demonstrated theoretical improvements from working at a higher sampling rate. Moorer pointed out that post-production processing, such as filtering, equalization, and compression, will result in less distortion in the audible band, as the errors are spread over twice the bandwidth—and half of that bandwidth is above 20 kHz.[8] Measurements discussed in Chapter 17 confirmed these conclusions. In addition, if the destination after processing is a high-resolution medium, then the master can be left at the higher sample rate and wordlength, avoiding another generation of potentially sound-veiling 16-bit dither and the consequences of low-pass filtering at the end of the process. Thus, consumers should not scoff at DVDs which have been digitally remastered from original 16-bit/44.1 kHz sources. They will be getting real, audiophile-quality sonic value in their remasters.

1 I was the recording engineer for the world's first 96 kHz/24-bit audio-only DVD.

2 According to Julian Dunn.

3 One argument by those who do not fully understand the nature of PCM is the "connect the dots" argument, which goes like this: 'We need more dots than only 2 to properly describe a 20 kHz sine wave.' But this is erroneous; only 2 dots (samples) are necessary to describe an undistorted 20 kHz sine wave; when reproduced through a DAC, the low-pass filtering smooths out the waveform and eliminates all the glitches.

4 In correspondence. JJ is the inventor of the science of perceptual coding, which led to coding developments such as mp3, Atrac, etc.

5 This is based on the length of the shortest organic filter in the human ear. Jim Johnston notes that the 50 kHz number nicely matches the original work with anti-aliasing filters done by Tom Stockham for the Soundstream project.

6 Craven, Peter (2003) Controlled Pre-response Antialias Filters for Use at 96 kHz and 192kHz, AES 114th Convention Preprint 5822.

7 A link to Lavry's papers can be found at digido.com/links/. Lavry also warns (in correspondence):

> Using an ADC designed for 192 kHz operation at 96 kHz is not the same as using an ADC intended to operate at 96 kHz! The 192 kHz design already has the accuracy tradeoffs; using it at 96 kHz does not remove the tradeoffs, it is in fact the same conversion with a X2 additional decimation.
>
> ...anyone that decimates or does sample rate conversion from 192 kHz to 96 kHz or 48 kHz or 44.1kHz [who] says they hear a particular sound from that original 192 kHz conversion, is in fact supporting the fact that what they hear resided all along under the new Nyquist.

A related reaction comes from Crispin Herrod-Taylor (in correspondence):

> Probably the only real way of proving whether 192 K sounds better than 96 K is to get a good ADC chip [with no compromise] running at 192K, and then add a DSP which has optimised filtering for the 192K and 96K.

8 Julian Dunn (in correspondence) clarifies:

> A 3 dB reduction in distortion results because the error products are spread amongst twice the bandwidth. This is true for **uncorrelated** quantization errors which fall evenly throughout the frequency range from dc to fs/2. And does not work for distortion products which will correlate with the signal [such as from compressors].

Nika Aldrich (in correspondence) qualifies:

> ...increasing the sampling frequency [simply] in order to increase dynamic range is an exercise in futility: the effect is swamped by other forms of noise anyway.

Jim Johnston (in correspondence) indicates:

> processing at higher rates is **required** for any non-linear processing, such as compression. These non-linear processes produce new frequency components, some at higher frequencies.

Conclusion: Other arguments aside, a high enough sampling rate for processing is required to avoid aliasing of these new frequency components (see Cranesong and Weiss FFTs in Chapter 17).

CHAPTER 21

# Jitter–Separating the Myths from the Mysteries

## I. Introduction

One of the hardest-to-explain phenomena in digital audio is **jitter**, because understanding the influence of jitter on our digital recordings means we have to reconsider much of what we learned from years of analog experience. In a classic Marx Brothers movie, Groucho's girlfriend catches him embracing another beautiful woman. In defense, Groucho quips, "Are you going to believe me, or your own eyes?" Let me apply this to audio and ask, "Are you going to believe the facts, or your own ears?" For in this digital audio world, sometimes we have to re-evaluate the evidence of our senses. Fortunately, this re-evaluation is based on well-established physical principles.

In 1980, because most sound system's digital converters and processors had low resolution, jitter errors were not regarded as a very high priority compared to nonlinearity, noise modulation, truncation, improper dithering, aliasing and other errors which created more audible problems. But today, where audio performance frequently reaches 20-bit level (the limit of most current "24-bit" converters), jitter problems are more evident. The symptoms of jitter mimic the symptoms of other converter problems—blurred, unfocused, harsh sound, reduced image stability, loss of depth, ambience, stereo image, soundstage, and space—though usually in such a subtle way that it can take time for even a critical ear to learn to identify them.

What causes these problems? Is our digital audio actually being affected by jitter in our clocks? The simple answer is: Sometimes yes, mostly no!

Should we believe our ears? It'll take a whole chapter to sort this one out.

## II. What is Jitter?

Digital audio is based upon the concept of sampling at regular time intervals. Keeping those intervals constant requires a consistent clock. If the frequency of the clock varies during A/D conversion, then since the waveform will be at the wrong amplitude at each sample point when the digital audio is played back, the audio will be permanently distorted. That's why it is critical to

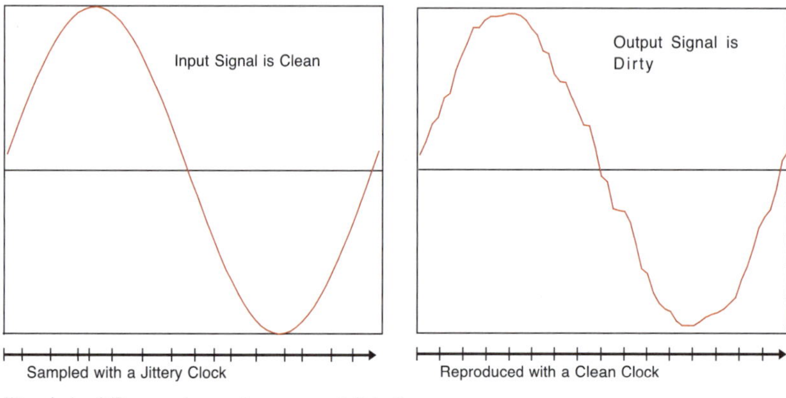


*Jitter during A/D conversion creates permanent distortion*

have a consistent clock during A/D conversion. Similarly, an inconsistent clock will yield distortion during D/A conversion. We call this inconsistency *jitter*. One period of a 44.1 kHz clock is 22.7 µs. Our tests seem to show that variations in that period as short as 10 ps may cause audible artifacts, depending on the quality of the reproduction system and our hearing acuity. As sample rate increases and wordlength expands, jitter must be proportionally lower to maintain sound quality, because jitter affects the absolute noise floor. Jitter produces sidebands (additional frequencies, or tones) that mask inner detail in a recording.

We can measure jitter in two places:

1) **interface jitter**, the jitter present in the interconnections between equipment.

2) **sampling jitter**, the jitter in the clock which drives the converter. Luckily, sampling jitter can be so reduced that it becomes inaudible: if a converter has excellent internal jitter rejection, then even high interface jitter may not result in audible sampling jitter. In this chapter, we are mostly concerned with sampling jitter, because interface jitter is rarely important unless it causes a breakdown in communication between devices.

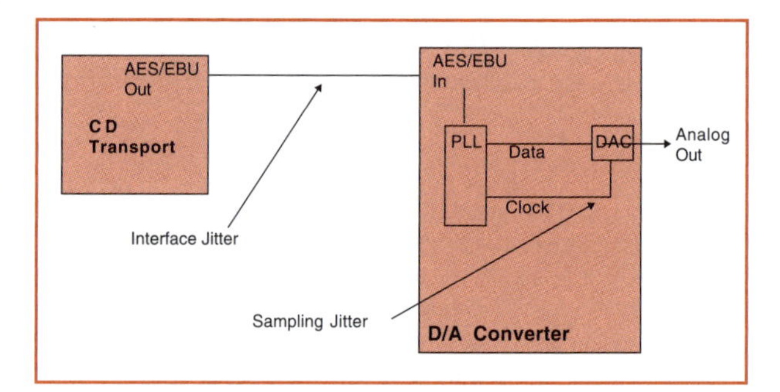


In the figure above, it is up to the PLL (Phase Locked Loop, to be explained) inside the DAC to create a very high frequency sampling clock which drives its components. If it is a superb PLL (very rare), none of the artifacts of incoming interface jitter will be transmitted to the sampling clock.

## III. Jitter, When it Matters, When it Doesn't

As I said in the first edition, if leaping to conclusions about jitter were an Olympic event, sound engineers would win the gold medal—an entire subculture continues to develop around digital cables and word clock generators in an attempt to achieve better sound reproduction. This has led some engineers to change cables everywhere they hear that such a replacement makes a difference, or to experiment with "stable" external clocks, each of which produces a different sound.[1] I don't blame them for trying, but the fact is that jitter problems must ultimately be solved **within** the converters themselves. No cable can remove the inherent jitter problems in the AES/EBU-SPDIF interface, because the embedded clock interacts with the data stream. Thus, external jitter reduction units will always be limited in their effectiveness because jitter may be increased at the output interface between the jitter reducer and the DAC.

Since engineers hear improvements with the better cables,[2] they conclude the cables will also improve their digital audio processors. But this is largely a misconception (shortly we'll describe the very few exceptions). **Audio processors process data, not clock**, so any sonic improvement is due to a cleaner clock being passed to the DAC, not to a difference in the data being processed. Believe the facts, not your own ears! The listening problem has an immediate solution—a better DAC!

**How to Lie With Measurements**

*"Traditional audio signal-to-noise ratio measurements have (almost) no relationship to the sound of a converter when it is receiving signal."*

Clock jitter can produce insidious audio artifacts in converters. Manufacturer's specifications often hide these artifacts because there is no established criterion for the effects of jitter on converters. For example, some ADCs (and a few DACs) report exceptional >120 dB signal-to-noise ratios, theoretically equivalent to >20-bit performance, but is this true in practice? These figures are obtained by the traditional method of calculating signal-to-noise ratios: first measuring a full-scale sine wave signal, then removing the signal and measuring the residual analog noise. But this method does not take into account *the noise modulation and distortion when a clock is jittery and the audio signal is complex (such as music)*, which accounts for some of the previously-unexplained sonic differences between converters. Most *signal-to-noise ratio* measurements quoted in manuals are therefore irrelevant, and most people have never heard true 20-bit performance, let alone 24.

**Digital Print-Through.** Ideally, the converter's PLL should completely reject incoming jitter with its clock-smoothing circuit, as inadequate *jitter attenuation* allows some of the incoming jitter to pass to the critical conversion clock. The worst sounding type of uneliminated jitter is **signal-dependent jitter**, caused by the designs of external interfaces such as AES/EBU and SPDIF. Although signal-dependent jitter is analogous to analog tape flutter, it behaves very much like analog tape print-through because it adds a blurred quality to the sound. Around 1975, analog tape manufacturer BASF demonstrated that an analog tape with lower print-through sounds cleaner and quieter than a tape with higher print-through even if the latter has a lower hiss level.[3]

Similarly, a converter which successfully rejects jitter can sound much cleaner than another with a lower static noise floor. As well as **signal-dependent effects** (which yield distortion from intermodulation between the sample rate and the audio signal), jitter can produce **random effects** (which translates to a higher random noise floor which can also be signal-dependent), and **discrete frequency effects** (such as other clocks in the box producing random tones and inter-modulation between the other clocks and the main sampling clock). Some of these effects are more benign to the ear than others, which is why it is so difficult to find a single criterion for evaluating jitter performance.

### Storage Media

*There is no jitter on a storage medium*—only the data is stored, not the clock (there is no clock on a compact disc). A new clock is generated on playback, and jitter may arise only when data is clocked out of the medium. Bits are usually stored in a very irregular fashion: on hard disks, the data may be out of order, non-contiguous, and widely spread. Data stored on CD (in EFM format) must be unscrambled and decoded during playback, but scattered storage has nothing to do with *jitter*, since time is not involved until the data is played back. To find the causes of playback jitter, we have to study the complete mechanism.

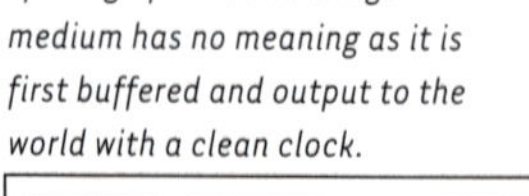

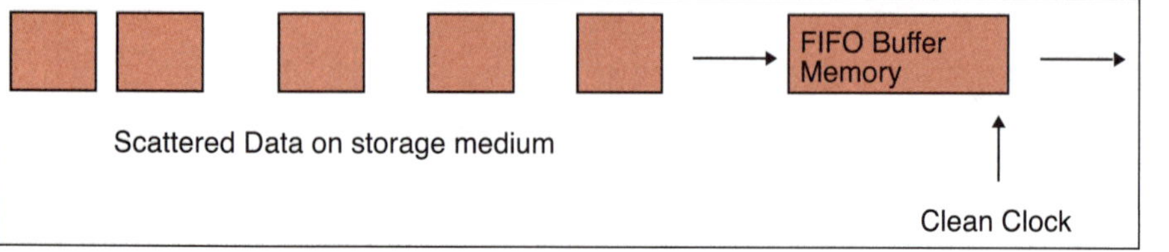


*Spacing of data on storage medium has no meaning as it is first buffered and output to the world with a clean clock.*

During playback, widely scattered data is collected into a buffer memory whose output is controlled by a steady clock (pictured below left). The quality of that clock and its driver circuitry is the origin of any interface jitter. Manufacturers differ widely in their abilities to keep outgoing clocks under control but all face the obstacle that clock stability is not important to the computer-based technology that we have adapted to digital audio. In fact, the standard computer hard disc interfaces are *asynchronous* (non-clocked), having a completely irregular output with enormous equivalent jitter. When such non-clocked interfaces are used, it is the task of following circuitry to make the data conform to a steady clock.

### Clock Stability Requirements for Converters

An ordinary crystal oscillator is sufficient for a computer that processes data, but audio converters require an extraordinarily stable master oscillator. To get 20-bit performance at 44.1 kHz requires oscillator stability (jitter) at or below 25 ps peak-to-peak.[4] One nanosecond (1000 ps) in the time domain equates to 1 GHz, which is why a critical converter's circuitry must be shielded and isolated from even the tiniest RFI or clock leakage that can enter via power supply, grounds, or emissions. It is now obvious why good-sounding converters are rare and expensive, and why the converters on most computer-cards do not sound very good: there are a lot of interference and power supply issues within a computer chassis.

## IV. How to Get the Best Performance from Converters

There are two ways to clock a converter:

a) **Internal Sync**, where a (hopefully) stable crystal clock located inside the converter (very close to the sampling clock pin of the converter chip for the best audio performance) directly drives the circuitry. This is not very costly in parts but does require good layout and power supply design.

b) **External Sync,** which as we have seen cannot be used directly, and requires a PLL, perhaps the fundamental culprit of jitter-induced converter artifacts. The PLL has to filter jitter caused by poor source clocks, by the AES/EBU line itself, or by interference along the cable which brings in the clock. Thus, the common use of unbalanced wordclock cables can produce ground loops in the clock signal itself. Since it is far more difficult and expensive to build a good PLL than a stable crystal oscillator, **only in an excellent converter design can jitter performance via PLL be as good as, or negligibly worse than, via internal clock.**[5]

Examples of External Sync:

i) **AES/EBU sync**, which is prone to *signal-related jitter*, also known as *program-modulated jitter* or *data-dependent jitter* as first illustrated by Chris Dunn and Malcolm Hawksford in their seminal paper.[6] Thus AES/EBU "black" will produce a cleaner clock than AES/EBU with signal, with a typical PLL. A "smart" PLL in a converter can still reduce this interface jitter to inaudibility.

*{ "Only in an excellent converter design can jitter performance via PLL be as good as, or negligibly worse than, via internal clock" }*

ii) **Wordclock sync**, which can yield extremely low jitter, because the PLL required is simpler. Despite this, only a small number of the converters I've tested have inaudible degradation due to jitter under wordclock, and even fewer under AES/EBU!

iii) **Superclock sync**. Superclock is a very high frequency clock, as much as 256 times the base sample rate. The object of superclock was to avoid a PLL entirely as the native clock in converters is already at this high rate. However, there is no such thing as a free lunch—it is very fragile, and manufacturers must still pay attention to jitter issues with superclock.

iv) **Other Interfaces. The embedded clock in** Firewire should not be used to drive a system. For lower jitter, a separate wordclock cable is needed or even better, the converter itself should be on internal sync. Let Firewire carry the data, but not the clock. The most meticulous manufacturers of firewire interfaces carefully separate data and clock issues.

### Single Box Solutions

In this image of a DAW (next page), the ADC and DAC are enclosed in a single box, so one clock drives them both (this applies to 2-channel or multi-channel converters).

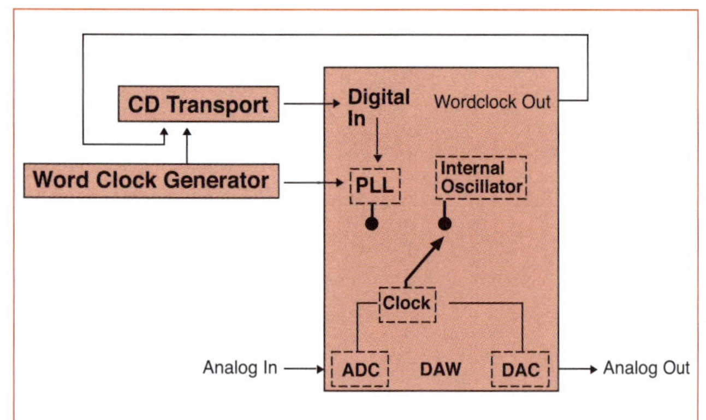


*ADC and DAC in a single box*

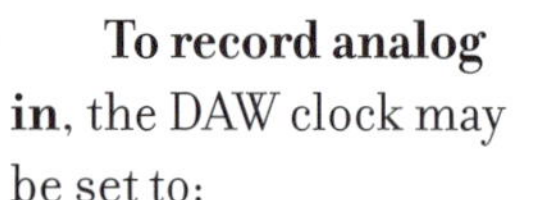

**To record analog in**, the DAW clock may be set to:

- internal oscillator
- external sync via wordclock
- external sync via AES/EBU or SPDIF

In theory and often in practice, the cleanest-sounding option for analog recording is to use internal clock.

**To record from a digital source**, the DAW clock and the source must be common. For example, from a CD or DVD transport, the DAW clock may be set to:

- internal oscillator, in which case the CD transport must receive external sync from the DAW's wordclock output.
- external sync via wordclock, in which case the CD transport may receive external sync from the DAW or the wordclock generator.
- external sync via AES/EBU or SPDIF, for example the CD transport pictured. This is the most common option for dubbing since most CD transports cannot be externally clocked.

Because jitter has no influence on data transfer, all three clocking options will work. In theory, and often in practice, the cleanest-sounding monitoring during the transfer is with internal clock.

## Multi Box Solutions

Jitter is harder to optimize with ADC and DAC in separate boxes, pictured below.

**For recording from analog sources**, the options are:

- ADC as master on internal oscillator, with DAC slaving via wordclock or directly by digital input
- Word Clock Generator driving all boxes

In theory and often in practice, the cleanest sound is with the ADC as master clock for recording and the DAC for playback. But standalone DACs usually do not have an internal sync option so the simplest playback solution is to use a jitter-immune DAC which is locked via its digital input.[7]

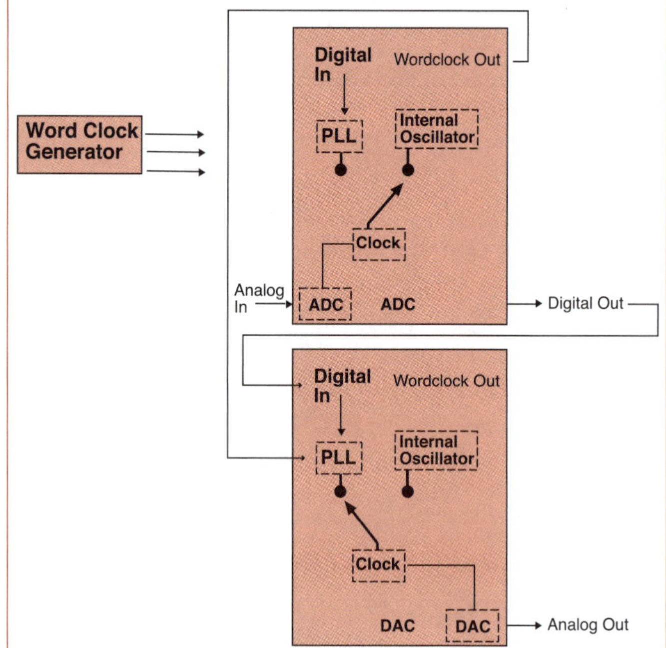


*Jitter is more difficult to control when ADC and DAC are in separate boxes.*

### Jitter Gain (Jitter Accumulation)

When digitally processing, the DAW or console can be the source of the unprocessed signal and capture the processed signal at the same time (pictured at right).

When chaining processors via AES/EBU, the interface jitter usually accumulates due to the limitations of the PLL in each box. So the interface jitter is greater at the output of Processor #3 than at the digital output of the Console. However, since the incoming signal is in sync with the console's output, the console's PLL circuit conforms that signal to the console's clock, and **incoming jitter is completely irrelevant**. Note that even if the console is running on external sync, the jitter gain of the processing chain still does not affect the console jitter, however the external clock does affect it and the DAC may not perform as well as when internally synced.

### Development of Jitter-Immune Converters

**DACs.** Recent advances in design have produced a few affordable DACs with both good-sounding analog circuitry and virtual immunity to incoming jitter. Instead of using a traditional PLL, these DACs have a stable internal crystal clock and a digital ASRC (Asynchronous Sample Rate Converter) for secondary PLL and anti-imaging filter. Although ASRCs add some distortion, they have greatly improved. Some jitter-immune DACs still employ traditional PLLs (that conceivably sound better) but they are relatively expensive.

**ADCs.** The state of the art in ADCs is also undergoing a revolution with some manufacturers experimenting with a topology that uses a crystal

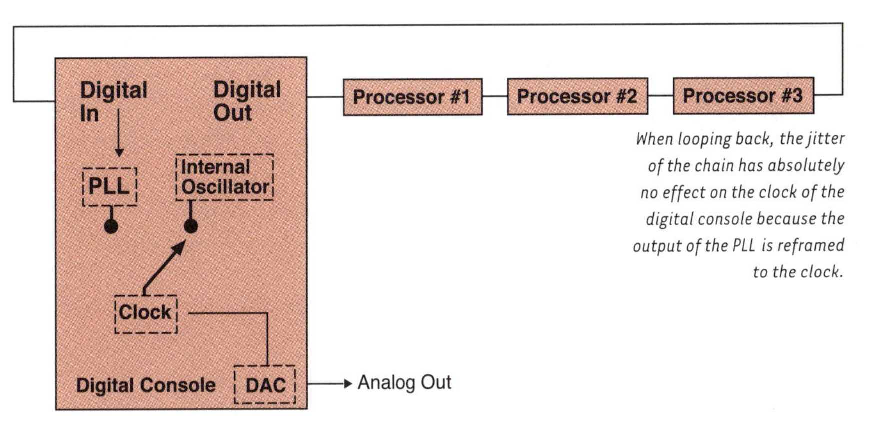


*When looping back, the jitter of the chain has absolutely no effect on the clock of the digital console because the output of the PLL is reframed to the clock.*

oscillator to drive the converter chip at a high sample rate for lowest distortion and best performance, which is then connected to an ASRC chip to synchronize the data to an external clock.[8] The ADC data is sample rate converted, and the ASRC functions as a downsampling filter. The jury is out on the sound quality of this topology, but preliminary impressions are that it sounds excellent. If this technology becomes established, then good-sounding, inexpensive, jitter-immune ADCs will be easily available and much of this chapter will be unnecessary! Until then though, and for the rest of this chapter I will be referring to "traditional-style" ADCs and DACs (which do not incorporate an ASRC).

To make a superior (traditional) ADC that produces only **inaudible** jitter effects requires time, research, and critical design implementation of PCB layout, grounding, internal clock distribution, and rigorous separation of digital and analog signals. The engineers who produced a superior converter model spent one **man-year** on the phase locked loop alone, and a further year on the converter details.

**The Miracle of the Blessed DAT Resurrection**
Always wanting to see how far equipment can be pushed, I decided to demonstrate the veracity of the laws of physics at an AES Convention. I built a special clock oscillator whose jitter could be altered, and connected it to a DAT machine, simulating extreme problems with clocking. The digital out of this machine was then connected to another DAT machine to make a dub.

While monitoring the record machine in record monitor, I increased the jitter of the source machine until the record machine exhibited serious distortion on its analog output; it sounded like a vocoder in overload.

**How important are digital cables** in a studio with perfect converters? For data transfer, mismatched impedances (e.g. 110 ohm to 75 ohm) cause no concern except at high sample rates, where they can cause poor signal transmission (and high interface jitter), glitches or dropouts. Balanced digital connections reduce RF radiation into sensitive analog stages.

### The Internet and Jitter

The internet has no clock. "Realtime" files played over the internet pass in irregular packets; they meet a clock for the first time when the computer gets ready to feed a DAC. The key to clean internet monitoring is to use a large enough buffer, followed by a crystal clock and a jitter-immune DAC.

## V. Mixing, Processing And Jitter

### Jitter does not affect the data...

...in an all-digital mix in digital consoles that do not contain ASRCs. After the initial A/D conversion, the data can pass from processor to processor, from medium to medium, regardless of clock jitter—just as long as the interface jitter is low enough to allow an error-free transfer. Similarly, clock jitter has no effect on the performance of most outboard digital processors, which are nearly all *state machines*. A state machine is defined as any type of processor which produces identical output for the same input data, which does not look at data timing or speed, but only at the state or recent history of the data. **In other words, most digital processors are completely immune to jitter.** We could make the clock completely irregular, or even slow it down to 1 sample per second, and eventually, the processor would output all the correct data words. For any processor which takes in a file and outputs another file, operating without a clock, *jitter* has no meaning.[9] Similarly, communication between plugins is asynchronous, happening in fractions of real time. Data is stored in a buffer at the end of the plugin chain for realtime output when it is clocked out of the buffer. Nor is jitter an issue with digital bounces, the data file gets stored sequentially regardless of clock. If audible differences exist between capture and source, it's likely an issue with automation not keeping up or a slow CPU not managing all the calculations in the required time. All current professional oversampling processors are also state machines because they use synchronous SRC which itself is a state machine.[10]

Although some digital pitch processors such as Autotune™ are not state machines (due to their randomizing algorithms), these too are not affected by jitter. They deal with each sample coming in, one at a time, regardless of the regularity of the clock feeding the box.

### Jitter affects the monitoring

**Jitter becomes meaningful in a digital mix only during monitoring**, when the data is clocked out of a DAC. This is where everyone gets confused. Let us emphasize: if high jitter during the monitoring **does seem** to affect the overall sound quality, it really only affects that individual listening experience, and has no effect on the data. This is what I call "ephemeral jitter". **Don't confuse the messenger with the message.[11] The message (the data) remains intact; so if it sounds degraded, blame the messenger (the clock inside the monitor DAC).**

If connections are improved and the sound gets better, it does not mean that the digital equalizers are suddenly performing better—just that a cleaner clock is getting to the DAC.

**Jitter affects the data during a digital mix only...**

- when signal leaves the digital realm to use outboard analog processors, hence superior converters and clocking must be used for outboard equipment feeds.

- when using a digital console containing ASRCs, which are not state machines. These affect the data, are sensitive to clock jitter, and are especially problematic in low cost consoles with compromised clocks.

**Analog Mixing**

Jitter performance is critical when mixing with an analog console and digital multitrack, however, the fact that some engineers hear an improvement with external clocks does not mean that this is the best or most efficient solution. Consider external wordclock to be an expensive add-on that cannot possibly maximize performance, since the output jitter of a PLL is worse than any good crystal.[12]

In the case of a cheap internal crystal oscillator, the external clock may improve sound, but you're pouring money on top of a bad converter investment. In the case of a medium-quality converter, an extraordinary external clock that costs nearly as much as the converter can improve the sound, but it would be far more economical to apply the cost of the external clock to a better converter running on internal sync. As Eelco Grimm (in correspondence) points out...

> if the external clock has lower jitter below the corner frequency of the internal PLL, then low frequency jitter performance can improve. Prism converters have a corner frequency below 200 Hz while typical converters' PLLs are above 2 kHz! So it is highly likely that a very good converter like a Prism will not be affected at all, or possibly degrade no matter what external clock you feed it.

So in all cases, replacing an existing converter with a superior converter running on internal sync will probably get the most bang for the buck.

Later in this chapter we present some measurements to help guide you, which you can duplicate with readily-available equipment and test signals. It's amazing how few manufacturers take advantage of these simple measurement techniques, or perhaps they're too embarrassed to publish the data.

## VI. Stop Leaping to Conclusions: Real World Examples

Here are some applications of the principles we've discussed.

**Example A: Digital Copying**

Engineer Betty would like to do some Digital Copying (cloning), from CD to Computer. First she notices that her CD player sounds better than her computer because as mentioned, the internal clocks of typical computer interfaces are not as clean as those in CD players. But she's more concerned that her

Anyone listening to this machine's output would conclude it was broken and that it was making a defective recording. But believe the facts, not your own ears, because on playback, there was no trace of the distortion; the playback sounded very clean. Thus demonstrating that digital dubs are not susceptible to jitter.

computer sounds **better on playback than on record!** What is going on here? The reason is that, as the theory says, her computer's internal oscillator is performing better than its PLL on external sync.

Digital copies really are perfect. Betty asks, "shouldn't I copy the file now that it sounds better?" But as this is a case of ephemeral jitter, Betty's data is just fine and should sound the same each time it is properly clocked.

#### Example B: Copying via SPDIF vs. AES/EBU

Engineer Don believes that AES/EBU sounds better than SPDIF through his DAC. So he decides he should make all digital dubs through AES/EBU. But the two interfaces are functionally equivalent, so the DAC he is monitoring through cannot be jitter-immune (there is no rule which of the two interfaces would have higher jitter). Even if he doesn't replace his DAC, he can safely make digital copies through either interface. He can prove their equivalence by doing a null test on two consecutive digital copies (see Chapter 17).

#### Example C: Clock Accuracy?

Ray was told that an accurate crystal clock would make his equipment sound better. This is only the case in an ADC, where the pitch would become incorrect when later played back with a different clock. But for an all-digital production studio **stability counts more than absolute accuracy**.

*"Don't confuse the messenger with the message"* — Andy Moorer

A crystal may produce 44,100 Hz **on the average**, but a jittery crystal oscillator or PLL deviates around that average. During playback in a totally digital studio, even if the crystal is several Hertz off, causing an audible pitch error, the end result will sound correct when later reproduced with a correct crystal. If I'm in a hurry, I can speed up the playback clock to 48 kHz or even faster if the equipment supports it, and still make a valid dub. This illustrates the fact that jitter cannot influence the accuracy of a dub: we can increase the source frequency to 10,000 times greater than the frequency deviation due to jitter and still make a perfect data copy! Dubbing is done on a sample by sample basis; the job of the clock is simply to deliver succeeding samples into the queue.[13]

## VII. Concern for the rest of the world...

Since much listening to music is done with inferior consumer DACs, it's very important that the masters we cut have the best possible sound. No jitter exists on a storage medium, but there is some (controversial) evidence that CDs sound superior when cut at low speeds, and that CDs cut with a jittery clock sound worse than those cut with a clean clock, especially when auditioned on a cheap CD player.[14] We theorize that certain mechanical parameters of the disc are altered by the cutting speed, making it more difficult for the CD player's servo mechanism which passes the varying load to the CD player's power supply and affects the stability of the master clock. Bruno Putzeys offers a possible explanation (which not all authorities agree with)...

Most cheap players use a crystal that's on the same chip as the servo/motor drive circuit. Changes in servo activity heat or cool the chip and the speed changes.

It only takes a few picoseconds to make an audible difference. Regardless of the theoretical reasons why this might be happening, remember that any audible difference between data-identical CDs is an ephemeral and correctable phenomenon. Time and again, when the clocking has been fixed, formerly audible differences disappear. Furthermore, one can "restore" the sound quality of an "inferior" CD by copying it back to a workstation and then outputting on a good writer at low speed. In this case the digital dub sounds better to me than the original! While this claim is unproven, I recommend that the CD production plant cut glass masters at a low speed. Bear in mind that a CD sent for manufacture will be ripped at the factory so it's only the data that's relevant. So bizarrely a high speed master is OK but potentially not a high speed reference.[15]

## VIII. Things That Go Bump In The Night

### Framing and Timing Errors: Wordclock to AES timing error

Although jitter is often the scapegoat for a motley of problems in digital audio, the fact is that 99% of the time, glitches, clicks, dropouts, noises and lockup problems, are caused by **framing problems**, not by jitter at all. Framing problems are caused by timing differences in critical signals and cannot be solved without software or hardware modifications. Below is an oscilloscope photo, at the top of which is the start of the AES preamble (which defines the beginning of the AES data word), and on the bottom, the point where wordclock changes from high to low.

To complicate matters, there is no standard that defines the timing of the wordclock transition (low to high or high to low) with the AES preamble. This variable is a timing difference of 180 degrees, or approximately 11 µs at 44.1 kHz, which is enough to cause glitches, or lose signal completely. Only one model of workstation has a menu choice that allows us to choose the wordclock phase, making it more compatible with products of various manufacturers. But the best solution is to use AES/EBU as a clock source, ensuring that all clocks will be in phase. However…

### AES to AES framing error

Digital audio is a small industry, still experiencing growing pains. And since some digital audio processors produce an AES output that is out of timing with their AES input, intolerant consoles and workstations have trouble locking to them. Once I was forced to insert a simple reverb unit via analog, because the digital console would not lock to it on a digital send/return path. The fault was caused by the console's intolerance to AES framing errors, aggravated by the reverb unit's output being slightly out of framing (timing), as seen in the figure on the next page. We can probably prove it's a framing

*This oscilloscope photo compares the timing of the start of the Channel A AES preamble against the start of wordclock at the output of a digital processor. This timing offset of 28% of the length of the AES frame is 3 points greater than the permissible tolerance in standard AES11 and would cause locking trouble to intolerant consoles or DAWs or other receivers.*

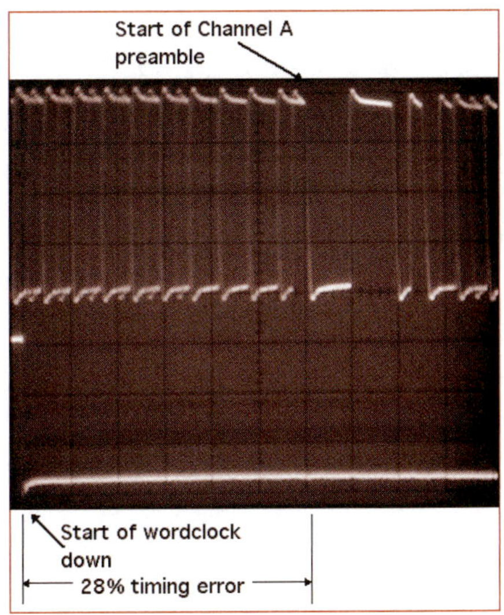

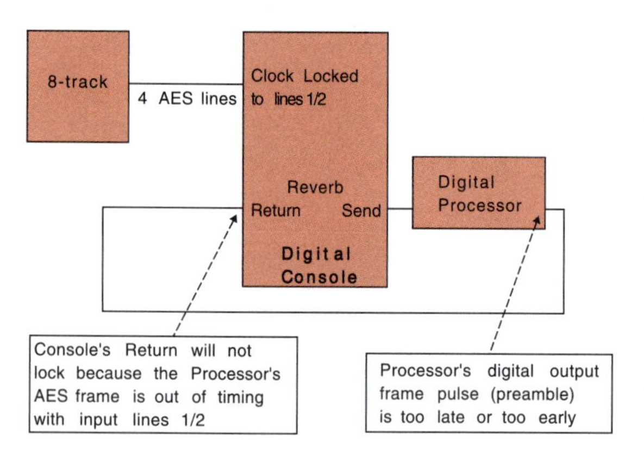


*How AES to AES framing error can cause locking problems*

problem without an oscilloscope: in this situation, set the external processor to run on its internal clock, and lock the console to the external processor on its reverb return. If the console will lock and pass audio from the external processor, then the previous problem was due to framing issues.

Framing errors are cumulative in a chain of processors if they are chained via AES/EBU (or SPDIF). If the framing error of each box is in the same direction, then the total error could be enough to cause locking problems in sensitive consoles and DAWs. We may be able to stabilize the system by locking the last processor in line to external sync (wordclock or AES). If the last processor in line is framing-tolerant on its AES input, then locking it to external sync will force its output to a known framing and hopefully to within the tolerance of the DAW.[16, 17]

## IX. How It Works

### Simple in Theory...

Most engineers don't need the heavy technical details of how equipment works, but usually a couple of nagging questions remain, like... What is a reclocking circuit? Why do we need a high-frequency clock?

**Reclocking Circuit.** The data inside typical audio processors is moved along by a clock pulse, traveling serially from chip to chip. This clock bus is distributed to all the critical chips inside the box. As we've seen, it doesn't matter if this clock is jittery, proper data still makes it to the next chip in line. But sometimes data needs to be *reclocked*, for instance when feeding a DAC. Pictured below is a simple reclocking circuit; on the left side is an incoming data word that's jittery; the data value is (conveniently) 10101010. This word passes, one bit at a time, into a logic circuit called a *D-type flip flop*, which is being fed a clean clock. Almost magically, the data neatly marches out of the flip flop, and in theory, all the jitter is gone and the data is ready to feed the DAC. This is why a PLL is often known as a **clock recovery** circuit. Notice how the clean clock's pulses permit the flip flop to properly "sample" each data value, but only if the clock pulse lands within the acceptance time of each incoming bit. In this illustration, the fourth (and eighth) data bit is in danger of being missed if it arrives a moment later, in which case the clean clock would land on the previous bit and the wrong data would be output, producing audible clicks or glitches.[18]

The figure (at right, next page) illustrates why a PLL is needed. If we are passing 24-bit audio bit by bit, then we need a high-frequency clock pulse that is 24 times the frequency of wordclock. Wordclock enters the device, and has to be multiplied up to the higher frequency to drive those bits around, known

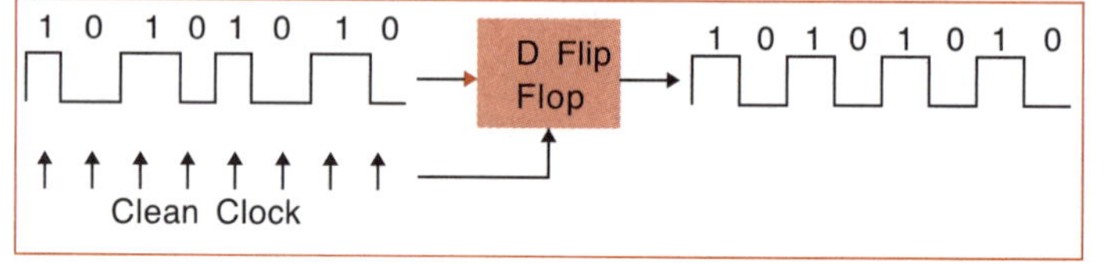


*A Simple Reclocking Circuit*

as the bitclock. It is easy to divide down without creating jitter, but very difficult to multiply up, and it's the job of the sophisticated circuitry of the PLL to create the higher frequency while reducing incoming jitter.[19] A PLL-based circuit is a sort of electrical flywheel: it tries to find a center, holding reasonably steady while still following the average frequency of the incoming source.

### ...Complicated In Practice

What makes these circuits so difficult to design well is that at high frequencies, leakage from any portion of the circuit can travel through back paths to contaminate the clean portion of the circuit. These paths include power supply and ground. Couple that with outside interference and ground loops, and you have created an analog designer's nightmare. 10 picoseconds error can make the difference between an 18 or 20-bit noise floor. Some manufacturers use a dual-PLL, where the first is an analog circuit, and the second a voltage-controlled crystal oscillator (VCXO), in an attempt to get the jitter down to that of a quartz crystal. Unfortunately, designs using VCXOs cannot varispeed because of their narrow frequency tolerance. It is difficult, yet possible to design a jitter-immune PLL that's as good as a crystal, has wide frequency tolerance and quick lockup.

## X. Jitter Measurements

Here we present some jitter measurements made on DACs. ADCs and DACs can be tested for the effects of jitter using a very high frequency sine-wave test signal, but the test signal must be very pure and frequency-stable, either crystal or digitally-generated. For these DAC tests, we used the J-Test signal invented by the late Julian Dunn, an independent consultant best-known for his work on the Prism brand of converters.[20] The 24-bit J-test signal was not available, so the 16-bit version was used; we'll have to ignore some artifacts that are part of the source signal. Here are a few guidelines: The lower the measured noise floor, the less jitter. We have not fully learned which jitter spikes are psychoacoustically important, but, as stated before, my listening tests show that jitter must be very low (close to the system noise) to be inaudible. Since test equipment varies, J-Test results will differ (use a high quality A/D on internal clock to capture the measurement), but relative rankings will likely remain.

In the color plates section, ***color figure C21-1*** shows, in red, the noise floor of my UltraAnalog ADC (used to sample the outputs of various DACs under test), and in blue, the artifacts of the 16-bit J-Test signal, which are at -132 to -135 dBFS.[21] This means if we appear to measure jitter in the device under test below -132, it may simply be due to artifacts of the test signal. It's more important to look at how the jitter artifacts affect the DAC's own noise floor and at what particular frequencies, than to calculate the actual jitter value in picoseconds.

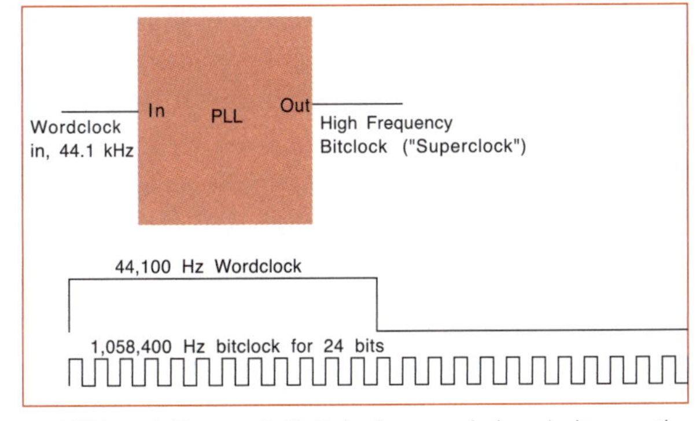


*A PLL is needed to generate the higher frequency clock required to move the individual bits from place to place.*

> *"Most digital processors are completely immune to jitter"*

***Color figure C21-2*** shows a considerable measured difference in jitter performance when an inexpensive consumer DAC is fed from two different sources. A cheap consumer CD player yields the highest output jitter, with the output of Sonic Solutions even less. If this were a linear display instead of semi-log, it would be more obvious that jitter usually produces paired artifacts around the center frequency, usually at equal deviation around the center. Compare the consumer DAC's performance to that of the excellent *jitter-immune* TC Electronic System 6000 DAC. When fed from either of two sources, the TC's jitter is effectively identical and just about as low as its quiescent noise floor!

***Color figure C21-3*** shows that sync mode hardly affects the TC's textbook-quality jitter performance, with extraordinary measurements in internal sync and very slight degradation when locked via AES/EBU. When slaved via AES/EBU, it produces very slightly more jitter (only the two discrete frequency blue lines circa -117 closest to the center frequency). When on internal sync (red trace), its jitter is nearly as low as the UltraAnalog's noise floor; most of the *grass* is the noisy 16-bit J-test signal itself. I may hear a slight degradation in sonic clarity, a smeared image and brightness when the TC is slaved to AES/EBU, which implies that the black-colored spikes at approximately -117 dBFS may be audibly significant to an extremely critical listener with the finest playback system.

The Weiss is the first DAC I have measured with no apparent trace of discrete frequency jitter in its output when locked via AES/EBU ***(Color Figure C21-4)***. Instead, its noise-floor rises with the test signal and the jitter "skirts" appear to widen; all incoming jitter has been converted to random noise. Or is the sonic improvement due to euphonic coloration (higher noise floor masking discrete jitter components, since we can no longer see the floor of the test signal itself)? This also raises concerns about potential converter noise modulation with signal, which may mask low level signals or reverberation. However, low-amplitude random noise is the most benign signature one could wish for and the Weiss DAC sounds great. I did not test the DAC on internal sync.

**In Conclusion:** We have penetrated jitter mysteries and dispelled jitter myths, which makes it a lot easier to construct a good-sounding studio.

1 Leading to the "Wordclock Du Jour" effect. And an erroneous audiophile magazine DAC review marveling at a DAC that's "good enough to reveal digital cable differences!" Fact is, a device that "reveals" an apparent difference must be considered **defective**.

2 There is only one right "kind" of digital cable, one whose impedance is a correct match for the circuit (e.g. 75 or 110 ohms). Some audiophile manufacturers have made so-called *digital* cables which are improper for the circuit, but since they affect the sound of a typical, consumer-grade DAC in some unpredictable way (usually adding jitter, not reducing it), consumers have been known to play with such cables to tune their systems. It's a losing battle, because the cable-induced jitter reduces resolution and colors the sound.

3 Analog tape print-through is a phenomenon where one layer of magnetic tape magnetically imparts some of its signal on the adjoining layer. After years of storage, it is possible to hear two or three repeating echoes in the tail decay of a song recorded on analog tape (which can be repaired by the "adding tails" technique explained in Chapter 7). But even when print-through does not provide a distinct echo, it is always there to some extent, affecting the clarity of the sound.

4 According to a simplified formula: Moses, Don (October 1992) Enclosure Detuning for 20-Bit Performance, *Journal of the AES preprint 3440*.

> The following expression utilizes Carlson's similar triangle analysis method and is useful for the case where: (1) the jitter deviation is small compared to the sampling interval, (2) distortion is measured at the zero-crossing of a sine wave, (3) the peak-to-peak amplitude is normalized to 1-V, and (4) the maximum slope is approximated as 2 x the information bandwidth:
> Resolution (in dB) = 20 log (time deviation x 2 x information bandwidth)
> For example, 25 ps of jitter, 20 kHz information bandwidth, yields:
> 20 log (25 ps x 2 x 20 kHz) = -120 dB, which provides 20-bit resolution.

In other words, if we double the sample rate to 88.2 kHz (the information bandwidth becomes ~40 kHz), the same amount of jitter reduces signal to noise ratio by 6 dB. For 20-bit performance, at 88.2 kHz, if we consider the information bandwidth goes to 40 kHz, the jitter would have to be halved, to less than 12 picoseconds. And for each 6 dB improvement or 1-bit increase in wordlength, the jitter must be halved again. Even if we limit the information bandwidth to 20 kHz, in order to get excellent performance with long wordlength, it boggles the mind the degree of engineering care required to produce a low-jitter converter.

5 It is possible to make the video clock slave to the converter using an *AES/EBU to video sync converter*, a product-design opportunity for fussy audio mastering engineers who want to run ADCs on internal clock.

6 Dunn, Chris & Hawksford, Malcolm (1992). Is The AES/EBU/SPDIF digital audio interface flawed? *Journal of the AES* preprint 3360.

7 In a multiconverter situation, which box should be the master? According to Ian Dennis, in Resolution Magazine Oct 2006:

> Using the best box in the studio is usually misguided, since good boxes are equally at home as master or slave whereas bad boxes are usually OK as master but perform poorly when slaved.

8 Crystal Semiconductor application note, A/D Conversion with Asynchronous Decimation Filter.

9 When transferring from file to file, the process deals with one sample after another. The bowling ball analogy: Throw a series of bowling balls down the alley, some white and some black. Although their timing is irregular, when they land back on the stand, the white and black are in the same order, so the output data is identical. **Digital processors (except for ASRCs) look at the samples (bowling balls), not at their time of arrival**.

10 Any processor which adds random dither is not a state machine in the strictest sense. All good synchronous SRCs employ internal dither to linearize the process. The randomizing effect of dither means that each output pass will produce slightly different data **at each instant**. However, **on the average**, the output stream is really the same, and if we could subtract the random dither from the output signal, each pass would be identical.

11 As illustrated in a demonstration of extreme jitter that I performed at the AES Convention, when the audio was so distorted by a jittery clock that it was unrecognizable, but the data remained intact. See the accompanying sidebar. Credit to Andy Moorer for coining the message/messenger dichotomy.

12 In the vast majority of converters manufactured today, "the jitter caused by wordclock is typically 15 times higher than when using a quartz based clock", according to the manual for the RME model ADI-8-DD format converter.

13 Those dyslexics in the audience will appreciate that I am taking slight liberty with this discussion for ease of understanding. Since the left hand end of the bitstream is the last to go into the flip flop, the "fourth bit" counted from left to right is actually the fifth bit to go in! This leads to the requirement that software has to decide whether to make the left or right end of the bitstream be the most significant bit. Intel and Motorola have been fighting over which end comes first for decades.

14 Note that the sonics have nothing to do with the soft error rate, a 1x write can exhibit higher soft error rate even if it sounds better to some ears. Anecdotal evidence: reports from the musical artists themselves led engineers at Sony Corporation to work on improving the jitter in their CD cutting systems. Without outside influence, some major artists had been reporting that their CD pressings did not sound as good as the reference CDRs they had received. Critical listeners making CDRs have heard superior sound with computer-based CD Recorders than with standalone CD recorders. In standalone recorders, the master clock driving the laser is slaved to the AES/EBU input, while computer-based recorders use a FIFO buffer and a crystal clock to drive the laser. To see how difficult this assertion is to prove, visit the Prism Sound weblink. Regardless, the issue is only for clients listening on cheap, jitter-susceptible DACs.

15 Thanks to Andy Jackson (in correspondence) for this reminder.

16 The best hardware solution I've found to framing issues is an external rackmount AES/EBU interface made by RME. Its framing tolerance is excellent; it cleans up framing issues, conforms all inputs to the same framing on its output, saving the day by making unrelated sources look like a single digital multitrack to my sensitive DAW.

Julian Dunn clarifies:

> Framing is a synchronization issue covered in AES11. These define the permitted output alignment error (+/-5% of a frame period) and the tolerance to input timing offset (+/-25% of a frame period) before the delay becomes uncertain. The specifications for the interface itself (AES3, IEC60958) do not allow a receiver's ability to decode data to depend on the relative alignment of clocks—as long as the dynamic variation is within the jitter tolerance spec. (about +/-4% of a frame period at low jitter frequencies).

17 If you're spending $30,000 and upward on a digital console, request the manufacturer to sign an agreement that the digital inputs and wordclock framing tolerances must meet or exceed the AES11 synchronization specs or the manufacturer will correct the problem at no charge. This amounts to a sad wakeup call to the manufacturers, but consumers should be entitled to interface real-world equipment to their consoles.

18 However, a crystal which is off the standard center frequency can cause locking problems, if using a low-jitter PLL with a narrow lock frequency range (which are also intolerant to varispeeding). But if the system components lock, then an off-standard crystal won't affect digital dubs at all. Some PLLs have a narrow and wide setting to deal with sources that are a bit off the standard. Switching to wide increases frequency tolerance, but also increases the PLL's jitter. Don't be concerned, as long as the PLL is not driving a converter. If a system has locking problems not due to framing errors, measure the source sample rate, and if it's off tolerance, trim the master crystal oscillator frequency.

19 Many bitclocks are 32x the wordclock, or greater, to allow for a longer internal wordlength. A typical PLL may generate a superclock which is 128, 256 or even 384 times the wordclock frequency, and is then divided down using a simple divider.

20 The J-Test is a special signal designed to aggravate a DAC's jitter. It contains a fundamental signal at 1/4 the sample rate, which is 11.025 kHz at 44.1 kHz SR and a low frequency component added to deliberately add data-jitter on the AES input. The test is particularly designed to pick out the interaction to the sample clock from the data on the AES or SPDIF interface that is used to derive the sample clock. When AES/EBU is not involved, it would be more practical to use a simple clean high frequency tone at 1FS.

Julian: "There are four 24-bit numbers in a sequence that is 192 samples long that repeats.

0xC00000, 0xC00000, 0x400000, 0x400000 (x 24 i.e. 96 samples)

0xBFFFFF, 0xBFFFFF, 0x3FFFFF, 0x3FFFFF (x 24)

| Hexadecimal | Binary |
|---|---|
| C00000 = | 1100 0000 0000 0000 0000 0000 |
| 400000 = | 0100 0000 0000 0000 0000 0000 |
| BFFFFF = | 1011 1111 1111 1111 1111 1111 |
| 3FFFFF = | 0011 1111 1111 1111 1111 1111" |

The 16-bit version of the J-Test signal is currently available on a CD from Audio Precision and on another test CD from Checkpoint Audio in the Netherlands. Further information can be found at the late Julian Dunn's website maintained in his honor.

21 The measured amplitude of the noise depends on the number of points (bins) in the FFT, the window which is used, and the quality of the measuring ADC, which is why each reviewer's results will be different. These measurements were taken with a 32K point FFT with an averaging time of about 2-4 seconds, and a Hanning window.

CHAPTER 22

# Technical Tips And Tricks

## Introduction

This chapter provides tips on how to maintain a digital audio studio.

## 1. Timecode and Wordclock in a Digital System

### Drifting drifting drifting

A common task in a digital audio studio is to slave a sequencer via timecode to a master DAW. If the sequencer is **not** recording or monitoring digital audio, it can be set to timecode sync. It will then slave (adjust its speed) to the incoming timecode, and **will always stay locked to timecode. It will not drift.**

However, unless the slave is locked to the same source as the incoming timecode, it will **drift out of sync**. This is because the slave machine takes a *timestamp* or *trigger* from the first valid timecode it sees. From that point on, the interface ignores incoming timecode and creates its own timecode locked to the digital audio clock. Also, make sure the slave is set to the same sample rate and timecode as the source.

A multiple sample rate system can be locked together in real time without causing timecode to drift as long as a real-time **synchronous** SRC is used. For example, when coming from a 48 kHz source DAW to a 44.1 kHz destination (slave) DAW. Even though the slave is set to a different sample rate, the frames are synchronous.

*"There must be only one clock master in any system"*

### Pull-ups

Things are far simpler without video, NTSC video in particular, as the issue of pull-ups occurs only in NTSC video countries. At timecode rates of 30 fps, there are exactly 1470 digital audio samples per 44.1 kHz frame, or 1600 samples per 48 kHz frame, so the audio rate is an **integer-multiple** of the timecode rate. But when NTSC video is involved, the timecode and video frame rate is 29.97 fps, which yields a non-integer number of audio samples per frame. To produce a wordclock at this rate, slaved to the NTSC video, a sophisticated wordclock generator creates a *pull-up* wordclock with the right ratio. As long as all digital audio devices are locked to that clock, then lip sync will not be lost. The digital audio output of a digital video deck can be substituted for a wordclock generator, as it will have the correct pull-up ratio.

### Wordclock Voltages

The problem with standards is there are so many of them! With Johnny-come-lately digital, no voltage standard was developed for wordclock, which produced a chaotic situation. Many of the earliest wordclock generators were based on video sync (blackburst) generators, which produce 4 volts peak to peak into a 75 ohm load (abbreviated $4\ V_{p\text{-}p}$). Later, wordclock generators appeared based on the video standard of 1 volt. Yet a third standard is based on TTL-level, with 2.5 volts terminated, and 4-5 volts unterminated! Chances are that any wordclock lock issue can be traced to these voltage incompatibilities. The best solution is to use generators that produce 4 volts and receivers that can accept anything between 1 to 5 volts, which are cross-compatible. We may be able to fix a disfunctional receiver, by removing the load termination resistor. If this doesn't work, then we need to measure amplitudes with an oscilloscope and get creative with the circuitry inside our distribution amplifiers, which may require removing the buildout resistor on one or more outputs in order to get a little more voltage. Caveat emptor.

### House Video and DARs Sync

It is not advisable to lock multiple digital audio devices via video sync. This is because neither wordclock nor video define the beginning of the digital audio frame or identify channels. Since there is no phase reference in video sync, using house video directly will produce a wordclock of the correct frequency but not the correct phase, so two video recorders containing digital audio will likely not be phase locked, which will cause unpredictable phase shift.

It is slightly preferable to use wordclock instead of video sync, but as mentioned in Chapter 21, this can result in channel reversals or latency differences. For this reason, do not split channels between audio devices which are slaved to wordclock. The only dependable sync reference for multiple digital audio devices is called **AES-11**, also known as **DARS** (digital audio reference signal), functionally equivalent to AES/EBU with muted audio, which maintains channel beginning markers (blocks) and channel-to-channel relationships. If video is the source, derive a master AES-11 sync signal from the video source. From that point on, all the digital audio devices will be referenced **to each other**, but there is no guarantee they will match the digital audio outputs of any other video recorders. When in doubt, test for phase shift between channels. To resolve this chaos, video and audio houses are migrating to a new SMPTE time standard known as ATR (absolute time reference).[1]

## II. Debugging AES/EBU and SPDIF Digital Interfaces

When the AES/EBU and SPDIF[2] interfaces were created, the idea of using standard microphone connectors and cables seemed like a godsend, but these were never intended to carry the high frequencies of digital audio (about 6 MHz bitrate for 48 kHz SR). So eventually we ended up with special RF-rated cables attached to our old-fashioned XLR connectors. However, AES/EBU cable makes excellent analog line and mike cable due to its high bandwidth and low capacitance.

### Software Issues

The most common issue is sample rate misidentification. When recording from an external digital source, most DAWs ignore or cannot read the incoming sample rate flag (metadata). Instead, the DAW assigns the flag of the new file to the rate of its session, regardless of the true rate of the source. This results in a mismatch if an external ADC is running at a different rate from the session. No one will hear a pitch error until the material is later played back with the correct clock. The cure is to fix the flag in the file header (metadata), which can be performed by **Soundhack** or **Sample Manager** (both Macintosh programs) without having to rewrite the entire file, very useful for batch processing errors of this type.

### Hardware Issues

Glitches or intermittent sound is most likely a hardware issue. We don't know how well a digital audio receiver is working until it stops or produces audible glitches, which is why digital audio receivers ought to include signal-quality indicators. The proper way to assess a hardware interface is to measure its objective performance by looking at the width of an eye pattern on an oscilloscope. Always use matched-impedance cabling, especially for long runs and high sample rates. When measured with a terminated load, the 110 ohm balanced signal should measure between 2 and 7 volts p-p, while the unbalanced signal should not be below 0.5 Vp-p.

Shields are unnecessary in the **balanced interface**, as can be illustrated by the success of Belden's Mediatwist™, consisting of four bonded-twisted pairs for up to 8 channels, which performs as well as the highest-grade coax. In fact, standard Cat 6 twisted pair Ethernet cable can carry four AES/EBU signals very well. The biggest problem with the **unbalanced consumer interface** is that its low voltage (0.5 $V_{p-p}$) does not give much margin for error with coax cable loss. These issues could have been avoided if the SPDIF interface protocol had specified 1 volt like the **AES-3ID** standard, which uses a BNC connector, popular with video houses.

**Improving the stability of the unbalanced interface**. The stability of the unbalanced interface can be improved by upgrading to **low-loss** 75 ohm cable, and/or by raising the output voltage from 0.5 volts to 2.5 volts, easily done by replacing the voltage divider at the transmitter with a single 75 ohm resistor, as in this next figure:

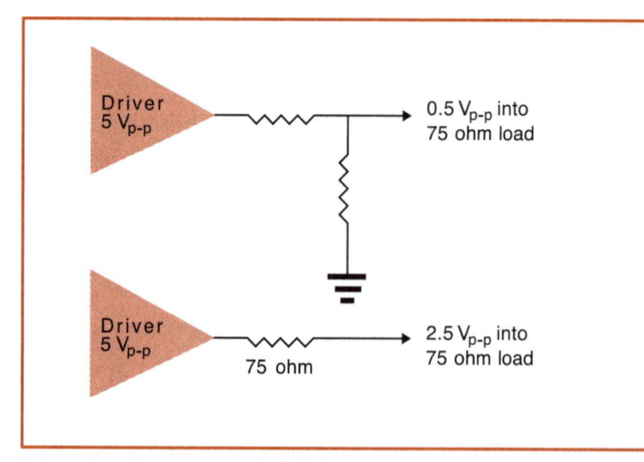


*Voltage Level improvement for SPDIF transmitter*

This modification to the transmission side works very well because it raises the noise margin of the receiving circuit at no significant cost or interference with other circuits. *Warning: modifying circuits usually voids the warranty.* Although the AES standard is between 2 and 7 volts, note that the same audio receiver chip is used for both AES/EBU and SPDIF decoding and it can accept from as low as 200 $mV_{p-p}$ to as high as 7 volts. Higher voltages are usually not a problem with SPDIF, but extremely low source voltages reduce noise margin and may introduce dropouts or glitches. The major difference between AES and SPDIF at the input is a change of connector and termination resistor between 75 and 110 ohms, as most SPDIF circuits already use input transformers.

## Converting Impedances

A mismatched impedance (as well as circuit imbalance) will result from putting an RCA connector on one end of an XLR cable without changing the source or load resistors. However, with short cable lengths and low sample rates, the mismatched signal may pass adequately.

An impedance-matching transformer is probably the best way of converting between 110 ohm balanced and 75 ohm unbalanced, but the following cheap and simple circuits can work perfectly. To convert a 75 ohm source on coax cable to feed a 110 ohm input, solder the coax cable onto the XLR male with a 237 ohm resistor across pins 2 and 3, as in the figure below. The 237 ohm in parallel with 110 ohms produces a 75 ohm termination; if the internal circuitry is not far from the connector, the impedance bump should be negligible. If this adapter passes 192 kHz sample rate cleanly it will surely work nicely at lower rates. We also recommend the voltage level mod, since SPDIF voltage level may be a bit low for a long cable run.

This next adapter (pictured next page) turns a 110 ohm source into a 75 ohm impedance by using a coax cable on a female XLR connector with a 237 ohm resistor across pins 2 and 3, which may work even better than an impedance-matching transformer because it maintains a higher voltage level.

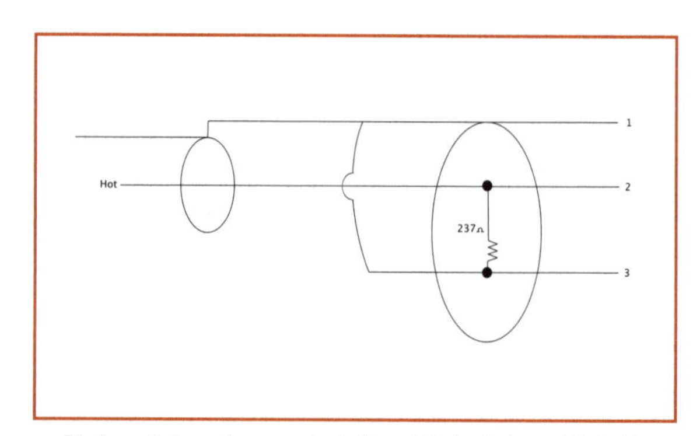


*75 ohm unbalanced source adapted to a 110 ohm balanced digital input*

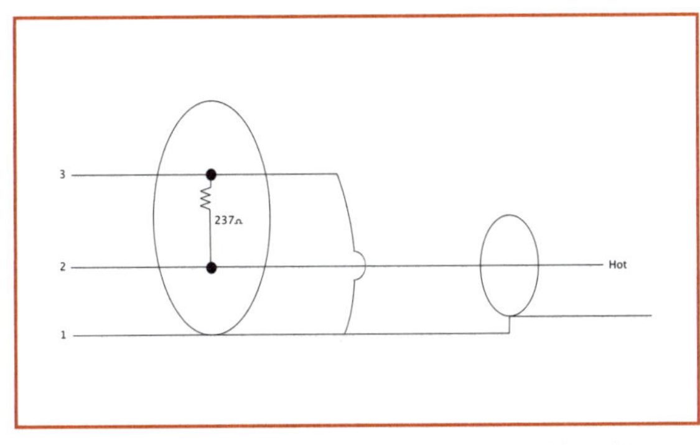

*110 ohm balanced source adapted to a 75 ohm unbalanced digital input*

Be aware that reversing pins 2 and 3 of an AES/EBU cable does not affect the audio in any way. Polarity reversal of the audio signal can only be accomplished with a digital processor or console or in the analog domain.

### Cable Lengths

**With copper cable**, the higher the sample rate, the shorter the tolerable cable length, because of the possibility of interfering reflections from the impedances and connectors at each end of the cable. The AES3 standard specifies usable lengths up to 100 meters at 48 kHz, which is possible with careful termination and high-bandwidth, matched-impedance cable. However, at 1/4 wavelength, reflections are at their worst, aggravating errors with cables that are close to 20 meters (66 feet) at 48 kHz, or 33 feet at 96 kHz. Neither the XLR nor the RCA connector was designed with exacting impedance specifications, so avoid passive hardware patchbays, splices, and extensions. Cable length has an insignificant effect on latency, since a cable length difference of over 200 meters would be required to exceed the AES11 framing tolerance.

**With optical cable**, the main concerns are bit integrity and interface jitter. As we explained in Chapter 21, jitter is only a consideration when delivering signal to a DAC or sync signal to an ADC. Plastic Toslink cable is rated for up to 5 meters (10 meters with some receivers) beyond which there is unacceptable signal loss. A legitimate test for cleanliness of an optical interface is *margin distance before dropout*. While observing the lock indicator on an AES receiver, or, simply listening to the audio, disconnect the cable from the input and slowly pull it outwards. The amount of distance before losing lock is an indicator of the margin of sensitivity and the strength of the optical signal. 1/8 inch to 1/4 inch or more is a good margin. Glass fiber has much less loss than plastic, and can transmit for thousands of feet.

### Two-wire 96 kHz and 192 kHz

Originally, the highest sample rate that could be carried on an AES/EBU interface was 48 kHz (slightly higher with varispeed). In order to double the sample rate with recorders that can only handle 48 kHz, a system called S/Mux was invented that places half the samples on one cable and half on another, each cable running at half the final rate. As of this writing, most copper interfaces can run 96 kHz single wire, but many 192 kHz interfaces require two wires, each carrying one channel. Keep good documentation and carefully mark cables to avoid reversed channels. Optical (ADAT format) can also run S/Mux—when the interface says 48 kHz/8 channels it's actually 96 kHz/4 channels.

## III. Nothing Special about PMCD

The term **PMCD** was invented by Sonic Solutions as a method of allowing glass masters to be cut directly from CDRs; the PQ codes were written into the lead-out portion of the PMCD master. However, this term is completely out of date, so if a client asks for a PMCD, it really refers to a standard CD-DA master.

## IV. The writable DVD confusion

DVD-RAM is a rewritable medium, claiming 100,000 re-write cycles, but it is not commonly used. DVD-RW can be read on more players, and the technology is limited to 1000 re-write cycles. Look for a player labeled *RW-compatible.* DVD-R (pronounced "DVDR", not "DVD minus R") can be played on most set top DVD players, yet have difficulty with older computer drives. DVD-R can only be written once, which is not a problem since blanks are very affordable. DVD+R is a variant invented in a format war that has received less success than DVD-R. Look for many more flavors and capacities evolving along with format wars.

1 Replacing video reference, ATR is a format for transmitting a universal date and time reference for the purposes of distributing synchronization information and the distribution of time. This is under the jurisdiction of the SMPTE group as well as the AES working group SC-02-05.

2 SPDIF stands for *Sony Philips Digital Interface*, which grew up into the IEC60958 standard, which supercedes IEC 958. Officially, type 1 is consumer with the consumer bitstream (protocol) on unbalanced RCA or Toslink optical connectors. Type 2 is professional, with the professional bitstream over XLR balanced connectors. There is also the AES-3ID standard, which transmits the professional bitstream over a 75-ohm BNC connector at 1 volt p-p.

"We'll FIX it in the shrinkwrap."

—Frank Zappa

# PART IV: IN CONCLUSION

"Every day in every way I'm getting better and better"

—Bob's Mom

CHAPTER 23

# Education, Education, Education

## What Have We Learned?

As we reach the end of our journey, it has become clear that the craft of **Mastering** requires attention to detail, technical and musical knowledge, and the ability to get along well with a wide range of people from artists to record company executives. In other words, be able to leap over tall buildings in a single bound! But nobody is infallible—we all make mistakes; the key is to correct them before the product goes out. That's why it's important to develop a system of **quality control**, which reduces the level of mistakes until they're below the radar.

## Mastering Your Own Project

Many readers are interested in trying their hand at mastering, or need to master a project that they can't afford to send out to a mastering house. If you find yourself in that situation, don't fret about not having optimum acoustics or even the right equipment, for talent is far more important than tools. If you have lots of time, follow good judgment and the advice of this book, you can overcome economic obstacles and produce a decent-sounding master. Since typical small studios have noisy fans, low-resolution monitors, interfering console and rack surfaces and uneven frequency response, approach the mastering session using a trial-and-error process. It will take much longer than in a proper studio with an experienced mastering engineer, but it can be accomplished.

Listen to the great-sounding pop recordings of yesteryear as examples to emulate. Bring the master in progress to different environments to check for translation. Audition the developing master on high

**From Ear to There**

Our listening audience is in a precarious situation—for the hearing of the average person, especially that of young people, has been deteriorating at an alarming rate! Some of the hearing loss can be blamed on the proliferation of earphone listening, since unisolated earphones cause the user to raise sound pressure to dangerous levels. To prevent ear damage, audio engineers should lobby for all earbuds to be isolated models. Expense is not an obstacle as it is possible to construct a low cost isolating earbud.

Life in a noisy city affects our hearing thresholds; use protection by wearing ear plugs on the subway and noisy streets. If you're drilling a metal hole, sawing a board or hammering a nail, wear ear protection.

quality headphones for noises and distortion. Try to obtain the aid and **perspective** of an experienced producer, or an objective professional whose ears you trust to ensure you haven't been so close to the material that you're missing something essential, especially if you are the artist. For if you know the lyrics by heart, then you are the wrong person to judge the vocal level! The requirement for perspective is the reason why mastering engineers avoid mastering their own mixes.

> Without collaboration, the music is not being given its full potential. There is a reason for the artist, the engineer and the producer because each one can worry about their own *thing* and they can collaborate on the final outcome. When music is done in a *virtual vacuum*, it does not sound as good.[1]

### Spread the Word

Audio is our passion, so let's spread the gospel of good sound. In Chapter 14, we learned how the loudness race is detrimental to sound quality. Show producers the square waveform of hypercompressed material. Make them aware that people don't like to listen to such fatiguing recordings—so the album loses critical word-of-mouth advertising. Show them that a decent amount of dynamic range produces an enjoyable, lively and clear album. Show them that hypercompression is incompatible with radio and lossy (mp3) encoding; recordings with **lower** intrinsic level sound **louder** and have the most impact on the radio (see Appendix 1).

Show the public that adjusting the volume control is normal practice; that we must turn it up and down in a noisy car. Guide them to the compressor button if they're annoyed by having to ride levels.

Spread the word about loudness normalization: engineers can supply algorithms to home server systems so there will no longer be an artificial loudness advantage to overcompressed material (see Chapter 14). Just as in the days of LPs, music parties in the home will be fun once again because when played on the normalized server, old and new recordings can coexist without making people jump to the volume control every 3 minutes.

Let's encourage car audio equipment manufacturers to incorporate a compressor as part of every car's system. Some sophisticated cars have automatic level controls tied with the speed and ambient level, which is a tremendous engineering advance. When cars have this equipment, fewer producers will demand overcompression.

### How Loud Should I Make It?

Not all producers and engineers understand the concept of the calibrated monitor level control (Chapter 14), but this simple guide can help them quantify their levels: The better-sounding record on the air and in the home will fall within the range of 3 to 6 dB[2] lower in average level than the most distorted CD in existence.

Most producers appreciate the advice of an experienced professional. We will not lose the job by suggesting to a producer that we have mastered the record to the maximum possible level without losing quality. If he prefers differently, then of course we turn it up, for we are a service industry. But we can demonstrate the sonic improvements that come from avoiding hypercompression. When our focus is quality and education, reproduced sound will get better.

*"Be prepared to be at least 6 dB lower than the current 'winner' if you want your record to sound acceptable!"*

Prolonged exposure to high-level music at clubs and concerts causes permanent hearing damage to the tiny, delicate parts of the ear. In a perfect world, club owners would be required to pass out ear plugs to customers, because alcohol dulls our senses so we don't notice that the music is painfully loud. Our job as audio professionals is to educate our audience about this very real threat, so that listeners do not turn into an endangered species.

When driving or flying a long distance to a mastering session, be sure to wear ear plugs, which greatly reduce travel fatigue. Try to arrive the previous night and get a local night's sleep before the session, which reduces threshold shift and improves perception during the mastering session.

1 Tom Bethel, Mastering Engineer's Webboard.

2 Or even lower!

3 J.Hall, email of 20 October 2005. "My rock band just put out a record that has exactly this in the liner notes." I've owned many stereos in my life and the volume knob was always the biggest.

*"This CD was mastered for sonic quality, not volume; for the best of both worlds, locate your volume knob and turn it clockwise until satisfied."*[3]

"Never doubt that a small group of thoughtful, committed citizens can change the world. Indeed, it is the only thing that ever has."

—Margaret Mead

About the Margaret Mead quotation on the previous page: The Institute for Intercultural Studies has been unable to locate exactly when and where this famous adage was first cited. They believe it probably came into circulation through a newspaper report of something said spontaneously and informally. They know that it was firmly rooted in Mead's professional work and that it reflected a conviction that she expressed often, in different contexts and phrasings.

# [ Appendix 1 ]

# Radio Ready: The Truth

## I. Introduction

Do you ever wonder what happens to your recording when it is played on the radio? Do you want to know how to get the most out of radio play? Radio, like all technology, is constantly changing. Digital radio (internet, satellite, and terrestrial) changes the way that our records sound. Though Section II deals with FM radio, its precautions are equally relevant to digital radio as we will see in Section III. In 2000, participants in the Mastering Webboard engaged in a friendly collaboration to find out what range of levels we are using. Engineer Tardon Feathered offered a rock and roll mix on his FTP site, which was then mastered by webboard engineers. Tardon produced a two-CD collection of these masters called **What Is Hot?** The intrinsic loudness of the cuts on this compilation ranges from extremely hot and highly distorted to very light, with a loudness difference of 9 dB!

Next, the Webboard participants felt it would be important to demonstrate what happens to these cuts when passed through radio processing. Enter Bob Orban, who volunteered to process the music with typical radio station presets. Tardon then prepared a compilation CD, comparing the songs before and after radio processing. This figure is a comparison of five mastered cuts before and after Orban processing.

Notice that regardless of the original level, after radio processing every source ends up with similar density: soft passages are raised radically, and loud passages are slammed to a maximum limit. Listening to this revealing CD, every track ends up at the same loudness, proving beyond a shadow of a doubt that there is no advantage to "compressing for the radio." In addition, the radio processing yields **negative** effects from extremely compressed tracks, distorting nearly every original, except for the softest track, which came in at about a K-14. The rightmost and most squashed source track was unlistenable after Orban processing. The radio processing also somewhat randomizes the stereo image and lowers the high end, but attempting to compensate for these losses in mastering only aggravates the distortion and does not help the clarity of the final product.

Please meet guest authors—Bob Orban and Frank Foti.[1] Both of them are considered to be the world's authority on radio processing. Bob is the

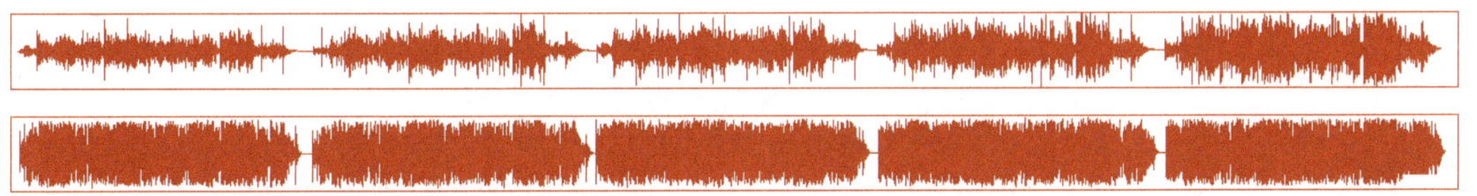

*At top, five mastered cuts of the same music, with increasing loudness and visual density. At bottom, the same cuts passed through the Orban radio processor.*

engineer and designer of the Optimod line of audio processors, while Frank is the creator and lead designer of the Omnia product line. Together, their products are used by nearly every FM radio station around the world. Here's what Orban and Foti say about *what goes on inside the box...*

## II. What Happens to My Recording When it's Played on the FM Radio? *by Robert Orban and Frank Foti*[2]

Few people in the record industry really know how a radio station processes their material before it hits the FM airwaves. This article serves to remove the many myths and misconceptions.

Every radio station uses a transmission audio processor in front of its transmitter. The processor's most important function is to control peak modulation to the legal requirements of the regulatory body in each station's nation. However, very few stations use a simple peak limiter for this function. Instead, they use more complex audio chains. These can accurately constrain peak modulation while significantly decreasing the peak-to-average ratio of the audio. This makes the station sound louder within the allowable peak modulation.

### Garbage In—Garbage Out

Manufacturers have tuned broadcast processors to process the clean, dynamic program material that the recording industry has typically released throughout its history. (The only significant exception that comes to mind is 45 rpm singles, which often were overtly distorted.) Because these processors have to deal with speech, commercials, and oldies in addition to current material, they can't be tuned exclusively for "hypercompressed," distorted CDs. Indeed, experience has shown that there's no way to tune them successfully for material which has arrived so degraded.

For 20 years, broadcast processor designers have known that achieving highest loudness consistent with maximum punch and cleanliness requires extremely clean source material. Orban has published application notes to help broadcast engineers clean up their signal paths. These notes emphasize that any clipping in the path before the processor will cause subtle degradation that the processor will often exaggerate severely. The notes promote adequate headroom and low distortion amplification to prevent clipping even when an operator drives the meters into the red.

About 1997, we started to notice CDs arriving at radio stations that had been pre-distorted in production or mastering to increase their loudness. For the first time, we started seeing frequently recurring flat topping caused by brute-force clipping in the production process. Broadcast processors react to pre-distorted CDs exactly the same way as they have reacted to accidentally clipped material for more than 20 years—they exaggerate the distortion. Because of phase rotation, the source clipping never increases on-air loudness—it just adds grunge.

The authors understand the reasoning behind the CD loudness wars. Just as radio stations wish to offer the loudest signal on the dial, it is evident that recording artists, producers, and even some record labels want to have a loud product that stands out against its competition in a CD changer or a music store's listening station.

In radio broadcasting this competition has existed since about 1975,[3] when radio stations used simple clipping to get louder, and this technique has now migrated to the music industry. The figure at right shows a section of a severely clipped waveform from a contemporary CD.

The area marked between the two pointers highlights the clipped portion. This is one of the roots of the problem; the other is excessive digital limiting in CDs that does not necessarily cause flat-topping, but still removes transient punch and impact from the sound.

The problem today is that sophisticated and powerful audio processing for the broadcast transmission system does not coexist well with a signal that has already been severely clipped. Unfortunately, with current pop CDs, the example shown is more the norm than the exception.

The attack and release characteristics of broadcast multiband compression were tuned to sound natural with source material having short-term peak-to-average ratios typical of vinyl or pre-1990 CDs. Excessive digital limiting of the source material radically reduces this short-term peak-to-average ratio and presents the broadcast processor with a new, synthetic type of source that the broadcast processor handles less gracefully and naturally than it handles older material. Instead of being punchy, the on-air sound produced from these hypercompressed sources is small and flat, without the dynamic contours that give music its dramatic impact. The on-air sound resembles musical wallpaper and makes the listener want to turn down the volume control to background levels.

There is a myth that broadcast processing will affect hypercompressed and clipped material less than it will more naturally produced material. This is true in only one aspect—if there is no long-term dynamic range coming in, then the broadcast processor's AGC[4] will not further reduce it. However, the broadcast processor will still operate on the short-term envelopes of hypercompressed material and will further reduce the peak-to-average ratio, degrading the sound even more.

Hypercompressed material does not sound louder on the air. It sounds more distorted, making the radio sound broken in extreme cases. It sounds small, busy, and flat. It does not feel good to the listener when turned up, so he or she hears it as background music. Hypercompression, when combined with "major-market" levels of broadcast processing, sucks the drama and life from music. In more extreme cases, it sounds overtly distorted and is likely to cause tune-outs by adults, particularly women.

*A severely-clipped waveform from a contemporary CD*

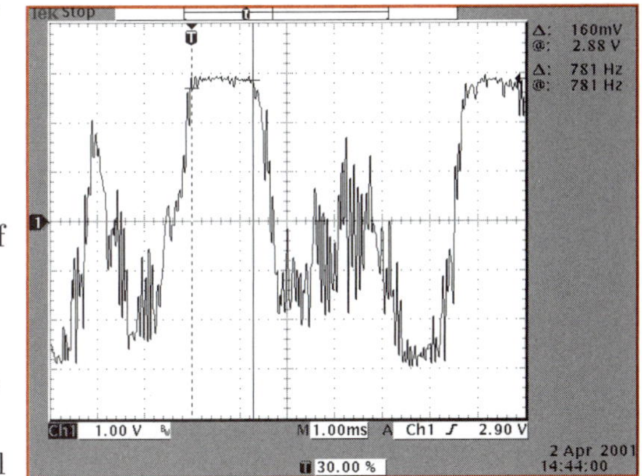


### A Typical Processing Chain—What Really Goes On When Your Recording is Broadcast

A typical chain consists of the following elements, in this order:

**Phase rotator**. The phase rotator is a chain of allpass filters (typically four poles, all at 200 Hz) whose group delay is very non-constant as a function of frequency. Many voice waveforms (particularly male voices) exhibit as much as 6 dB asymmetry. The phase rotator makes voice waveforms more symmetrical and can sometimes reduce the peak-to-average ratio of voice by 3-4 dB. Because this processing is linear (it adds no new frequencies to the spectrum, so it doesn't sound raspy or fuzzy) it's the closest thing to a "free lunch" that one gets in the world of transmission processing.

There are a few prices to pay. In the good old days when source material wasn't grossly clipped, the main price was a very subtle reduction in transparency and definition in music. This was widely accepted as a valid trade-off to achieve greatly reduced speech distortion, because the phase rotator's effects on music are unlikely to be heard on typical consumer radios, like car radios, boomboxes, "Walkman"-style portables, and table radios.

However, with the rise of the clipped CD, things have changed. The phase rotator radically changes the shape of its input waveform without changing its frequency balance: If you measured the frequency response of the phase rotator, it would measure "flat" unless you also measured phase response, in which case you would say that the "magnitude response" was flat and the phase response was highly non-linear with frequency. The practical effect of this non-linear phase response is that flat tops in the original signal can end up *anywhere* in the waveform after processing. It's common to see them go right through a zero crossing. They end up looking like little smooth sections of the waveform where all the detail is missing—a bit like a scar from a severe burn. This is an apt metaphor for their audible effect, because they no longer help reduce the peak-to-average ratio of the waveform. Instead, their only effect is to add unnecessary grungy distortion. Thanks to phase rotation, any clipping in the source material causes nothing but added distortion without increasing on-air loudness at all.

**AGC.** The next stage is usually an average-responding AGC. By recording studio standards, this AGC is required to operate over a very wide dynamic range—typically in the range of 25 dB. Its function is to compensate for operator errors (in live production environments) and for varying average levels (in automated environments). Average levels vary mainly because the peak to average ratio of CDs themselves has varied so much from about 1990 on. Peak-normalizing hard disk recordings (to use all available headroom) has the undesirable side effect of causing gross variations in average levels. Indeed, 1:1 transfers (which are also common) will also exhibit this variation, which can be as large as 15 dB![5]

The price to be paid is simple: the AGC will eliminate long-term dynamics in your recording. Virtually all radio station program directors want their stations to stay loud always, eliminating the risk that someone tuning the radio to their station will either miss the station completely or will think that it's weak and can't be received satisfactorily. Radio people often call this effect "dropping off the dial."

AGCs can be either single band or multiband. If they are multiband, it's rare to use more than two bands because AGCs operate slowly, so "spectral gain intermodulation" (such as bass' pumping the midrange) is not as big a potential problem as it is for later compression stages, which operate more quickly.

AGCs are always gated in competent processors. This means that their gain essentially freezes if the input drops below a preset threshold, preventing noise suck-up despite the large amount of gain reduction.

**Stereo Enhancement**. Not all processors implement stereo enhancement, and those that do may implement it somewhere other than after the AGC. (In fact, stand-alone stereo enhancers are often placed in the program line in front of the transmission processor.)

The common purpose of stereo enhancement is to make the signal stand out dramatically when the car radio listener punches the tuning button. It's a technique to make the sound bigger and more dramatic. Overdone, it can remix the recording. Assuming that stereo reverb, with considerable L–R energy, was used in the original mix, stereo enhancement, for example, can change the amount of reverb applied to a center-channel vocalist. The moral? When mixing for broadcast, err on the "dry" side, because some stations' processors will bring the reverb more to the foreground.[6]

Because each manufacturer uses a different technique for stereo enhancement, it's impossible to generalize about it. The only universal constraints are the need for strict mono compatibility (because FM radio is frequently received in mono, even on "stereo" radios, due to signal-quality-trigged mono blend circuitry), and the requirement that the stereo difference signal (L–R) not be enhanced excessively. Excessive enhancement always increases multipath distortion (because the part of the FM stereo signal that carries the L–R information is more vulnerable to multipath). Excessive enhancement will also reduce the loudness of the transmission (because of the "interleaving" properties of the FM stereo composite waveform, which we won't further discuss).

These constraints mean that recording-studio-style stereo enhancement is often incompatible with FM broadcast, particularly if it significantly increases average L–R levels. In the days of vinyl, a similar constraint existed because of the need to prevent the cutter head from lifting off the lacquer, but with CDs, this constraint no longer exists. Nevertheless, any mix intended for airplay will yield the lowest distortion and highest loudness at the receiver if its L–R/L+R ratio is low. Ironically, mono is loudest and cleanest!

**Equalization.** Equalization may be as simple as a fixed-frequency bass boost, or as complex as a multi-stage parametric equalizer. EQ has two purposes in a broadcast processor. The first is to establish a signature for a given radio station that brands the station by creating a "house sound." The second purpose is to compensate for the frequency contouring caused by the subsequent multiband dynamics processing and high frequency limiting. These may create an overall spectral coloration that can be corrected or augmented by carefully chosen fixed EQ before the multiband dynamics stages.

**Multiband Compression and Limiting.** Depending on the manufacturer, this may occur in one or two stages. If it occurs in two stages, the multiband compressor and limiter can have different crossovers and even different numbers of bands. If it occurs in one stage, the compressor and limiter functions can "talk" to each other, optimizing their interaction. Both design approaches can yield good sound and each has its own set of tradeoffs.

Usually using anywhere between four and six bands, the multiband compressor/limiter reduces dynamic range and increases audio density to achieve competitive loudness and dial impact. It's common for each band to be gated at low levels to prevent noise rush-up, and manufacturers often have proprietary algorithms for doing this while minimizing the audible side effects of the gating.

An advanced processor may have dozens of setup controls to tune just the multiband compressor/limiter. Drive and output gain controls for the various compressors, attack and release time controls, thresholds, and sometimes crossover frequencies are adjustable, depending on the processor design. Each of these controls has its own effect on the sound, and an operator needs extensive experience if he or she is to tune a broadcast multiband compressor so that it sounds good on a wide variety of program material without constant readjustment. In broadcast there's no mastering engineer available to optimize the processing for each new source!

**Pre-Emphasis and HF Limiting.** FM radio is pre-emphasized at 50 microseconds or 75 microseconds, depending on the country. Pre-emphasis is a 6 dB/octave high frequency boost that's 3 dB up at 2.1 kHz (75 µs) or 3.2 kHz (50 µs). With 75 µs pre-emphasis, 15 kHz is up 17 dB!

Depending on the processor's manufacturer, pre-emphasis may be applied before or after the multiband compressor/limiter. The important thing for mixers and mastering engineers to understand is that putting lots of energy above 5 kHz creates significant problems for any broadcast processor because the pre-emphasis will greatly increase this energy. To prevent loudness loss, the processor applies high frequency limiting to these boosted high frequencies. HF limiting may cause the sound to become dull, distorted, or both, in various combinations. One of the most important differences between competing processors is how effectively a given processor performs HF limiting to minimize audible side effects. In state-of-the-art processors, HF limiting is usually performed partially by HF gain reduction and partially by distortion-cancelled clipping.

**Clipping.** In most processors, the clipping stage is the primary means of peak limiting. It's crucial to broadcast processor performance. Because of the FM pre-emphasis, simple clipping doesn't work well at all. It produces difference-frequency IM distortion, which the de-emphasis in the radio then exaggerates. (The de-emphasis is flat below 2-3 kHz, but rolls off at 6 dB/octave thereafter, effectively exaggerating energy below 2-3 kHz.) The result is particularly offensive on cymbals and sibilance ("essses" become "efffs").

In the late seventies, Bob Orban invented distortion-cancelled clipping. This manipulates the distortion spectrum added by the clipper's action. In FM, it typically removes the clipper-induced distortion below 2 kHz (the flat part of the receiver's frequency response). This typically adds about 1 dB to the peak level emerging from the clipper, but, in exchange, allows the clipper to be driven much harder than would otherwise be possible.

Provided that it doesn't introduce audibly offensive distortion, distortion-cancelled clipping is a very effective means of peak limiting because it affects only the peaks that actually exceed the clipping threshold and not surrounding material. Accordingly, clipping does not cause pumping, which gain reduction can do, particularly when gain reduction operates on pre-emphasized material. Clipping also causes minimal HF loss by comparison to HF limiting that uses gain reduction. For these reasons, most FM broadcast processors use the maximum practical amount of clipping that's consistent with acceptably low audible distortion.

Real-world clipping systems can get very complicated because of the requirement to strictly band-limit the clipped signal to less than 19 kHz despite the harmonics that clipping adds to the signal. (Bandlimiting prevents aliasing between the stereo main and subchannel, protects subcarriers located above 55 kHz in the FM stereo composite baseband, and protects the stereo pilot tone at 19 kHz). Linearly filtering the clipped signal to remove energy above 15 kHz causes large overshoots (up to 6 dB in worst case) because of a combination of spectral truncation and time dispersion in the filter. Even a phase-linear lowpass filter (practical only in DSP realizations) causes up to 2 dB overshoot. Therefore, state-of-the-art processors use complex overshoot compensation schemes to reduce peaks without significantly adding out-of-band spectrum.

Some chains also apply composite clipping or limiting to the output of the stereo encoder, which encodes the left and right channels into the multiplex signal that drives the transmitter. It's actually the peak level of this signal that government broadcasting authorities regulate. Composite clipping or limiting has long been a controversial technique, but the latest generation of composite clippers or limiters has greatly reduced interference problems characteristic of earlier technology.

### Conclusions

Broadcast processing is complex and sophisticated, and was tuned for the recordings produced using practices typical of the recording industry during almost all of its history. In this historical context, hypercompression is a short-term anomaly and does not coexist well with the "competitive" processing that most pop-music radio stations use. We therefore recommend that record companies provide broadcasters with radio mixes. These can have all of the equalization, slow compression, and other effects that producers and mastering engineers use artistically to achieve a desired "sound". **What these radio mixes should not have is fast digital limiting and clipping. Leave the short-term envelopes unsquashed. Let the broadcast processor do its work. The result will be just as loud on-air as hypercompressed material, but will have far more punch, clarity, and life.**

> *"The main obstacle to good digital radio sound is low bit rate."*

A second recommendation to the record industry is to employ studio or mastering processing that provides the desired sonic effect, but without the undesired extreme distortion from clipping. The alternative to brute-force clipping is digital look-ahead limiting, which is already widely available to the recording industry from a number of different manufacturers (including the authors' companies). This processing creates lower modulation distortion and avoids blatant flat-topping of waveforms, so is substantially more compatible with broadcast processing. Nevertheless, even digital limiting can have a deleterious effect on sound quality by reducing the peak-to-average ratio of the signal to the point that the broadcast processor responds to it in an unnatural way, so it should be used conservatively. Ultimately, the only way to tell how one's production processing will interact with a broadcast processor is to actually apply the processed signal to a real-world broadcast processor and to listen to its output, preferably through a typical consumer radio.

## III. Digital Radio (BK)

Digital radio stations may or may not use such strong processing—but they all share this main obstacle—**low bit rate**. Satellite radio uses a variable bitrate codec; all channels share the total bit budget, with music channels being allocated the highest rate, but the average bitrate is still pithy compared to high fidelity standards. The bitrate used by satellite radio is a proprietary secret, but after some detective work I have determined it is significantly below 96 kbps on the average, perhaps lower than 64 k!

Codecs are designed to work with normally dynamic music, which is not very dense. The available bits are allocated to the frequency bins that have audio activity. But clipped or hypercompressed material spreads distortion across the audio spectrum and fills in all the frequency bins, making the codec ineffective, using up all the available bit bandwidth! The broadcasters can't afford to loan extra bits from adjacent channels, so the music channel starts sounding worse and worse. To make the recording semi-presentable on the air, operators are likely to severely filter or lower the level of this material. In other words, clipped and hypercompressed material does not broadcast any better on digital radio than it does on FM.

A good way to demonstrate this effect is to encode masters through a low bitrate mp3. Clipped or hypercompressed masters will sound viciously distorted, especially the S channel. Observe the spectragram of ***Figure C5-1*** in the Color Plates.

1 Robert Orban, Orban Inc. (A CRL Company). Frank Foti, Omnia Audio

2 Edited and adapted from a 2001 AES presentation.

3 Bob Ludwig (in correspondence) mentions that competition in radio broadcasting was already happening in the late 1960's, noting WABC "color radio" added EMT plate to everything to increase average density.

4 Automatic Gain Control. A compressor that brings up low-level passages. See Chapter 11.

5 No wonder CD changers are a predicament. See Chapter 14.

6 On the other hand, the other radio processing, especially the compression, reduces depth, plus, distant reception areas tend to lose separation so, improving the stereo image in mastering may not be such a bad thing.

# [ Appendix 2 ]

# The Tower of Babel Audio File Formats

### Platforms, Extensions and Resource Forks

Macintosh files are divided into two parts, the **data fork** (which is the main part and which is transferable to a PC), and **resource fork.** The Mac OSX plan is to eventually obsolete the resource fork and determine the file type strictly from the extension (e.g. .aif, .wav).

When transferring files between platforms, the **WAV**, and **AIFF** file types (described below) are the most universal. All the information about the file is contained within its header, which, unlike the resource fork, is interchangeable between the Mac and PC.

**Split and Interleaved**. A **mono** or **split** file contains one channel, as opposed to an **interleaved** file, which can contain multiple channels. We prefer to receive interleaved files wherever possible, because it is easier to group them and prevent interchannel time-slippage.

**Barbabatch**, **Sample Manager** and **Soundhack** on the Mac can batch convert between file formats, sample rate convert, change levels, redither, etc. On the PC, **Wavelab 6** is probably the most powerful file conversion tool as it can import raw files with no extension and ignore the header, allowing nearly any manipulation.

## Linear PCM

There are two popular linear (by nature lossless) audio file formats in current use: **AIFF**, and **WAVE**. The broadcast wave (BWF) variant of WAV commonly uses an extension of wav, but applications use the header to tell if the file is BWF. Acceptable extensions are: aif, aiff, wav, and bwf. Data is stored in *chunks*, and manufacturers can write proprietary chunks. The byte order of linear file formats is either **big-Endian (msb)** first, which is the Motorola standard, or **little-Endian (lsb)** first, the Intel standard. If a program accidentally misreads the wrong end of a file, the result will be nearly full-scale white noise. Reversing the ends wastes one instruction cycle, so manufacturers are sometimes fussy about which file format they prefer.

### AIFF

**Audio Interchange File Format** supports standard bit resolutions in multiples of 8, up to 32 bits fixed point, although most AIFFs are 16 or 24 bits. Byte order is **big-Endian** first **(Motorola).** While most professional PC programs can read and write AIFFs, this format was created for use on Macintosh computers, but it does not require a resource fork. There is reportedly a floating-point

AIFF file type, but as of this writing, the high-end mastering programs using AIFF insist on **fixed-point notation**. Sample rates up to 192 kHz and beyond are supported. AIFFs have no official provision for time-stamping nor for dealing with the 32-bit length limitation (see below) which will likely obsolete AIFF in due time.

A variation of AIFF is called **AIFC** (short for AIFF-C), which employs optional lossy data reduction (coding) and can use floating point notation. I have not seen AIFC supported by a high-end mastering program, but I have seen the AIFC file type accidentally applied to a standard AIFF by Mac programs such as Quicktime.

### WAVE and BWF

The **WAVE** file format, developed by Microsoft, is the most popular audio format, using a standard extension of **.wav**. It supports a variety of bit resolutions (both fixed and floating point), sample rates, and channels of audio. At the time of this writing, if interchanging WAVs with Macintosh computers, convert 32-bit floating point to 24-bit fixed, as it is unlikely the Mac will read the floating point. Byte order is **little-Endian** first **(Intel)**.

**Chunks** can be manufacturer-specific. The format has grown in a somewhat disorganized manner, and now supports many variant and sometimes unstandardized types of chunks. But the high-end programs seem to be successful ignoring the chunks that don't make sense to them!

The EBU has added a "broadcast wave extension" chunk to the basic wave format, called **Broadcast Wave (BWF)** which continues to use the WAV file name extension. The EBU has legislated this format to be a standard of interchange, so it has taken over from standard WAV in all applications. It contains the minimum metadata information that is considered necessary for all broadcast applications, such as unique source identifiers, origination station data, time stamping, etc. Although the standard WAV has some provision for time-stamping in one of the standard SMPTE timecode formats supported by some PC and Mac programs, the BWF timestamp format is standardized.

**Length limitations.** The audio in WAVE files is held in chunks defined with 32-bit integers. In February 2007, EBU Recommendation R111-2007 allows for multichannel BWF files together with channel identifiers for surround sound. For a quad 24-bit file at 96 kHz, the longest possible duration is just over an hour. In 5.1 surround at 192 kHz/24-bit, the limit descends to some 20 minutes. A way to get around the time limit was to record multiple mono BWFs, each no longer than 4 GB. However, RF64 the extended file format, adds a <link> chunk allowing a cluster of files stacked together to get a longer playing time than available with the 4 GB limit for one individual WAV file in 32-bit environments. RF64 is suitable for multichannel, stereo or mono.

### Sound Designer II

SDII (.sd2) format was invented by Digidesign for use on the Mac and is very incompatible with the PC because the metadata is kept in a resource fork. Since Digidesign has officially **obsoleted** SDII, we highly recommend running all sessions in WAV (BWF) and converting older files to WAV. Barbabatch on the Mac should preserve the SDII timestamp information when converting.

## Lossless Coded

**FLAC** is a very popular lossless coded (master-quality) format that can save space and download time. It is the functional equivalent of **zip** in the audio world except that FLAC can be played and decoded in real time. DAWs should begin to support FLAC because a lossless coded file is guaranteed to play accurately due to the built-in checksums. Considerable time would be saved as it is not necessary to zip a FLAC file for transmission. If the EFM encoder on the Laser Beam Recorder at the plant could decode FLAC files, then audio integrity would be guaranteed from the mastering studio to the disc cutter.

Many semi-pro applications (such as **iTunes**) and hardware (such as the Slim Devices **Transporter**) can read metadata from a file's ID3 tag, including genre, artist, year of introduction, and replay gain. Unfortunately, this "consumer-oriented" tag is incompatible with the broadcast WAV version of metadata and so it is hit or miss whether these applications will read the tags on WAV files. To avoid the sound losses of mp3, another positive use of FLAC is to play master-quality audio from a server in a home or studio environment, while displaying and working with metadata. **Media Monkey** on the PC, a very sophisticated freeware program, allows lossless conversion to and from FLAC and manual entry of all the metadata. Sadly, iTunes does not support FLAC, preferring its own proprietary **Apple Lossless** format.

## Lossy Coded

**AAC** (used in iTunes) and **mp3** are **lossy coded file formats**, that is, some audio information has been sacrificed in the effort to save space and increase transmission speed. Once sound has been encoded into a lossy format, sound quality can never be restored, which is why it's called a *lossy format!* Since these have become widespread and mislabeled *CD Quality* we sometimes get them as original sources! This violates the source-quality rule. Whenever possible, we ask to have these replaced with higher-quality, earlier-generation sources, or the sound quality will obviously suffer, especially after mastering processing.

## Metafile Formats

Metafile formats are designed to interchange all the information needed to reconstruct a project. Unfortunately, some manufacturers are reluctant to adopt another's format, so this valuable effort has not made enough progress.

### AES-31

The **AES-31** file Interchange standard was developed by the SC-06-01 AES standards committee jointly with several manufacturers. The goal is to interchange basic projects, timestamp and crossfade information as well as audio files. There has been some success but as of this writing the format is not supported by Digidesign.

### OMF

The **Open Media Format** was produced by Digidesign to interchange Pro Tools Session and audio data with other workstations. In multi-vendor environments, success rate may vary!

[ Appendix 3 ]

# Preparing Tapes and Files for Mastering

One major theme in this book has been the mastering engineer's comprehensive attention to sequencing, spacing (a.k.a. *assembly*), leveling, clean-up and processing. The better-prepared the mix tape or file, the better we all will look. Make the best mix you can, then let the mastering engineer do the rest of the magic, including the "heads, tails, fade-ins and fadeouts," for if something is cut off or faded prematurely, it will be lost. Don't be tempted to fade even if there is a noise, because we can create natural-sounding endings on tunes that everyone thought had to be faded, as described in Chapter 7. Mix engineers can also include a "fade example," which we can use if this proves to be the best choice. Given freedom, the mastering engineer can produce a seamless, flowing record album from the "loose parts" sent by the mix engineer. Leaving the tunes loose also permits the mastering house the most flexibility to change the order of the album (if necessary), or produce segues in the most artistic fashion.

In the last century, the most common formats we received for music mastering were linear, e.g. analog and digital tape and standalone CDR (which is linear for writing, but random-access for reading). But now the most popular formats are completely random access (file-based). Here's how to satisfy the needs of the mastering engineer when submitting finished mixes on the medium of your choice.

### Communication

Mastering is a collaborative process, even if you cannot attend the mastering session; the mastering engineer's job is to realize your desires and if possible to go beyond your wildest dreams! The mix engineer should discuss the music and his goals and involve the mastering engineer at an early stage; if in the neighborhood, bring over a sample to hear on the high-resolution, wide-range mastering monitors. Listen: Does it sound like music? Does it live and breathe? Do the climaxes sound like climaxes? Do the choruses have a bit more energy than the verses (the usual natural case)? Is the bass drum to bass ratio right? Is the sound as spacious and deep as you want it to be? How does it sound on several alternative systems? When the mastering date arrives, don't hesitate to provide or suggest a CD of similar music that appeals to you, yet leave your mind open to the creativity of the mastering engineer. When the mastering session is over, the mastering engineer will provide a reference disc that the mix engineer will check and if desired, suggest revisions or improvements.

## Mix Logs

The logs that accompany mixes are very important. Logs keep a project from being delayed as we don't have to chase down the catalog number or other essential information on the mastering day. Some engineers forget that **a disc of files has no order.**[1] So all logs should indicate the full title of each song, the corresponding file name on disc, and the order the song is to appear on the final medium, plus comments about fades, noises, any problem or concern, or special requests. ("Please leave that ugly laugh in between songs 2 and 3, I think it's funny.") Appendix 5 contains example logs.

**Stems, Splits and Alternate Mixes (e.g. Vocal Up/Down).** By all means provide alternate mixes or synchronized stems if possible. See Chapter 16.

## Linear Media (Analog tape, CD-DA)

Don't bother to reorder linear media. Leave the tunes out of order, leave the outtakes and alternate mixes (which may prove useful), and mark all *keeper* takes in the log. Don't bother to space the tunes on linear media other than leaving enough time to cue and to use leaders or program IDs to identify the cuts.

When mixing to a digital recorder, always record two at once (make data-identical mixes labeled "A" and "B"), and hold onto that safety—never send the only copy in the mail. Track IDs on CD-DA mixes do not have to be exactly placed, but they guide us to loading the proper tune.

## Level Check

As described in Chapter 5, mix with conservative levels, which is not a problem with 24-bit media. Print the mix with peak levels well under the top and no OVERs! I recommend -3 dBFS maximum. Roger Nichols reminds mix engineers using DAWs to visit each plug-in, reset the clip indicator and check the mix. If there's a clip, then redo the mix to avoid internal clipping, which can cause pops and snaps that nobody seems to notice until it gets to mastering.[2]

## Preserve Data Integrity

In general, send the earliest generation, unprocessed material to the mastering house. If you must edit, keep everything at unity gain if at all possible (do not normalize), even if the material is peaking low, as explained in Chapter 5. The same goes for temptation to equalize, compress, limit or process a mix after it has been made. If you do post-process, please send both versions to the mastering house, because we may have a better process with our tools, or we may combine it with other processes and reduce cumulative DSP.

## Maximum CD Program Length

Every plant specifies a maximum acceptable length, and some charge more for CDs over approximately 77 minutes. The final CD Master tape, including songs, spaces between songs, and reverberant decay at the ends of songs, must not exceed the limit, which at one popular plant is 79:38. The mastering engineer can determine the exact time after the master is assembled. DVD program lengths vary because of the data coding and must be determined at the time of authoring.

## Labeling tapes or discs. Which is the Master?

Don't forget to put a name and phone number on the source media in case it gets separated from the documentation! The **sources** for an album are **NOT** the master; the album (production) **MASTER** is the final, PQ'd, equalized, edited, assembled, and prepared tape or disc that needs no further audio work, and is ready for replication.[3] Please label the source media: **Submaster, Work Tape, Mix, Final Mix, Session Tape, Edited-Mix, Compiled-mix**, or **Equalized Mix**, to name several possibilities. This will avoid confusion in the future when looking through the tape library for the one and only *real* (production) master.[4]

### Analog tape Preparation

Begin and end the reel with some "bumper," followed by leader. If possible, put leader between songs (except for live concerts and recordings edited with room tone). Tape should be slow wound, tail out. Label each reel as recommended in Appendix 5. Indicate tape speed, record level for 0 VU in nw/m, record EQ (NAB or IEC), track configuration, whether it is mono, stereo or multichannel. Indicate if noise reduction is used and include the noise reduction alignment tone. Include alignment tones 30 seconds (or longer) each, at 0 VU, without noise reduction, minimum 1 kHz, 10 kHz, 15 kHz, and 100 Hz plus (highly recommended) 45 Hz and 5 kHz. Also highly recommended is a tone sweep (glide) from 20 Hz through 500 Hz. The tones must be recorded by the same tape recorder that recorded the music, and ideally, recorded through the same console and cables that were used to make the mix. Many mastering engineers prefer having the tones at the tail of a reel or on a separate reel. Store all tapes flat wound, tail out.

Many historic analog tapes do not include proper tones and sometimes it is not possible to put tones on new masters. If it was not possible to lay down tones on the session, then we will use sophisticated methods to guarantee azimuth and equalization accuracy.

*"The source tapes/files for an album are NOT the Production MASTER."*

### Give Handles

For live concerts and many other forms of music, it's useful to include *handles*, that is *raw footage* on either side of the intended music. This can include out takes, unfaded applause, breaths, coughs, noises, speech between tracks, etc. Handles are especially useful when a track might have to be noise reduced, for the noise sample we need can sometimes only be found just before the downbeat.

### What Sample Rate?

Until around 2000, I recommended that mix engineers try to work at 44.1 kHz if possible for CD, considering the then poor state of typical sample rate converters. This is no longer necessary nor desirable; high quality sample rate converters can convert between 96 kHz and 44.1 kHz with high integrity, as described in Chapter 20. The best recommendations are for mix engineers to work at the highest practical sample rate and longest available wordlength.

However, if you are mixing digitally, **do not sample rate convert yourselves**, but remain at the same sample rate as the multitrack. If you are mixing with an analog console, there is a marginal advantage to using a higher sample rate for the mixdown recorder than the multitrack. For example, even if mixing analog with a multitrack at 48 kHz, you will get slightly better results with a mixdown recorder at 96 kHz.

**Random Access Media: Preparing Files**

File discs require attention to detail, as a poorly-prepared CD-ROM or DVD-R can waste a tremendous amount of time at the mastering house. Make sure the mastering house will accept the file types you want to send. Here are some critical do's:

• **Leave blank sound at the head of the file**, in other words, start the first music at least 1 second into the file, **not at zero time** (this is to prevent glitches that often occur at the file start).

• For stereo and multichannel, high resolution, linear-format, **interleaved files** are preferred. Mono or split files are also acceptable. **No MP3's, please!**

• Try to do one project at a **single sample rate**. It takes considerable extra work to deal with multiple sample rates in a project and often involves a compromise as we must rate-convert some files to yield a common rate for the mastering. But if for some reason your project includes different rates, do not convert them yourself, carefully mark (log) the rate of the files for our information.

• Give each file a meaningful name related to the song title, like **Love Me Do**, not some meaningless serial number.

• Choose a high-quality name-brand blank. To my experience, Taiyo Yuden, the oldest manufacturer, continues to make the most compatible and reliable blank discs.

• For lowest error rate, test error rates with Plextools Pro to find the max. 8x to 16x is typically safe with today's media.

• Write a **Fixed** disc, i.e. a **closed session.** To verify the disc has been fixed, pop it into a PC or Mac CD reader (not a writer) and make sure it can read the file names.

• **DO NOT USE PAPER LABELS!** Stick-on paper labels may look impressive, but they increase error rate and they are dangerous at high rotational speeds. Labels have become partly or completely unglued and tear off in the reader, which is not a pretty sight! Also, do not label the disc with a ball-point pen, but with a soft marker, on the protected (overcoated) part of the top surface.[5] While I personally believe that the coating on professionally over-coated CDRs is sufficient protection from scratches and organic solvents (as in an aromatic Sharpie-brand marker), the most conservative mastering engineers recommend using water-based markers for labeling. Perhaps someone will do a long-term study measuring errors on CDRs with a coated-marked surface.

• Find out if the mastering house can accept 32-bit floating point files, and in which of several competing 32-bit formats. The most commonly used is compatible with Nuendo and Samplitude. When in doubt, write to **fixed-point 24-bit files** (also known as *Integer Format*).

• Use any standard sample rate up to 96 kHz. Verify the mastering house can use files with a higher rate before cutting.

• **File names** should not include hyphens (-), use an underline instead. Do not use the / or \ character. For best multi-platform compatibility, stay away from spaces and use alphas, numerics and underlines only.[6]

• **We love receiving files that include the intended track number in their name**. It is very useful to include the intended track number at the beginning of the file name (using two digits), which makes it much easier to assemble them in the album order. For example: **01 I Need Somebody, 09 I Got Rhythm, 10 She's So Fine.**

• **Avoid extra periods (dots) in Mac file names on Mac discs** because they might be transferred to PC and be confused with extensions; use one and only one dot in front of the 3-letter extension.

• DVD-R is a little more compatible than DVD+R.

### Split Files

Interleaved files are less subject to accidents since all the channels are guaranteed to start at the same point. For multichannel, include a note indicating the channel order used, e.g., L, R, C, LFE, SL, SR or L, R, SL, SR, C, LFE. If you must send split files, use a standard nomenclature to distinguish the channels, e.g. **Do It_L, Do It_R, Do It_SL, Do It_SR, Do It_C, Do It_LFE.** Letter abbreviations are preferable to ambiguous channel numbers.

### When You Get Your Master Back

If the master is sent back to you instead of directly to the CD plant, **don't handle it or play it.** Play the ref, not the master![2]

1 There is no track order on a non-linear, file-based medium. Often, clients ask me, "put the master in the order that's on the CD-ROM," but they forget the only order on the CD-ROM is the alphabetical directory of files.

2 Thanks to Roger Nichols for the nudge to put these recommendations in the Appendix.

3 Andre Subotin on the Mastering Webboard reminds us that there may be several true Masters, each of which we must clearly label, e.g. Production Master for Cassette; Master for foreign countries; etc.

4 Visit the NARAS Master Recording Delivery Recommendations, linked at digido.com.

5 Thanks to Clete Baker and Mike McMillan on the Mastering Webboard for clarifications on these points.

6 Thanks to Clete Baker on the Mastering Webboard for reminding me of this essential!

## [ Appendix 4 ]

# Premastering for Vinyl

To make the best premaster for vinyl, do nothing special, work with conservative levels and good headroom as the premaster has no effect on the level of the vinyl cut. A little better sound quality can be obtained by creating a data disk with 24-bit files rather than an audio CD. In this case we depend on the vinyl cutting engineer to decide the space between the songs, or we can include the desired space at the tail of each file and he puts them in order in a DAW and then outputs to the cutter. Duration is the only thing to be concerned about, especially when there is a lot of bass on the record. A ten-minute side is usually no problem even when there is heavy bass. It's technically possible to put half an hour on an LP side, but almost inevitably with loss of level, stereo separation and/or bass.

We should not apply a process because we think it's necessary for the cutting. When mastering for vinyl, our job is to get the SOUND we want to hear in the finished product onto the source medium.

Let the cutting engineer expert do the final processing for vinyl, which usually includes narrowing the separation at the bass end to protect the groove excursion, and some high-frequency limiting to protect the cutterhead. Let him do what he has to do for loudness and other idiosyncrasies of the medium. He will try to come as close as possible to our sound within the limits of the LP medium.

# [ Appendix 5 ]

# Logs and Labels for Tapes, Discs and Boxes

### Labeling Source Media

**I don't dare put an unlabeled disc down on my mastering desk, for it will immediately be lost in a crowd! Please do** put the following minimal information on every piece of source media, in case it gets separated from the box:

- Artist
- Album Title [or working title]
- Contact Name, phone number
- Tape or reel number
- Date [important to help separate out revisions]

### Labeling Those Boxes

The box label contains much more information than what's written on the reel or disc itself.

### Analog Tape Boxes: An example label

Some studios have preprinted labels with checkboxes for each option.

**Mix tape, Unedited, songs head leadered [or other descriptive]**

Artist: ______________________

Album Title: ______________________

Record Label: ______________________

Reel number: ____________ of ________

Catalog Number: ______________________

Studio, Address, Contact Phone #: __________

Engineer: ______________________

Assistant: ______________________

Producer: ______________________

Date: ______________________

Format, EQ, Speed, Level: *[e.g. 1/2" 2-track AES stereo, no noise reduction, 30 IPS, 0 VU = 320 nW/M, or 0 VU = 250 nW/M + 2 dB]*

Test Tones @ Head _____ @ Tail _____ consisting of ______________________ Hz at 0 VU

**Name of Song or**
**Track Length**
**Comments** *[e.g. "vocal up" or "needs fadeout" or "leave countoff at the beginning"*

Name of next song, etc.

Further comments can be written in a letter that accompanies all the media.

### Discs: Example Label

There is not enough room on a CDR or DVD-R surface to write everything we want to know. Some studios have prescreened discs with checkboxes. At minimum, the top surface of the disc itself should include:

**Mixes, Unedited** *[or submaster or other descriptive]*
Artist: ______
Album title: ______
Record Label: ______
Disc and File Format: *[e.g. ISO-9660 or HFS, or Masterlink, Stereo AIFF Files, 48 kHz/24 Bit]*
Disc # ____ of ____ Date: ______
[date is very critical]
Plus, if possible:
Contact name and Phone #: ______
Catalog Number: ______

Since there is not enough room to list all information on the disc itself, be sure to include the remaining information on the box, jewel box, and/or printed log (pictured opposite page) which accompanies the media. If possible, the log can be duplicated in a READ_ME.doc file which resides on the disc, so it will never be lost.

### Discs, Jewel Box or Paper cover label

Instead of using up several jewel boxes, some studios cleverly put CDRs inside a taped and folded printout of the disc's directory, which covers all the names of the tunes inside the disc. When shipping, put these paper-covered discs in a foam-lined hard-box to prevent scratching or breakage.

### Printed Log/letter

Accompany the discs or tapes with a printed log/letter to the mastering engineer. This is where you can also include all your comments and thoughts on the eventual mastering. You can put this in the form of a letter, which includes your story and feelings about the album and its sound. Some comments (especially the need for a fade!) may be superfluous, but put down anything you are concerned about.

Don't forget to include:

Artist: ______
Album title: ______
Record Label: ______
Disc, File Format, Sample Rate, Wordlength: *[e.g. ISO-9660 or HFS, or Masterlink, Stereo AIFF Files, 48 kHz/24 Bit]*
Contact name and Phone #: ______
Contact Address: ______
Catalog Number: ______

Mix engineer's log, with notes for fades and spacing. Logs can also be in the form of a letter to the mastering engineer which include this information.

| Title/File Name | CD track Order | Length (approx.) | ABS time/DAT or CD Program ID (not relevant if this is a disc of files) | ISRC | Comments *[e.g. by engineer, producer or artist]* |
|---|---|---|---|---|---|
| I Wanna Make You Happy/ 05_makehappy.wav | 5 | 4:02 | | ES6080132805 | This song needs a fadeout. Try starting circa 3:45 and be out by 4:00 from the downbeat so as not to hear the snickering! Please include the sticks at the beginning. |
| Love Me Do/ 02_lovemedo.wav | 2 | 2:55 | | ES6080132802 | This is an obvious tribute to the Beatles. The more Beatle-like you can make the mastering, the happier I will be! |
| Why Me?/ 04_yme.wav | 4 | 5:02 | | ES6080132804 | This is the only ballad on the album. The artist is not happy with her intonation entering the last chorus. Is there anything you can do about this? |

[ Appendix 6 ]

# Conversions

| Chart 1: Tape Fluxivity in dB and nanowebers per meter (nW/M) | | | | | |
|---|---|---|---|---|---|
| **Level dB** | **Reference 185** | **Reference 200** | **Reference 250** | **Reference 320** | **Reference 400** |
| 9 | 521 | 564 | 705 | 902 | 1127 |
| 8 | 465 | 502 | 628 | 804 | 1005 |
| 7 | 414 | 448 | 560 | 716 | 895 |
| 6 | 370 | 400 | 500 | 640 | 800 |
| 5 | 329 | 356 | 445 | 569 | 711 |
| 4 | 293 | 317 | 396 | 507 | 634 |
| 3 | 261 | 283 | 353 | 452 | 565 |
| 2 | 233 | 252 | 315 | 403 | 504 |
| 1 | 208 | 224 | 281 | 359 | 449 |
| 0 | 185 | 200 | 250 | 320 | 400 |

| Chart 2: dBu (reference 0.775 volts) converted to voltage | |
|---|---|
| **dBu** | **Volts** |
| 40 | 77.500 |
| 35 | 43.581 |
| 24 | 12.283 |
| 20 | 7.750 |
| 18 | 6.156 |
| 14 | 3.884 |
| 12 | 3.085 |
| 8 | 1.947 |
| 6 | 1.546 |
| 4 | 1.228 |
| 2 | 0.976 |
| 1 | 0.870 |
| 0 | 0.775 |
| -10 | 0.245 |
| -20 | 0.078 |
| -60 | 0.001 |

| Chart 3: Q to Bandwidth Conversions | | | |
|---|---|---|---|
| **BW** | **Q** | **Q** | **BW** |
| 0.20 | 7.21 | 0.6 | 2.19 |
| 0.30 | 4.80 | 0.7 | 2 |
| 0.40 | 3.60 | 0.8 | 1.7 |
| 0.50 | 2.87 | 0.9 | 1.53 |
| 0.60 | 2.39 | 1 | 1.39 |
| 0.70 | 2.04 | 1.2 | 1.17 |
| 0.80 | 1.78 | 1.4 | 1.01 |
| 0.90 | 1.58 | 1.5 | 0.94 |
| 1.00 | 1.41 | 1.7 | 0.84 |
| 1.20 | 1.17 | 1.8 | 0.79 |
| 1.40 | 0.99 | 2 | 0.71 |
| 1.60 | 0.86 | 3 | 0.48 |
| 1.80 | 0.75 | 4 | 0.36 |
| 2.00 | 0.67 | 5 | 0.29 |
| 2.50 | 0.51 | 6 | 0.24 |
| 3.00 | 0.40 | 8 | 0.18 |

*Please visit www.digido.com/links/ for a link to an online calculator together with the mathematical explanation.*

## [ Appendix 7 ]

# I Feel The Need For Speed

| Medium | Theoretical Speed MB/sec | Practical Speed MB/sec | Theoretical Speed Mb/sec | Practical Speed Mb/sec | Theoretical # mono tracks | Practical # mono tracks |
|---|---|---|---|---|---|---|
| **CD player (1x speed)** | 0.17 | N.A. | 1.35 | N.A. | 1 | N.A. |
| **DSL 384 kbps** | 0.05 | 0.04 | 0.38 | 0.30 | 0 | 0 |
| **ISDN** | 0.19 | 0.17 | 1.54 | 1.36 | 1 | 1 |
| **10 base T** | 1.25 | 1.12 | 10.00 | 8.96 | 5 | 4 |
| **100 base T** | 12.50 | 10.00 | 100.00 | 80.00 | 46 | 36 |
| **1000 base T (Gigabit Ethernet)** | 125.00 | 56.00 | 1,000.00 | 448.00 | 455 | 204 |
| **USB 1.0 slow** | 0.19 | N.A. | 1.50 | N.A. | 1 | N.A. |
| **USB 1.0 fast** | 1.50 | N.A. | 12.00 | N.A. | 5 | N.A. |
| **USB 2.0** | 60.00 | 29.50 | 480.00 | 236.00 | 218 | 107 |
| **Firewire 400** | 50.00 | 38.50 | 400.00 | 308.00 | 182 | 140 |
| **Firewire 800** | 100.00 | 57.60 | 800.00 | 460.80 | 364 | 210 |
| **ATA/133** | 133.00 | 62.60 | 1,064.00 | 500.80 | 484 | 228 |
| **SATA/300** | 384.00 | 63.20 | 3,072.00 | 505.60 | 1,398 | 230 |
| | | | | | | |
| **File system benchmark tests by SI Software Sandra Lite** | | | | | | |
| **OSX server (G4/SATA)** | N.A. | 9.00 | N.A. | 72.00 | N.A. | 33 |
| **NAS** | N.A. | 15.00 | N.A. | 120.00 | N.A. | 55 |
| **Local Drive (C:)** | N.A. | 33.00 | N.A. | 264.00 | N.A. | 120 |
| **Linux server (Software RAID 5)** | N.A. | 56.00 | N.A. | 448.00 | N.A. | 204 |
| | | | | | | |
| **File copying** | | | | | | |
| **NAS to Linux server** | N.A. | 6.73 | N.A. | 53.84 | N.A. | 25 |
| **Linux server to Local Drive (C:)** | N.A. | 43.01 | N.A. | 344.08 | N.A. | 157 |
| **NAS to Local Drive (C:)** | N.A. | 11.65 | N.A. | 93.20 | N.A. | 42 |
| **Local Drive (C:) to Linux server** | N.A. | 29.90 | N.A. | 239.20 | N.A. | 109 |

*MB = Megabytes, Mb = Megabits (8 bits/byte), based on 1 kB = 1024 B, mono tracks are 96 kHz/24-bit. Practical speeds were measured on PC. Mac benchmarks may vary.*

*For serving audio, RAID 5 systems are the best, most reliable solution via Gigabit Ethernet. A small studio may choose a Unix-based NAS, which is off the shelf, plug and play and easily administered. As you can see, it has moderate speed. For multi-track or multiple studios, a Linux-based server is extremely fast and economical, with Apple's XSAN being an expensive alternative. Benchmarks were measured with SI Software Sandra Lite, and copy tests with a stopwatch. The chart shows the Linux server is faster than a locally mounted single drive! Three clients could simultaneously pull 30 MB/sec over the LAN while copying a 600 MB file, theoretically equivalent to 109 simultaneous 96 kHz/24-bit tracks to each client but your mileage may vary.*

# [ Appendix 8 ]

# I Feel The Need For Capacity

| Year | Type of Storage | Capacity GB | Total Cost US Dollars | Cost per GB | # Mono Tracks / Hr | Facts |
|---|---|---|---|---|---|---|
| 1980 | Data General | 0.297 | $35,000 | $117,845 | 0.31 | Size: 2 feet x 3 feet x 3-1/2 feet high! |
| 1990 | SCSI Hard Disc | 0.60 | $750 | $1,250 | 0.62 | CD one hour 635,040,000 bytes |
| 2002 | IDE Hard Disc | 80 | $137 | $1.71 | 83 | Street price |
| 2012 ? | Colossal's Nanotechnology-based Hard Disc | 1,000,000 | $100 | $0.0001 | 1035631 | New technology that will be able to store about 125 TB / sq. inch |

*GB = Gigabytes, based on 1 GB = 1024 MB, mono tracks are 96 kHz/24-bit.*

Prior to 1990, I was making CD masters with linear editing using the Sony 3/4" editing systems. In 1990 I set up my first nonlinear mastering workstation, purchasing the highest capacity hard discs available, a pair of 600 MB SCSI hard discs, that cost $1500 retail, or $1.25 per MB. Fortunately, as our needs have gone up in 10 years, capacity has increased geometrically and cost has gone down. It's not out of line to expect typical storage capacity to tentuple in 10 years.

# [ Appendix 9 ]

# Footnotes on the K-System

### The VU Meter's Actual Ballistics

The **VU** meter's ballistics were analyzed as early as 1940. According to **A New Standard Volume Indicator and Reference Level**, Proceedings of the I.R.E., January, 1940, the mechanical VU meter used a

> copper-oxide full-wave rectifier which, combined with electrical damping, had a defined averaging response according to the formula **i = k * e to the p** equivalent to the actual performance of the instrument for normal deflections. (In the equation **i** is the instantaneous current in the instrument coil and **e** is the instantaneous potential applied to the volume indicator).... A number of the new volume indicators were found to have exponents of about 1.2. Therefore, their characteristics are intermediate between linear (**p** = 1) and square-law or root-mean-square (**p** = 2) characteristic.

### How 85 dB Became 83, Reverted to 85 (but was Really 83) and Remains Mired in Politics and Inexact Measurement Methods

The K-System is a cousin to the *theatre standard*, **Proposed SMPTE Recommended Practice: Relative and Absolute Sound Pressure Levels for Motion-Picture Multichannel Sound Systems.** This SMPTE Document RP 200 defines the calibration method in detail but has flip flopped its measurement technique in several revisions since 1995.

In the 1970s the value had been quoted as *85 dB SPL at 0 VU* but as the measurement methods became more sophisticated, this value proved to be in error because of several problems:

- variations in VU meters, as much as 2 dB
- variations in the amplitude of pink noise depending on cable length and bandwidth of measurement
- variations in the crest factor and style of the pink noise generator
- bandwidth of the loudspeaker system under calibration

A measurement instrument with mean response (simple averaging), such as a VU meter, will always respond differently to different styles of pink noise, even if their energy is the same. Only a true RMS meter responds correctly to energy level.

Several fudges were made to the generators and specifications so that Hollywood engineers could continue to use mean-responding meters and yield the historic **85** number. As long as the calibration remains a closed system, with specified pink noise generators and meters, it can be consistent, and consistency is the name of the game. When the respective measurement methods are examined, "83" and "85" are only 1 dB apart.

According to K.K. Proffitt, Dolby's wideband pink noise, specified as -20 dBFS, is actually -19 dBFS RMS, in order to fudge the mean-responding meter to read 0 VU. It is preferable to use the filtered pink noise especially since Dolby's filtered pink noise appears to give the same results as Holman's. If either Dolby or Holman's filtered pink noise is used, calibrate the large theatre to 85 dB SPL for closest conformance with RP200.[1]

### Why The K-System Divorced from the Theatre Controversy

Even if the K-System exactly followed the 85 dB theatre calibration, it would still be a bad choice! This is because small rooms sound 1 or 2 dB louder than large rooms due to increased transient response, lower reverberation and proximity and directivity of the loudspeakers. Which is why a lower number, 83, sounds about as loud in the small room. We also specify RMS meters for more accuracy and consistency regardless of the shape of the noise. See Chapter 15 for alignment techniques.

### Detailed Specifications of the K-System Meters

**General**: All meters have three switchable scales: K-20 with 20 dB headroom above 0 dB, K-14 with 14 dB, and K-12 with 12 dB. The K/RMS meter version (flat response) is the only required meter—to allow RMS noise measurements, system calibration, and program measurement with an averaging meter that closely resembles a *slow* VU meter. The other K-System versions measure loudness by various known psychoacoustic methods (e.g. LEQ and Zwicker).

**Scales and frequency response:** A tri-color scale has green below 0 dB, amber to +4 dB, and red above that to the top of scale. The peak section of the meters always has a flat frequency response, while the averaging section varies depending on which version is loaded. For example: Regardless of the sampling rate, meter version K-20/RMS is band-limited to 22 kHz, with a flat frequency response from 20 Hz-20 kHz +/- 0.1 dB, the averaging section uses an RMS detector, and 0 dB is 20 dB below full scale.

**Other loudness-determining methods are optional**. A true loudness meter will be monophonic and reflect the head-related transfer function. Regardless of the frequency response or methodology of the loudness method, reference 0 dB of all meters is calibrated such that 20 Hz-20 kHz pink noise at 0 dB reads 83 dB SPL on each channel, C weighted, slow. Psychoacousticians designing loudness algorithms recognize that the two measurements, SPL and loudness are *not* interchangeable and take the appropriate steps to calibrate the K-System loudness meter 0 dB, so that it equates with a standard SPL meter at that one critical point with the standard pink noise signal.

**Scale gradations**: The scale is linear-decibel from the top of scale to at least -24 dB, with marks at 1 dB increments except the top 2 decibels have additional marks at 1/2 dB intervals. Below -24 dB, the scale is non-linear to accommodate required marks at -30, -40, -50, -60. Optional additional marks may be included through and beyond -70. Both the peak and averaging sections are calibrated with sine wave to ride on the same numeric scale. Optional (recommended): A *10x* expanded scale mode, 0.1 dB per step, for calibration with test tone.

**Peak section of the meter:** The peak section represents the true, flat (1 sample) peak level, regardless of which averaging meter is used. An additional pointer above the moving peak represents the highest peak in the previous 10 seconds. Designers can add an oversampling peak movement as long as it is clearly marked and identified, especially since all our emphasis on loudness judgment is based on the averaging section and its scale. A peak hold/release button on the meter changes this pointer to an infinite high peak hold until released. The meter has a fast rise time (a.k.a. *integration time)* of one digital sample, and a slow fall time, ~3 seconds to fall 26 dB. An adjustable and resettable OVER counter is highly recommended, counting the number of contiguous samples that reach full scale.

**Averaging section:** An additional pointer above the moving average level represents the highest average level in the last ten seconds. An *average hold/release* button on the meter changes this pointer to an infinite *highest average* hold until released. The RMS calculation should average at least 1024 samples to avoid an oscillating RMS readout with low frequency sine waves, but keep a reasonable latency time. If it is desired to measure extreme low frequency tones with this meter, the RMS calculation can optionally be increased to include more samples, but at the expense of latency.

**Ballistics**: This is only relevant to the RMS meter, as the "ballistics" of the true loudness versions will be determined by the algorithm. After RMS calculation, the meter *ballistics* are calculated, with a specified integration time of 600 ms to reach 99% of final reading (this is half as fast as a VU meter). The fall time is identical to the integration time. Rise and fall times should be exponential (log).

1 Thanks to Tomlinson Holman (in correspondence) for explaining the historical source of the measurement errors, and to K.K. Proffitt for the most currently acceptable "fudge".

## [ Appendix 10 ]

# Recommended Reading, CDs for Equipment Testing and Ear Training


**Books**

Burroughs, Lou (1974) *Microphones: Design and Application*, Sagamore Publishing Company, Inc. Out of Print. A classic audio work, the first book to publish the 3 to 1 rule with frequency measurements of the anomalies.

Holman, Tomlinson (2000) *5.1 Surround Sound: Up and Running*, Focal Press. Also includes guides on the problems of locating speakers near consoles.

Howard, David M. & Angus, James (2001) *Acoustics and Psychoacoustics*, Focal Press. Includes good discussion of the time/frequency relationship of filtering.

Kefauver, Alan P. (1999) *Fundamentals of Digital Audio*, A-R Editions, Madison, WI

Kirk, Ross & Hunt, Andy (1999) *Digital Sound processing for Music and Multimedia*, Focal Press, Boston.

Nisbett, Alec (2003) *The Sound Studio: audio techniques for radio, television, film and recording*, 7th Edition. Focal Press. A classic work with practical techniques which will never go out of style. I started with the 1962 edition!

Owsinski, Bobby (2000) *Mastering Engineer's Handbook*, ISBN# 0-87288-741-3. A collection of interviews with mastering engineers.

Pohlman, Ken (2000) *Principles of Digital Audio*, McGraw Hill.

Watkinson, John (1988, regularly revised) *The Art of Digital Audio*, Focal Press, ISBN 0 240 51320 7. The definitive industry *bible*. This is where you must first go for in-depth information on how digital audio works and the specifications of much of today's digital audio equipment and interfaces.

**Articles in Print**

Blesser, A. & Locanthi, B., (1986) *The Application of Narrow-Band Dither Operating at the Nyquist Frequency in Digital Systems to Provide Improved Signal-to-Noise Ratio over Conventional Dithering*, AES 81st Convention, Preprint 2416.

Cabot, Richard C. (1989) *Measuring AES/EBU Digital Audio Interfaces*, AES 87th Convention Preprint 2819 (I-8).

Gerzon, M.A., Craven, P.G., Stuart, J.R., & Wilson, R.J. (1993) *Psychoacoustic Noise-Shaped Improvements to CD and other Linear Digital Media*, AES 94th Convention, Preprint #3501.

Lipshitz, S.P. & Vanderkooy, J. (1989) *Digital Dither: Signal Processing with Resolution Far Below the Least Significant Bit.* AES 7th International Conference-Audio In Digital Times, Toronto, 87-96.

Muncy, Neil, Whitlock, Bill et al (1995) Collection of definitive articles on grounding, shielding, power supply, EMI, RFI. Journal of the AES Vol. 43 Number 6, a special excerpt printing.

Nielsen, Soren & Lund, Thomas (2000) *0 dBFS+ Levels in Digital Mastering*. AES 109th Convention, Preprint #5251.

Stuart, J.R. & Wilson, R.J. (1991) A Search for Efficient Dither for DSP Applications, AES 92nd Convention, Preprint #3334.

Stuart, J.R. (1993) *Noise: Methods for Estimating Detectability and Threshold*, 94th AES Convention Preprint #3477.

### Compact Discs

*Auralia, Complete Ear Training software for musicians*, Rising Software, Australia.

Grimm, Eelco, (2001) *Checkpoint Audio Professional Audio Test Reference*, Contekst Publishers, Netherlands, ISBN 90-806111-1-5. Test and listening CD including J-Test, Bonger Test, and unique distortion and listening tests. Written in Dutch with no English translation (as of 2002).

Moulton, David, *David Moulton's Audio Lecture Series, Golden Ears audio ear-training self-study course*, KIQ Productions (Golden Ears).

*Various compilation and test CDs*, Chesky Records.

### Articles On the Internet

Please go to www.digido.com/links/ for online articles.

## [ Appendix 11 ]

# Eric James Biography

An awful lot has changed for Eric James in the five years since he first worked with Bob on the original edition of this book. He is still English, still inordinately fond of chamber music and acoustic jazz, but he is very happily no longer resident in Hong Kong, and rather more sadly, no longer in his mid-forties. Officially he is both the Director of URM Audio - a classical music editing/mastering/ SADiE training facility based in Suffolk, U.K.—and Preceptor in Philosophy at Corpus Christi College, Cambridge, where he teaches the work of the very musical Nietzsche and Wittgenstein (on whom he has written a book to be published later this year). Unofficially he is having a ball—quite possibly the best time of his life. In both worlds, and especially where they overlap, he gets to work with some of the nicest and most talented people you could ever imagine. Although URM Audio began life as a bit prissily purist, it is now a happily open-minded music production company which lets Eric indulge his interest in the application of classical fine-editing techniques to all forms of acoustic music (especially in vocals) and, so far even more successfully, the application of standard subtle mastering techniques to classical chamber and instrumental music.

Over the past five years his daughter, Jamie Martha Perry, has grown into a sparkling and talkative six-year-old, and the fact that she now lives thousands of miles closer is the best change of all.

**www.urmaudio.co.uk**

## [ Appendix 12 ]

# Bob Katz Biography

From his earliest years, Bob has been as curious as a *Katz*. He voraciously reads audio books, service manuals, product spec sheets, license plates, and bumper stickers. But his favorite reads are Science Fiction writers Spider Robinson and Frederick Pohl, which may explain Bob's punny personality. In his teens he dabbled in hypnotism and magic, but was a bit klutzy to turn that into a career. Bob is an animal lover—all dogs and cats love him back.

Coming from a family of medical doctors, musicians and composers, Bob gravitated to the B flat clarinet at the age of ten; his aunt, a viola teacher, gave Bob his first lesson in solfège and transposition. At the age of 13, he rebuilt his first tape recorder. After wiring the house for sound, he was forced by his parents to remove the microphones he had secreted throughout the house. Clearly destined for a career in audio, by high school he had begun an amateur recording career, plus studying the sciences and linguistics, practicing French and Spanish and looking for female pen pals on three continents. Perhaps out of default he was voted *most versatile* in his class. Eventually his language skills would reach the point where he can give seminars in any of three languages.

An enthusiastic young man with a passion for good sound, Bob developed a reputation as an audiophile around Hartford, Connecticut town. The local audio stores regularly invited him over, for Bob is never short of opinions. One day he was asked to audition a new pair of speakers with the designer present. After hearing a few notes, Bob ran out of the store covering his ears! Over the years, he has learned to be more diplomatic, but his opinions continue to be defined by a love for the art of audio.

In college he played in an *ad hoc* Dixieland ensemble, and the treat of his performance life was soloing *Sweet Georgia Brown* before the homecoming football crowd. Two years at Wesleyan University were followed by two more at the University of Hartford, studying Communication and Theatre, but he spent less time in the classroom and more at the college radio station, where he became recording director. A fan of the Firesign Theatre, Bob used to write and edit humorous radio ads, and he became a DJ, manning a free-form-progressive rock radio show titled *The Katz Meow*, and doing a stint on the commercial rock station.

Bob taught himself analog and digital electronics, and was influenced by a number of creative designers. In Hartford, Bob's mentor was Steve Washburn, an EE who invented a way to nearly double the power-handling of a Hartley 24" woofer and also constructed Bob's first custom-built portable audio console. Just out of college, Bob became **(1972)** Audio Supervisor of the *Connecticut Public Television Network*, producing every type of program from game shows to documentaries, music and sports, and he learned to mix all kinds of music live. When he wasn't working television, he was on location, recording music groups direct to 2-track.

In **1972**, Bob wrote his first article for **dB** magazine, describing a set of mike heaters he developed to warm his AKG microphones and keep them from sputtering due to changes in humidity. This spiked a *heated* controversy as Stephen Temmer of Gotham Audio wrote a response stating that "Neumann microphones are never affected by humidity" but Bob's experience was supported by some others and in those pre-internet times the controversy remained of modest proportions. Hooked by the writing bug, Bob is a natural-born teacher who puts himself in the mind of the learner. He is a columnist for Resolution Magazine and has written many articles and reviews in publications such as dB, RE/P, Mix, AudioMedia, JAES, PAR, and Stereophile.

**In 1977** he moved from Connecticut to New York City, and began a recording career in records, radio, TV, and film as well as building and designing recording studios and custom recording equipment. Long before the advent of the home PC, Bob taught himself several computer languages, and sold an assembly-language program that was used in an embedded system at a brokerage firm. During the primitive time before cell phones, the voice of *Matilda* became well known. Matilda answered Bob's phone and forwarded calls to any place Bob happened to be. Visitors to Bob's house were dismayed to discover that sultry-voiced Matilda was not flesh and blood but rather a 6502-based controller, DTMF encoders, decoders and other gear. Matilda's true identity remains a mystery today.

**From 1978-79**, he taught at the Institute of Audio Research, supervised the rebuild of their audio console and studios and began a friendship with IAR's founder Al Grundy, mentor and influence. Other New York era influences include Ray Rayburn and acousticians Francis Daniel and Doug Jones. **In the 80's**, one of his clients was the spoken-word label, *Caedmon Records*, where he recorded actors including Lillian Gish, Ben Kingsley, Lynn Redgrave and Christopher Plummer.

An active member of the *New York Audio Society*, Bob was the ultimate audiophile. This led to a full-page interview/article in the *Village Voice* called *Sex With The Proper Stereo*, a story about Bob's railroad apartment on East 90th with the empty refrigerator in the kitchen and mysterious monoliths in the living room.

But the refrigerator was not empty for long. In **1984**, Bob was doing sound for a motion picture in Venezuela and met multi-lingual Mary Kent, production assistant. After the filming, Bob invited Mary to come to New York for a vacation that became a permanent engagement! One day new girlfriend Mary came home and turned on the stereo system in the wrong order, blowing up the Krell amplifier and one of the Symdex woofers producing sparks and blue smoke. When Bob arrived home, he calmed her down— "Don't worry, Mary, your love for me means more than any stereo system." Bob and Mary have been together ever since (Mary jokes that she's really in love with the stereo system).

One day Bob received a call from musician David Chesky, who had read the *Voice* article and was looking for an audiophile recording engineer. In **1988** this led to a long and pleasant association with Chesky Records, which became the premiere audiophile record label. Bob specializes in minimalist miking techniques (no overdubs) for capturing jazz and other music that commonly is multimiked. His recordings are musically balanced, exciting and intimate while retaining dynamics, depth and space. **In 1989** he built the first working model of the DBX/UltraAnalog 128x oversampling A/D converter, and produced the world's first oversampled commercial recordings. Over the years, the converter was refined, until by **1996** Bob found a commercial model that performed slightly better. Bob has recorded about 150 records for Chesky, including his second Grammy-winner, and **in 1997** the world's first commercial 96 kHz/24 bit audio DVD (on DVD-Video).

This obsession with good sound has developed into Bob's passion: *Audio Mastering*. Daily, he applies his specialized techniques to bring the exciting sound qualities of live music to every form recorded today. **In 1990** he founded **Digital Domain**, which masters music from pop, rock, and rap to audiophile classical. Besides mastering, Digital Domain provides complete services to independent labels and clients, graphic design and replication. Mary, who became Bob's wife, is an accomplished photographer and graphic artist, the visual half of the Digital Domain team and more than two-thirds of the charm. **In 1996**, Bob and Mary moved the company from New York to Orlando, adding numerous Florida-based artists and labels to the international clientele.

In the **90s**, Bob invented three commercial products, found in mastering rooms around the world.

The first product, the *FCN-1 Format Converter*, was dubbed by Roger Nichols the *Swiss-Army knife of digital audio*. Then came the *VSP model P and S* Digital Audio Control Centers, which received a Class A rating in *Stereophile* Magazine. These devices perform jitter reduction, routing, and sample rate conversion.

Bob has delivered lectures and seminars to the Audio Engineering Society at the conventions and sections and chaired AES workshops. He has been Convention Workshops Chairman, Facilities Chairman and served as Chairman of the AES New York Section. In **1991**, Bob began the **Digido** website, the second audio URL to make the World Wide Web, an educationally-oriented site which has grown to be a premium source for audio information. Thousands of pages around the globe have linked to www.digido.com.

Bob's first 21st century invention has received a U.S. Patent. He designed and introduced an entire new category of audio processor, the **Ambience Recovery Processor**, which uses psychoacoustics to extract and enhance the existing depth, space, and definition of recordings. Z-Systems of Florida and Weiss Audio of Switzerland have licensed Bob's K-Stereo™ and K-Surround™ processes.

Bob has mastered CDs for labels including EMI, BMG, Fania, Virgin, Warner (WEA), Sony Music, Walt Disney, Boa, Arbors, Apple Jazz, Laser's Edge, and Sage Arts. He enjoys the Celtic music of Scotland, Ireland, Spain and North America, Latin and other world-music, Jazz, Folk, Bluegrass, Progressive Rock/Fusion, Classical, Alternative-Rock, and many other forms. Clients include a performance artist and poet from Iceland; several Celtic and rock groups from Spain; the popular music of India; top rock groups from Mexico and New Zealand; progressive rock and fusion artists from North America, France, Switzerland, Sweden and Portugal; Latin-Jazz, Merengue and Salsa from the U.S., Cuba, and Puerto Rico; Samba/pop from Brazil; tango and pop music from Argentina and Colombia, classical/pop from China, and a Moroccan group called *Mo' Rockin'*.

Bob mastered *Olga Viva, Viva Olga*, by the charismatic Olga Tañon, which received the **Grammy** for Best Merengue Album, **2000**. *Portraits of Cuba*, by virtuoso Paquito D'Rivera, received the **Grammy** for Best Latin Jazz Performance, **1996**. *The Words of Gandhi*, by Ben Kingsley, with music by Ravi Shankar, received the **Grammy** for Best spoken word, **1984**. In **2001** and **2002**, the Parents' Choice Foundation bestowed its highest honor twice on albums Bob mastered, giving the **Gold Award** to children's CDs, *Ants In My Pants*, and *Old Mr. Mackle Hackle*, by inventive artist Gunnar Madsen. The Fox Family's album reached #1 on the Bluegrass charts. African drummer Babatunde Olatunji's *Love Drum Talk*, **1997**, was Grammy-nominated.

Bob's recordings have received *disc of the month* in *Stereophile* and other magazines numerous times. Reviews include: "best audiophile album ever made" (McCoy Tyner: *New York Reunion* reviewed in *Stereophile*). "If you care about recorded sound as I do, you care about the engineers who get sound recorded right. Especially you appreciate a man like Bob Katz who captures jazz as it should be caught." (Bucky Pizzarelli, *My Blue Heaven* reviewed in the San Diego

Voice & Viewpoint). "Disc of the month. Performance 10, Sound 10" (David Chesky: *New York Chorinhos*, in *CD Review*). "The best modern-instrument orchestral recording I have heard, and I don't know of many that really come close." (Bob's remastering of Dvorák: Symphony 9, reviewed in *Stereophile*).

Some of the great artists Bob is privileged to have recorded and/or mastered include: Juan de Marcos González and the Afro-Cuban All Stars, Monty Alexander, Carl Allen, Jay Anderson, Lenny Andrade, Michael Andrew, Issa Babayago, Ray Barretto, Lucecita Benitez, Berkshire String Quartet, Ruben Blades, Gordon Bok, Luis Bonfa, Boys of the Lough, Bill Bruford, Irene Cara, Ron Carter, Brandi Carlile, Cyrus Chestnut, George Coleman, Willie Colon, Larry Coryell, Celia Cruz, Joe Cuba, Eddie Daniels, Los Dan Den, Dave Dobbyn, Paquito D'Rivera, Arturo Delmoni, Garry Dial, Dr. John, Toulouse Engelhardt, Robin Eubanks, George Faber, John Faddis, Fania All Stars, David Finck, Tommy Flanagan, Foghat, Fox Family, Johnny Frigo, Ian Gillan, Dizzy Gillespie, Whoopi Goldberg, Ricky Gonzalez, Bill Goodwin, Arlo Guthrie, Steve Hackett, Lionel Hampton, Larry Harlow, Emmy Lou Harris, Tom Harrell, Hartford Symphony, Jimmy Heath, Levon Helm, Vincent Herring, Conrad Herwig, Jon Hicks, Billy Higgins, Milt Hinton, Fred Hirsch, Freddie Hubbard, Garth Hudson, David Hykes Harmonic Choir, Dick Hyman, Ahmad Jamal, Antonio Carlos Jobim, Clifford Jordan, Sara K., Connie Kay, Kentucky Colonels, Lee Konitz, Hector Lavoe, Hubert Laws, Peggy Lee, Chuck Loeb, Joe Lovano, La Lupe, Patti Lupone, Gunnar Madsen, Jimmy Madison, Taj Mahal, Sean Malone, Manhattan String Quartet, Herbie Mann, Michael Manring, Marley's Ghost, Winton Marsalis, Dave McKenna, Jackie McLean, Jim McNeely, Milladoiro, Mississippi Charles Bevels, Los Mocosos, Max Morath, Paul Motian, Necrophagist, New England Conservatory Ragtime Ensemble, New York Renaissance Band, Sinead O'Connor, Johnny Pacheco, Eddie Palmieri, Van Dyke Parks, Gene Parsons, Gram Parsons, Danilo Perez, Itzhak Perlman, Billy Peterson, Ricky Peterson, Bucky Pizzarelli, John Pizzarelli, Chris Potter, Tito Puente, Kenny Rankin, Richie Ray and Bobby Cruz, Mike Renzi, Rincon Ramblers, Sam Rivers, Red Rodney, Rodrigo Romani, Michael Rose, Phil Rosenthal, Rey Ruiz, Mongo Santamaria, Horace Silver, Paul Simon, Lew Soloff, George 'Harmonica' Smith, Soneros Del Barrio, Spanish Harlem Orchestra, Janos Starker, Olga Tañon, Ben Taylor, Livingston Taylor, Clark Terry, Thad Jones/Mel Lewis Big Band, Steve Turre, Stanley Turrentine, McCoy Tyner, Jay Ungar, U.S. Coast Guard Band, U.S. Marine Band, Amadito Valdez, Kenny Washington, Peter Washington, Doc Watson and Son, Clarence White, Widespread Jazz Orchestra, Robert Pete Williams, Larry Willis, and Phil Woods.

—by Mary Kent (who knows him best)

# [ Appendix 13 ]

# Glossary

## A

**AES/EBU** The name of the digital audio interface jointly conceived by the Audio Engineering Society and the European Broadcasting Union. See Chapter 22.

**ADC** Analog-to-digital convertor, a circuit that converts continuous signals, coming from the analog domain, into discrete digital numbers.

**AGC** Automatic Gain Control. Compression that brings up low-level passages. See Chapter 11.

**AIFF** (along with **WAVE, BWF, SD2, MP3**), a type of audio file format. See Appendix 2.

**ALIASING** An alias is a beat note or difference frequency between the audio content and the sample rate, a form of intermodulation distortion. Proper filtering should eliminate aliases, but see Chapter 20. Note in an ADC, the higher the sample rate, the less chance of aliasing products being created against the normal audio content, but aliasing distortions could still arise from RF interference.

**ASRC** Asynchronous sample rate converter. A converter from one sample rate to another which can work with a wide relationship of input to output frequency, and thus can deal with varispeeded rates. Filter coefficients are continuously variable, computed on the fly. However, this may not yield the lowest distortion. See Chapter 19.

**AUTHORING** The process of recording source material (audio, video or other data) onto the release format, which may include adding menus, and interactive content. To author, one has to know all the rules and specifications of the format, e.g. DVD, Blu-Ray. See Chapter 1.

**AVG** Average.

**A-WEIGHTING** See Weighting.

## B

**BIT-TRANSPARENT** The output is a perfect digital clone of the input, including the source wordlength.

**BITRATE** Speed in number of bits per second. As distinguished from wordlength. In linear PCM, bitrate is the product of the wordlength (sample size) and the sample rate.

## C

**CODEC** (Coder-Decoder), is an algorithm that performs encoding (recording) and decoding (playback) on a digital data stream or signal. There are both lossy and lossless codecs. WAV, AIFF are examples of lossless codecs while mp3 and ACC are lossy. The sound quality of a lossy codec is dependent on the algorithm and bitrate.

**COMB FILTERING** Comb filtering of a signal is introduced when combining an audio signal and a delayed replica to the same output channel. See Chapter 3.

**COMPACT DISC DIGITAL AUDIO (CD-DA)** A 16-bit stereo 5" disc standard jointly developed by Sony and Philips in 1980. This is also known as the Red Book standard. See also **Red Book** for other forms of the compact disc such as the CD-ROM, which carries files.

**COMPRESSION RATIO** The ratio between input and output level of a compressor above the threshold point. See Chapter 10.

**CONVOLUTION** Technique used in DSP design that is a mathematical way of combining two signals to form a third signal. The two signals consist of the input and the impulsive response which allows the third (the output) to be calculated.

**CLIPPING** A form of distortion that happens when an amplifier is asked to create a signal greater than its maximum capacity. When trying to go above the maximum capacity of the amplifier, the signal is said to be "clipped". In digital, the maximum capacity is known to be be 0 dBFS, and any overs will cause distortion to appear on its outputs. See Chapter 5.

**CREST FACTOR** The difference between the peak amplitude of a waveform and its RMS value.

## D

**DAC** Digital-to-analog convertor, is a circuit that converts discrete digital numbers, into continuous signals (a voltage) in the analog domain.

**DAT** Digital Audio Tape Recorder. Short for RDAT, which stands for Digital Audio tape recorder with rotating heads. There was an SDAT (stationary head) standard, but this was never released.

**DAW** Digital Audio Workstation. Usually a computer with dedicated hardware and software for editing and processing digital audio.

**DB** Decibels. A logarithmic measure of audio level. See chapter 5.

**DBFS** 0 dBFS means "0 dB reference full scale," as on a digital meter. Full scale is 0 dB and the meter reads negatively below that. Full scale is the highest signal which can be recorded. Positive going signals with a value of 32767 or negative with a value of -32768 (at 16-bit) are at the maximum. Levels below those are translated to decibels, with 0 dBFS being full scale. For example, -10 dBFS is a level 10 dB below full scale.

**DITHER** A process that linearizes digital audio by adding a random noise signal at the point just before wordlength truncation. Dither is required for clean digital audio recording and processing. After dithering, the wordlength can be safely truncated or shortened, but truncation without dithering results in quantization distortion. See Chapter 4.

**DSD** (Direct Stream Digital) is the audio format used on the SACD (Super Audio Compact Disc), a rival format to the DVD-A. DSD, as opposed to multibit PCM, carries audio information using one-bit encoding.

**DVD-A** DVD originally stood for **Digital Video Disc**, but it has been redubbed **Digital Versatile Disc** as it can support computer, audio, and video formats. The -A suffix defines the multichannel audio disc standard that supports a wide range of PCM sample rates and wordlengths, and limited (still) graphics.

**DVD-V** A video and audio disc standard that supports multichannel digital audio sample rates up to 48 kHz/24-bit, and 2-channel digital audio at 96 kHz and 192 kHz, but there is usually not enough room on the disc to fit high-quality video and high resolution audio at the same time. When MPEG video takes up much of the space on the disc, usually coded (data-reduced) formats such as DTS or Dolby Digital carry the multichannel audio track.

**DYNAMIC RANGE** The range in decibels between the highest level which can be encoded and the lowest level which can be heard. Since this is a perceptual, or ear-based determination, it is an approximate number. In a properly dithered system, available dynamic range can be greater than its measured signal-to-noise ratio. See Chapters 4 and 5.

**DSP** (Digital Signal Processing), is the processing of a stream of information by performing numerical calculations.

## E

**EDL** Edit decision list. Also known as **Playlist**. Instead of cutting the actual audio, an EDL is a list of instructions of where and how to cut and reproduce the audio when played back. Thus, many different versions or playbacks of the same audio can be reproduced from the audio files. An EDL is to audio as a Word Processor is to words.

**E-E** Pronounced "E to E." Electronics to electronics. For example, when a tape recorder is put into record, its output monitors its input directly. This mode is known as E-E.

**EFM** (Eight-to-Fourteen Modulation), used for channel encoding on CDs. EFM breaks the data (PCM audio in case of the CD, together with data bits) into 8-bit blocks (bytes). Each 8-bit block is translated into a corresponding 14-bit word using a lookup table. These 14-bit words have a minimum of two and a maximum of ten zeroes. A binary one is stored as a change from a land (zero) to a pit (one) or a pit (one) to a land (zero), while a binary zero is indicated by no change.

**EMPHASIS** In an effort to improve the already excellent signal-to-noise ratio of the Compact disc, CDs (as well as digital tapes) can be recorded with emphasis. If it is decided to use emphasis, the recording is made with a calibrated high frequency boost (called Emphasis), and during playback, a corresponding high frequency rolloff (called Deemphasis) is applied. Thus, in theory, signal-to-noise ratio is improved, though in practice the loss of high frequency headroom may reduce any audible improvement. Most CDs made today do not use emphasis.

**EQUAL LOUDNESS CONTOURS** A measure of sound pressure over the frequency spectrum, for which a listener perceives a constant loudness. This was first measured by Fletcher and Munson in 1933, whereafter Robinson and Dadson did a new experimental determination in 1956, which was more accurate and become the ISO 226 standard.

## F

**FIR VS. IIR** FIR stands for finite impulse response and IIR for infinite impulse response. These are types of filters which can be implemented in equalizers. See Chapter 8.

**FIREWIRE** The name of a high-speed bi-directional serial interface originally developed by Apple computer, but then officially adopted as standard IEEE 1394, for use with digital audio, video, hard drives, controllers, etc.

**FIXED-POINT VS. FLOATING POINT** Fixed-point notation uses a finite binary number whose range in 16-bit is 96 dB, in 24-bit is 144 dB. But floating point notation can represent thousands of dB of dynamic range through use of exponents. Assuming a skilled designer, 48-bit fixed point is more than adequate for all standard DSP chores, though some feel that at least 40-bit float is necessary for complex calculations such as equalization and sample rate conversion.

**FLETCHER-MUNSON CURVE** See Equal Loudness Contours.

**FRAMES** There are two commonly used "frame" standards in CD work, with different lengths: 75 CD Frames in a second, as opposed to 30 SMPTE frames per second. Modern PQ lists are usually expressed in CD Frames, but the older 1630 systems used SMPTE frames, which have less timing resolution.

## G

**GAIN, LOUDNESS, VOLUME AND LEVEL** Distinctive terms each with their own meaning, carefully distinguished in Chapter 5.

**GLASS MASTER** Glass Mastering is the process of transferring the CD master to a physical image of the pits on a coated glass substrate. See Chapter 1.

## H

**HEADROOM** See **Nominal Level.**

**"HOT" CD** A recording with a high intrinsic loudness (explained below). See also Chapter 5.

**HYPERCOMPRESSION** Compression applied for the sake of increasing intrinsic loudness but without caring about the decrease in sound quality. See Chapter 14.

## I

**INTRINSIC LOUDNESS** The loudness of a program before the level is adjusted using the monitor control. See Chapter 5.

## J

**ISRC** International Standard Recording Code, defined by the RIAA as a unique code for each track on the CD. See Chapter 1.

**JITTER** Timing variations in the digital audio clock, producing distortions. See Chapter 21.

## K

**KHZ** Abbreviation for kiloHertz, meaning audio frequency in thousands of cycles per second. Commonly this usage also applies to sample rate, which can cause confusion. When there might be confusion, we add the term sample rate where appropriate.

**K-SYSTEM** An integrated system of metering and monitoring devised by the author (Ch. 14).

**K-STEREO™, K-SURROUND™** Patented processes for extracting and enhancing the already existing ambience of recordings.

## L

**LSB** Least significant bit. See Chapter 4.

**LATENCY COMPENSATION** Latency compensation within DAWs means that all outputs of every bus are sample aligned, no matter how many plugins are inserted on that bus.

## M

**MADI** (Multichannel Audio Digital Interface), is a communication protocol commonly used in digital audio, that provides serial data transmission over coaxial or fibre-optic cables and supports up to 64 channels, with sample rates up to 96 kHz and bit depths of up to 24 bits per channel.

**MLP** Meridian Lossless Packing, now part of Dolby HD, a data-reduction technique which made it possible to fit as many as 6 high quality channels of digital audio at 96 kHz SR on a DVD-A disc.

**METADATA** Metadata is data about data. It is used to facilitate the understanding use and management of data. MP3 uses ID3 metadata, which stores the title, artist, album, etc., with the audio itself.

**MICROSECOND** (µs) is one millionth of a second.

**MP3** (MPEG-1 Layer 3), popular audio encoding format that uses a lossy compression algorithm.

**MOORE'S LAW** The empirical observation made in 1965 that the number of transistors on an integrated circuit for minimum component cost doubles every 24 months.

## N

**NORMALIZATION** An automatic process available in most DAWs, whereby the gain of all program material is adjusted so the peak level will just arrive at 0 dBFS. Technically this should be called **peak-normalization**. There are many esthetic and technical reasons to avoid peak-normalization. **Loudness-normalization** is the more correct approach, see Chapter 14.

**NOMINAL LEVEL** The average or RMS level at which an audio device is designed to operate. **Headroom** is defined as the level allowed for peaks above the nominal level. See chapter 5.

**NYQUIST** Dr. Nyquist, while working for Bell Labs, discovered the sampling theorem, where he states that a sampled waveform contains ALL the information without any distortion, when the sampling rate exceeds twice the highest frequency contained by the sampled waveform. This theory forms the basis for all PCM-based systems. There can be no information in a sampled system above the Nyquist **Frequency**, which is 1/2 of the sample rate.

## P

**PANDORA'S BOX** In Greek mythology, Pandora was the woman who opened a box releasing all the evils of mankind, leaving only hope inside once she closed it.

**PICOSECOND** (ps) is one millionth of one millionth of one second, or 10-12 second. See Chapter 21.

**PLUG-IN** An extra process which can be inserted into a DAW. Some plug-ins utilize the power of an external DSP card, while others, called native plug-ins, utilize the computer's CPU.

**PQ CODING** The Compact disc contains a number of subcode areas, each area is named with a letter, from P to W, with information on track number, timing, and so on. See Chapter 1.

**POLARITY** The quality of having two oppositely charged poles, one positive and one negative. Changing polarity means reversing the positive charged pole with the negative, and vice versa. Analog sound travels through electronics as AC current, where polarity inversion can be applied by switching the positive and the negative wire. See Chapter 3.

## Q

**QUALITY CONTROL** (QC), Form of checking and ensuring the product that is being delivered is error-free. See Chapter 1.

## R

**RED BOOK** defines the standards for the audio CD as defined by Sony and Philips. No ordinary individual has a copy of the Red Book. The real Red book can only be found at authorized Compact Disc replication plants. The **Blue Book** defines enhanced CDs with audio and ROM material. **Yellow Book** defines CD-ROMs. **Green Book** defines compact disc Interactive. **White Book** defines the Video CD. **Orange Book** defines CD-R or Recordable CDs.

**REFERENCE** Reference CD. See Chapter 1.

**RMS** Root-Mean-Square. A method of averaging levels which computes the equivalent power of the material. For all naturally-occurring music, an RMS-responding meter will read several dB below the actual peak level of the music at any moment in time.

**RESOLUTION** We use the term resolution to indicate whether a source signal of a given level will be represented in the output. This can be expressed as a number of equivalent bits. See chapter 4 and 5.

## S

**SACD** See DSD.

**SDIF-2** Sony Digital Interface-2. The stereo version uses 3 cables, one for each channel and one for wordclock, thus avoiding the interaction between clock and data that causes interface jitter in the competing AES/EBU or SPDIF interfaces.

**SEGUE** A crossfade between two different types of music, pronounced seg-way, from the Italian seguire meaning to follow.

**SEQUENCING** Putting an album together and spacing it, not to be confused with MIDI sequencing. See Chapter 7.

**SNR** The abbreviation we use in this book for Signal-to-Noise Ratio. SNR of a digital system is the decibel ratio between the highest level which can be encoded (0 dBFS) and the dither noise. Since the noise can be measured with different weightings, SNR is simply a number we can use to compare, but may have little relationship to the actual range the ear hears. Dynamic range represents more closely what the ear hears, but it's difficult to define precisely the absolute lowest levels we can hear in any particular digital system. See Chapters 4 and 5.

**SPDIF** Shorthand for Sony-Philips Digital Interface. Standard IEC-958 and IEC-60958 defines this interface, usually found on an RCA (coaxial) connector.

**SRC** (also abbreviated SFC) Sample Rate Converter, or Sample Frequency Converter. See Chapters 20 and 21. A **Synchronous SRC** uses fixed filter coefficients, can only convert between certain fixed rates, e.g. 44.1, 48, 88.2 and 96 kHz, and cannot accept varispeeded sources. See also **ASRC**.

**STEMS** Individual components in a mix, commonly used in mastering when there are problems in the full mix. See Chapter 16.

**STATE MACHINE** A state machine is defined as any type of processor which produces identical output for the same input data, and which does not look at data timing or speed, but only at the state or recent history of the data. Most digital processors are state machines and thus are completely immune to jitter. (See Chapter 21).

## T

**TRUNCATION** Reduction of wordlength by cutting off the lower bits. If dithering was not performed first, then simple wordlength truncation causes distortion.

**TRANSPARENT** A device which is transparent sounds as clean on its output as the source.

**THD** (Total Harmonic Distortion), is a measurement of the harmonic distortion present, mostly used to test quality of equipment.

**TWO'S COMPLEMENT** is a specific format of binary notation that allows multiplication and addition of bits, and positive and negative numbers. Zero is the midrange, and the MSB forms the sign bit where 1 = negative. This is analogous to adding analog signals in an operational amplifier.

## V

**VCA** (Voltage Controller Amplifier), is an audio amplifier whose gain is controlled by a control voltage.

## W

**WOW AND FLUTTER** Speed variations in recordings, commonly caused by imperfections in analog tape recorders.

## Z

**ZIP** A file format that is popular for its data compression and archival purposes, to enable reliable and faster up and downloads. ZIP can store one or more files in a single archive.

# Index

## D

## E

**N**

## T



## U



## V



## W



## Y



## Z

# Afterword

## How This Book Was Written and Edited

From the moment the first edition of **Mastering Audio** went to press, I started collecting and compiling notes based on audience reactions from seminars and meetings. I also had numerous reader email and web forum questions and suggestions, and copious notes on how I wanted to reorganize and present the new areas in the second edition.

Then, using a professional utility called CopyFlow Gold, I was able to transfer the final version of the first edition back from QuarkXPress to Word for a complete rewrite. CFG also enabled a smooth return to Quark for final layout and proof.

After rewrite, writer and editor (who have not yet met face to face!) exchanged drafts of each chapter using FTP. We continued to use Word's **Track Changes** and **Comments** features to interact, annotate and comment the text, and view each others' changes. Near the last draft, I passed the copy of critical technical chapters via email to Dick Pierce, who suggested his changes (more refinement than corrections) which I integrated into the manuscript.

Thanks to the many readers and listening audience who provided great feedback and inspired new ideas. I am especially pleased that this second edition has taken on its own evocative vitality.

Bob Katz, August 2007, Orlando, Florida